D0928086

Sound as Popular Culture

Sound as Popular Culture

A Research Companion

edited by Jens Gerrit Papenburg and Holger Schulze

The MIT Press
Cambridge, Massachusetts
London, England

© 2016 Massachusetts Institute of Technology

Printed with the support of the Deutsche Forschungsgemeinschaft (DFG)

All rights reserved. No part of this book may be reproduced in any form by any electronic or mechanical means (including photocopying, recording, or information storage and retrieval) without permission in writing from the publisher.

This book was set in Stone Sans and Stone Serif by Toppan Best-set Premedia Limited. Printed and bound in the United States of America.

Library of Congress Cataloging-in-Publication Data

Names: Papenburg, Jens Gerrit, editor. | Schulze, Holger, editor.
Title: Sound as popular culture : a research companion / Jens Gerrit
    Papenburg and Holger Schulze, eds.
Description: Cambridge, MA : The MIT Press, [2015] | Includes bibliographical
    references and index.
Identifiers: LCCN 2015038277 | ISBN 9780262033909 (hardcover : alk. paper)
Subjects: LCSH: Sound in mass media. | Sound–Social aspects.
Classification: LCC P96.S66 S68 2015 | DDC 302.2301–dc23 LC record available at
    http://lccn.loc.gov/2015038277

10 9 8 7 6 5 4 3 2 1

To the memory of the life and research of Maria Hanáček

Contents

Preface

Our work on this volume started in 2009. It all began with a phone conversation, in which we, Jens Gerrit Papenburg and Holger Schulze, discussed our mutual interest in working together on some form of collective research project concerning the state of contemporary research on sound in popular and media culture—a project that could be conducted at the cross section of a variety of disciplines, scholarly and professional practices, aesthetic genres, and everyday experiences. From this starting point, and in collaboration with our great and cherished colleague Maria Hanáček, we developed a project proposal to bring both young and experienced scholars working in these fields in Germany together with scholars from all over Europe, the United States, and Canada.

As soon as the national German Research Foundation (DFG) had granted us the funding for our project as a research network, we started inviting a group of twenty-two researchers to take part in a series of international workshops: at the Berlin University of the Arts and the House of World Cultures Berlin (host: Holger Schulze), at the London College of Music (host: Simon Zagorski-Thomas), at the Academy of Fine Arts Vienna (host: Diedrich Diederichsen), at the Humboldt University Berlin (host: Peter Wicke), at the Leuphana University Lüneburg (host: Rolf Großmann), and at the Istituto Europeo di Design Milano (host: Franco Fabbri). Between 2010 and 2013, we held and heard presentations, working papers, preliminary concepts, revisions, and discussions concerning the definition and the history of sound, both the everyday and the studio aspects of sound, and regarding anthropological, technological, and design approaches to sound as part of contemporary popular culture in the twentieth and early twenty-first centuries. Our thanks go to the network's members and guests: Karin Bijsterveld, Susanne Binas-Preisendörfer, Jochen Bonz, Michael Bull, Claudia Bullerjahn, Thomas Burkhalter, Mark J. Butler, Diedrich Diederichsen, Veit Erlmann, Franco Fabbri, Golo Föllmer, Marta García Quiñones, Rolf Großmann, Thomas Hecken, Anahid Kassabian, Carla J. Maier (neé Müller-Schulzke), Carlo Nardi, Thomas Schopp, Jonathan Sterne, Paul Théberge, Peter Wicke, and Simon Zagorski-Thomas. Thanks also to everyone who contributed a lecture or performance, or who took part in one of the discussions at our workshops, in particular Ilaria Bertuetti, Steve Goodman, Simon Grab, Antye Greie, Sarah-Indriyati Hardjowirogo, Julian Henriques, Daisuke Ishida, Matteo Lo Valvo, Morten Michelsen, Marco Montiel-Soto, Christina Nemec, Peppino Ortoleva, Franz Pomassl, Michael Spahr, Sutsche & Fello, Georg Weckwerth, and Stefano Zorzanello.

Over the course of five years, this network's manifold research approaches became increasingly interlinked and interwoven—and at the same time, the over-arching field of research developed rapidly, demonstrated by the large number of new anthologies, new journals, and foundations of new research associations: the European Sound Studies Association (ESSA) in 2012 as well as the long overdue (re) foundation of a German-speaking branch of the International Association for the Study of Popular Music (IASPM-DACH) in 2012. The ESSA association was not only supported by members of our research network taking part in a series of workshops organized by our colleagues from the impressive Nordic Research Network of Sound Studies (NORSOUND), but also by our members co-organizing its first international conference in 2013 in Berlin. The refoundation of IASPM-DACH, mainly initiated by Maria Hanáček, started with a promising workshop in Bern (Switzerland) in 2013 and a groundbreaking first conference in Siegen (Germany) in 2014; we feel honored that Susanne Binas-Preisendörfer, IASPM-DACH's first president, is also a member of our research group. The volume that you are holding in your hands (or that you are scrolling through on your screen) is the final outcome of our collective explor-atory expeditions into sound as popular culture. May it serve in teaching, in research, and in inspiring and inciting a variety of discussions of critical and fruitful new approaches to research in the near and far future.

We would like to thank all of the colleagues and friends who supported us, who encouraged us, and who worked continuously with us to find the best approach, the best structure, and the best research issues to discuss in this volume. A special thank you to Jonathan Sterne, who read and commented thoroughly on an early draft of the book proposal; to Anahid Kassabian, who supported us by helping us find the best destination for the book proposal; and to Marta García Quiñones and Rolf Großmann, who discussed the first version of the volume's manuscript with us in depth. A very special thank you also to Jessica Ring for her excellent translations, edit-ing, and helpful comments. Jess did for our book what a brilliant mastering engineer does for a record production. Thanks to Stephanie Schulze for formatting the chapters of this volume according to the publisher's guidelines, as well as to Burkhard Mei-schein for his support. Many thanks to our research network's assistants, Ann-Kathrin Eickhoff and Andreas Lipowsky: you did not just support us with the organization of the six workshops, you also contributed inspiring intellectual input to the proj-ect. Thanks to Paul Snowden, whose logo design shaped the research project's public image and to Doug Sery from the MIT Press for his support, as well as to the anonymous reviewers of the book.

We would also like to thank Marcus Gammel from Deutschlandradio Berlin for his ongoing support and inspiration in the production and airing of three hour-long radio features about this research network's work between 2011 and 2014. Moreover, we would like to thank Gisela Nauck and Roland Posner for their similar support in conceptualizing and producing two special issues for the journals *Positionen: Texte zur aktuellen Musik* (2011) and the *Zeitschrift für Semiotik* (2012). And, most importantly, we would like to thank our families for their support and patience. Holger sends his thanks and love to his little tribe, Maren, Nanouk, Rasmus and Pirkka; and Jens sends his thanks and love to Anna, Anima, and Ophelia.

During our half-decade-long journey into intensive research work and many inspiring, illuminative, and transformative discussions on all of the unforeseeable, subtle details in sound cultures, in the fall of 2013, we had to mourn the loss of our cherished colleague, third main collaborator, and friend: Maria Hanáček. Maria died much too young. We will miss her commitment, her scholarly curiosity, her competence, and her friendly and open personality. From the beginning, Maria was a crucial driving force—of the third part of this volume in particular—as well as more generally concerning research approaches that focus on music technology and studio production, including the relevant discourses of practitioners in that context: instrumentalists, sound engineers, session musicians, sound designers, and marketing experts. As the two remaining editors, we asked ourselves during that forced hiatus: how could we bring this work to a convincing end, work that we had initiated, moved forward, executed, and developed as a team of three? How could we honor the heritage of a young and promising scholar whose ideas and interests can now be found in people's memories and in fragmentary manuscripts, rather than in well-edited published books and chapters? We tried to finish this volume in the spirit of Maria's research, of her rigorous and vigorous interest in theory as well as in practice. We know one thing for sure: Maria would have liked our research project to conclude in this way, namely with strong international collaboration, a collaboration that developed step by step over the course of our six workshops and continued on after MIT Press became interested in the project.

Jens Gerrit Papenburg and Holger Schulze, Berlin, January 2015

Introduction: Sound as Popular Culture

Jens Gerrit Papenburg and Holger Schulze

Researching sound as popular culture is the study of sound as both an integral and constitutive part of culture. The investigation of sound as popular culture would be more aptly described as a study *through* sound than a study *about* sound. An examination of sound as popular culture explores epistemologies reaching beyond the dualisms of subject–object and text–context. Those three research goals guide the direction of this volume: in the first section ("I. Outlining a Non-Discipline") through fundamental reflections on a series of ontologies and epistemologies of sound, as well as on specific terminologies and research methods; in the second ("II. Formations of Listening") by analyzing historical and contemporary case studies on how listening and listening's relation to sound was and is conceptualized in various sound cultures; and finally ("III. Producing Sonic Artifacts"), by describing, questioning, and reconceptualizing how practices of sound generation are applied by practitioners in the diverse fields in which sounds are produced, mastered, distorted, processed, or enhanced since the mid-twentieth century.

The Sound of Consultants and Clouds

We are watching a video, listening to a monologue. A sound expert is standing in front of an audience—is he a designer? A consultant? A musician? All of the above and more? He is equipped with all of the state-of-the-art miking and amplifying technology that is necessary these days for an internationally accessible lecture series. He is quite tall, and appears calm, sincere, and dedicated. He says: "Over the next five minutes, my intention is to transform your relationship with sound" (Treasure 2009, 0:16). His speech is well articulated but still easygoing, adapting the rhythm of professional speaking to the strict time frame of this TED Talk, the world-famous lecture series on "Technology, Entertainment, Design."[1] The sound of his voice is not only attuned to the auditorium in which he is performing; it also transmits well and efficiently over the laptop speakers through which people can experience the lecture online. Over the course of the video clip, a wide range of everyday sounds (traffic sounds, alarm sounds, jackhammers, so-called brand sounds), "natural" soundscapes, and, of course, music is intermittently emitted and mixed in with the speaker's voice. In addition, the audience intervenes acoustically as a laughing crowd from time to time, affirming the entertaining elements provided by the speaker. He is on a mission, and his whole appearance—bald head, earnest stare, and relaxed attire with loosely cut yet fitted pants and shirt—apparently coincides with the audience's idea of what an intellectual creator who might provide them with novel insights should be. He is a successful advocate for listening and for raising general awareness about the audible. His manner of presentation is fully in accordance with the internationalized American style of presenting academic content, namely in that superficially optimistic, and as such deeply paternalistic, way that has been so intensely promoted for decades now through the *Californian Ideology* (Barbrook and Cameron 1996): an ideology that comes along as an assemblage of concepts stemming from post-1960s countercultural bohemia, networked digital technologies, "free market" fantasies, and hip(pie) entrepreneurship. Our speaker is presenting and performing a culturally and historically unique form of the practices and knowledge of consultants in the twenty-first century. And it does not come as a great surprise that a "society of consulting" (Pias and Vehlken 2010, 9), with its consultants for diverse areas, such as business, marriage, finance, fashion, academia, drugs, has also produced—rather sooner than later—a consultant for sound.[2]

The video then proceeds to run through some of the last decades' more general assumptions in sound theory. The speaker starts off by reproducing some of the most common yet very specifically exemplified claims about how sound resonates within individual bodies ("Sounds are affecting your hormone secretions all the time, but also your breathing, your heart rate … and your brain waves" (Treasure 2009, 0:50). Some of the examples he presents reflect approaches by sound research pioneers such as R. Murray Schafer; but all of the examples present these fundamental claims as absolute and irrefutable truths. Contrary to the sound research pioneers' "sentimental" ecologist approach (Hosokawa 1984), as was developed in the ecology movements of the 1960s and 1970s, the speaker seems to be in a rather good mood, looking confidently ahead to the potentials of a sonic future. He is speaking the truth: the truth about sound. This supposed truth is revealed to us through technical diagrams and statistics that promise an easily manageable chart of "outcomes" that are quite convenient for Westernized consumer culture and employee culture: *physiological, psychological, cognitive*, and *behavioral*. The terms chosen by the consultant to speak about sound stem from a vocabulary most commonly associated these days with the language of science and economics: *bandwidth, processing, activity, productivity, model*. Sound increases the productivity or effectiveness of its listeners—and according to this speaker, that is its only conceivable significance. Instead of describing sound as a culturally and historically specific phenomenon and concept explicitly, as has become common practice in sound research over the last years, this speaker is interested in offering a "better" life to—dare we say: Westernized?—Subjects through sound.

In any case, the assumptions about sound—exemplified here by the speaker Julian Treasure, chairman of the Sound Agency, a consultant company specialized in designing sound for corporations and brands—do not stray completely from contemporary sound scholarship. Treasure's assumptions do indeed selectively represent some elements of an approach toward material culture and sensory studies that is currently relevant in sonic theory. As such, the consultant does actually manage to promote a transfer of infobits from sound research. However, the manner in which it is done and the mission that lurks behind the speaker's litany stem from the aforementioned knowledge and practices of consultants. As such, it is riddled with implications, mannerisms, and biases that are not at all part of current research—research that has quite different implications, mannerisms, biases: the need to present hermetic, expert knowledge; and the need for an intricate, sometimes even incommensurable analysis and narcissistic argument (as in the case of this introduction perhaps). In contrast, a sound consultant's performance is conducted in a strongly populist and educating manner—albeit pedagogically reductionist—which seems to be the behavior required for the TED lecture series' most successful presentations—a series that has served as the Olympics of Californian Ideology. Rather than being about ambitiously presented "objective" truth (which even in accordance with the strictest positivism would imply its future falsification as well as its purely tentative nature), the consultants' sound knowledge is a strategic knowledge; even if—and this is the trick—it promotes general and non-individualistic "objectivity," it follows a pragmatic and quite personal interest and ambition.

To clarify the most striking elements and basic assumptions of this objectivistic analysis of the consultant's presentation we can take a closer look at two especially disconcerting statements. The first refers to a common notion about the general relevance of music: "Music is the most powerful form of sound that we know that affects our emotional state" (Treasure 2009, 1:20). This statement might be described as meaningless from a research perspective, and more an example of proselytizing than of actual scholarship. Yet precritical claims such as this one are actually included in any educated citizen's common knowledge about cultural practices in music today. In such cases, musical anthropology is boldly extended to a musical education, even a musical eschatology, which we can call an "audiopietism" (Schulze 2007), or even a kind of "theology of sound" (Sterne 2011). In this context, shifting one's focus to sound and music is understood as an improvement strategy for Subjects, one which has even be linked to a mystic hope of universal salvation and redemption scenarios. However, the consultant's bias toward promoting the general relevance of sound and music for contemporary culture becomes even more apparent in another remarkable statement from this presentation: "recognition + association = power" (Treasure 2009, 3:50). To exemplify this, he provides two examples from the canon of contemporary popular culture's sound design: the iconic starting chord in the Beatles' *A Hard Day's Night* (ibid., 3:54) and the ascending semitone of John Williams's famous shark motif for *Jaws* (ibid., 4:00). Although there is no mention of this point, it is obviously necessary to recognize a specific sound as being a part of popular culture in the first place in order to associate anything with it—only then is it possible that this association will have some powerful impact on the listener. The physical effect of sound on the body is culturally defined through a reference to the material and sensory dimensions of popular culture. Sound thus becomes a materiality through which the visceral and affective potentials of popular culture can be shaped.

In these examples, the speaker repeatedly refers to findings that were proven by empirical research and presents them as unfalsifiable; but is that not a contradiction in itself? As Jonathan Sterne noted: "Claims about the transhistorical and transcultural character of the senses often derive their support from culturally and historically specific evidence—limited evidence at that" (Sterne 2003, 18). At first glance, this paradox is quite pervasive in the field of popular culture in regard to sound theory and it hints at popular culture's vitalist and decidedly presentist approach, even if such a vitalist und presentist potential is itself constructed. Sound, in particular (and the ambitious theoretical reflections upon it), is indeed a genuine part of popular culture in the twenty-first century. Both the claim that there is a dire need for a more subtle and refined consideration of sound and the call for more research into the domain of individual listening habits and specific listening environments are widely thought to be statements of high distinction denoting a refined life experience. The consultant therefore ends his talk with the following mesmerizing mission statement for a heightened sonic awareness in everyday life: "If you're listening consciously, you can take control of the sound around you. It's good for your health. It's good for your productivity" (Treasure 2009, 5:20). Such an improvement of the Subject through a shift toward sound resonates strongly with audiopietism and the theology of sound mentioned above. However, along with the consultant's audiopietism comes a particular specificity: It rejects any metaphysical qualities or transcendent utopias. Thus, becoming a "better" Subject through a greater sensitivity for sound signifies something very concrete in the case of sound consultancy and Californian Ideology, namely, becoming a more successful, free, and creative entrepreneur.

Although that presentist approach leads to an energetic, focused, and quite convincing argument, right here, right now, the downside is that it brings a subjective fallacy into play if it is falsely accepted as an example of valid research: this narrow research field's rather limited evidence and the individual lifetime experience of one sole speaker are extrapolated to be of general relevance to, and even insight into, every possible anthropological, historical, or cultural situation—now and forever. An extremely strange assumption. However, the consultant's claims about the "general" effects of sound should not be misunderstood as academic knowledge and criticized as such, but rather, first and foremost, they should be treated as consultant knowledge. And as that form of knowledge, the consultant's sound claims serve a strategic function. Their aim is to create a convincing and vital presence and performance in the TED Talk format, thereby putting the topic "sound" on the agenda for a public that is presumed to understand that "scientific knowledge" is synonymous with "objective" knowledge. Although the consultant does not specifically discuss culture in the presentation, culture is in fact implicitly present throughout his entire talk. In other words: even if he does not address popular culture, the presentation *itself* constitutes a very characteristic part of contemporary popular culture, one which can be associated with strategies of "creative" self-improvement and self-marketing. This consultant performance has proven as a useful starting point for an analysis of particular aspects of popular culture and to outline this volume's epistemological interests. However, the discourse it offers is itself not really helpful for the development of a challenging concept of popular culture.

The TED Talk is not the only platform where we observe sound being incorporated into the Californian Ideology, into practices and business models, and into the Subject's strategies for improvement that correspond to those values. Moreover, its inclusion became particularly obvious when the radical transformation of the media industry through the tech industry's file-sharing platforms in the late 1990s was regarded as a model for a more general industry change, covering many top-selling industrial sectors. The film industry is one obvious example, but there is also the electric power industry—both of which have been affected by the so-called sharing-economy first practiced in the music sector. Furthermore, this shift towards an incorporation of sound can be deduced from the increased interest of leading tech companies' in audio companies—such as Apple's acquisition of Beats Electronics and Beats Music in May 2014, or Eric Schmidt's extensive visit to the headquarters of Native Instruments' Berlin in October 2014. Sound is therefore present and productive in the deterritorialized Californian Ideology. It does not make a significant difference if sound's promises of salvation are true or false. The only thing that matters in this context is whether the articulation of sound and salvation promises functionality and produces real practices and worlds.

Currently, audio streaming services are exploring and capitalizing on the typical sound consultant's plea for an improvement of the (listening) Subject through the control of the sounds in her or his environment.[3] If we listen in this way carefully enough, we might capture another auditory phenomenon, a phenomenon that has been formed by a combination of digitalization and *ubiquitous listening* (Kassabian 2013): the constant playing of music using various streaming services with differing but converging business models and pricing structures. As people navigate their own individual way through the day, these services accompany them, to an even greater extent than previous tools for ubiquitous, mobile, and personal stereos: they are easier to carry around on your daily routes than a *KLH model 11 portable record player* (introduced in 1962), a *Sony Walkman* (introduced in 1979), a *Sony Discman* (introduced in 1984), or an *Apple iPod* (introduced in 2001). With its presumed immaterialization, the commodification of ubiquitous listening rose to an unforeseen level. And this immaterialization is a delusion: it is nothing but the clever obfuscation of its highly refined and in part irresponsibly inexpensive material carriers via all of the cables, interfaces, software suites, server farms, mastering and postproduction companies, and all of the satellites and the space launch programs behind them. This rapid development of a globalized commodification of ubiquitous sonic and musical experience is currently one of the core drivers of global sound culture and entrepreneurial culture.

These days, the distribution, sharing, and consumption of sound in its manifold guises— (co-)organized by numerous audio platforms (such as SoundCloud), streaming services (such as Spotify), online music stores (such as Beatport, iTunes, and Amazon), but also by the vast amount of other hosting, streaming, and sharing services that are considered illegal according to most current legislation and jurisdiction—is a core part of what we call popular culture. Historically speaking, it is quite astonishing how quickly quasi-universal access to such a variety of sound productions has been made possible through the rapid expansion of worldwide digital data networks and their related business models. Ubiquitous music has thus become a major paradigm of twenty-first century popular culture: turning that music into your life soundtrack is presumably one of the fundamental selling points for any contemporary form of music consumption. Apparently, we seem to require this soundtrack for our everyday lives. It drives our turntables, it fuels us; it is our most effective click track for performance, desire, digestion, and rest. Businesses such as SoundCloud operate, capitalize, and monetize on popular culture's need for it. Their business model focuses mainly on the storage of sound data, providing online software interfaces for sound transmission and reproduction. In addition, they are involved in big data harvesting, data which is then sold to numerous clients in the industry and in the state administration, as well as—we must assume nowadays, unfortunately—a tacit data trade to globalized, networked secret intelligence services, such as the "Five Eyes." SoundCloud's intention is to provide a software environment that fosters, animates, and encourages the consumption and exchange of sound productions. This might sound like mere marketing lingo, but it does indeed quite accurately describe the performative activity that a company like SoundCloud simultaneously inspires and exploits. The principle characteristics of the sound productions in popular culture also play into this: the relative brevity of the pop song, the established commodified form of sounds, the public media personas related to its products, and the easily accessible options for data transfer and cultural exchange provided by a pop soundtrack make it the perfect currency in current popular culture. The undisturbed circulation of capital and profits are only lightly disguised in these products, and a *pop capitalism* profits best from globalization. In the case of SoundCloud, in the beginning it actually combined Stockholm's innovation-friendly research and development environment with the recruitment-friendly, highly skilled, and comparably low-salary market of freelance artists, designers, programmers, and information architects working in Berlin. Finally, they engaged attorneys in London, making good use of the strong, capital-friendly business laws there. That meant that all of the different aspects of the company's initial development—its foundation, employee recruitment, and establishing of legal offices—were scattered all over this continent, a quite common business practice in globalized capitalism: the actual market forces operate beyond the laws of any one nation—the consumers and employees, however, are still forced to live in accordance with those particular national laws for the most part, with that specific local lifestyle, costs of living, and taxation level. The everyday culture of digital pop is a culture that involves playing cleverly between the various legislative systems, workday time zones, and urban as well as other social environments. For the companies running contemporary popular culture, nations and specific cultural traditions are primarily cultural artifacts to be capitalized on; it is as if, for them, they are simply hindrances to their business that only exist on paper. And their consumers—meaning all of us, including you—enjoy this immensely. Companies such as SoundCloud, but also Apple and Amazon, are using pop sound productions as their most attractive elements, teasing us to turn them into indispensable sources of joy, inspiration, and energy in our lives. We love paying for them—with money, but with our spare time and valuable person-based data as well.

While the kind of consultant sound knowledge that prevailed in the TED Talk discussed above is also at play here, what is more effective in this case is the knowledge of the users, listeners, producers, social networking aficionados, and the notorious *prosumer* (Toffler 1980), although the concept of the "productive" consumer has recently turned from a desired utopia into a quite often worn-out reality. At the core of early twenty-first-century popular culture is a mutual interpenetration of digital business models, employee culture, and sound production and performances. The knowledge of how to listen in culturally specific ways and how to perform and produce sounds in effective and capitalizable ways is what drives these businesses and their clients—us. The "Empire" of which Hardt and Negri (2001) spoke is actually a sonic one: a sonic empire. The sensory culture of listening techniques and of sound practices as well as the empire's highly refined material artifacts, its interventions, architectures, and mannerisms have yet to be described at the intersection of popular music studies, sound studies, and cultural studies.

Studying Popular Culture through Sound

In all their diversity, the chapters of this volume promote sound as a subject through which popular culture can be analyzed in an innovative way. In that respect, this book is *about* the manifold sounds of popular culture. From everyday sounds, such as the sound images of radio stations (Schopp's "Records on the Radio" and Föllmer's "From Stationality to Radio Aesthetics") to the mediated screams of a baby monitor (Mihm's "Baby Monitor"), from the sounds of computer games (Grimshaw's "Computer Game Sound") to audio lectures (Nardi's "Critical Listening"), from the noises and sounds of special mass events in places such as soccer stadiums (Bonz's "Soccer Stadium as Soundscape") and dance clubs (Butler's "Listening Orientation"), to rather idiosyncratic vernacular practices such as playing records backward to search for hidden messages (Smith's "Over-Hearing")—sound and the discourses, technologies and practices with which it is correlated are part and parcel of popular culture's many domains.

Even if the sounds of popular *culture* include the sounds of popular *music*, the former cannot be reduced to the latter. Sound can be aestheticized, but does not have to be aestheticized as music to become a subject for sound as popular culture. That is, this volume is linked to popular music studies to the extent that these studies deal primarily with the *sound* of popular music, and not first and foremost with, say, cultural contexts, visual worlds, lyrical content, or the music industry's institutions (Diederichsen's "Sound/Music"; Fabbri's "Concepts of Fidelity" and "Syd's Theme"; Seay's "Sonic Signatures"; and Zagorski-Thomas's "Sonic Cartoons").[4] However, in contrast to popular music studies, the contributions in this volume also analyze sounds beyond music, between contingency and design. That statement links this volume, once again, to the academic field of sound studies as it has developed since the late 1990s, with an interest in sounds as culturally and historically specific phenomena that correlate with general culture theoretical questions how specific concepts of sound and listening are involved in the constitution of subjectivity, knowledge, or modernity (Burkhalter's "Sound Studies across Continents"; Schulze's "Sonic Epistemology"; Maier's "Sonic Modernities"; García Quiñones's "On the Modern Listener"; Burkhalter's "World Music 2.0"; and Bonz's "Distorted Voices").[5] Even when the sound of popular culture resonates and is reflected in the monographs, anthologies, and articles from the domain of sound studies, there are two prevailing trends: the study of sound in the more "exclusive" expert spheres, such as labs, test sites, studios, and clinics, or in the "everyday" sphere. These two spheres do indeed overlap with popular culture, but popular culture is not reducible to those spheres. Moreover, and we can identify this as a third trend, prominent publications in the sound studies field deal with sound in the context of avant-gardist sound art, which is usually conceptualized in strict differentiation to the popular. Therefore, the conceptualization of sound as popular culture still identifies a research gap in the academic field of sound studies. Contrary to approaches from the perspective of the history of science, the study of sound as popular culture is not restricted to an examination of expert environments, forms of training, or knowledge; it does not seek to follow the trend toward scientification and refinement, searching for a means to further elevate some particular discourse or research area. Unlike approaches that follow the category of sound as an artistic entity, an examination of sound as popular culture is not limited to aestheticized and specifically theorized forms of sound production and their simultaneously transformed listening practices and aesthetic discourses. And in contrast to established and often monolithic approaches in absolute favor of analyzing everyday practices, an analysis of sound as popular culture respects the impacts of non-everyday contexts on sound—and their effects on social differences (Nardi's "Sound and Racial Politics"; Erlmann's "Invention of the Listener"; Diederichsen's "Existential Orientation").[6]

While popular culture is still an underrated object of study in sound studies, sound in turn remains inadequately examined in the field of cultural studies that examines popular culture (based on its inclusive and broad concept of culture).[7] Surprisingly, some of the anthologies and monographs coming out of this academic field are still exploring popular culture as a comparatively silent culture: "Popular culture is chiefly marked by four characteristics: visualization, commodification, entertainment and technology" (Betts and Bly 2013, 4). Thus, in regard to such publications, there is clearly an urgent need to tune or to "retune" popular culture. This tuning of popular culture is the optimization of popular culture's epistemological potential. We, as researchers, are attuning ourselves to popular culture through an analysis of its sounds. As opposed to the famous intention to holistically, eschatologically explore the tuning of the world (Schafer 1977), this attunement, with its research subjects and its research methods, makes a great effort to remain quite purposefully in the realm of popular culture. In that sense, in this volume cultural studies is tuned to a level at which sound studies and popular music studies can be recognized as part of the same discourse. In the process, the cultural practices concerning sound should no longer be seen as mere semiotic or signifying processes that refer to some outerworldly model of reference to super-signs, but as thoroughly material, physical, perceptual, and sensory processes that integrate a multitude of cultural traditions and forms of knowledge, for example, musical or technological, and which form an actual new discourse. This volume is guided by this approach in order to focus on the historical and cultural specificity of sound by referring to specific practices, technologies, and discourses (Maier's "Sound Practice"; Bijsterveld's "Ethnography and Archival Research"; Mrozek's "Historicization in Pop Culture"; Birdsall's "Sound and Media Studies"; Kassabian's "Listening and Digital Technologies"; Théberge's "Listening as Gesture and Movement"; Schulze's "Corporeal Listening"). Moreover, by using sound as a subject to analyze popular culture, this book is not only *about* sound. In fact, it goes one step further—it is a study *through* sound. The chapters of this book therefore provide an analysis of crucial aspects of popular culture, its echoes and recordings from the past, its resonant presents but also—by listening to the contradictions, breaks, and discontinuities—its possible futures.

The basic goal of this book is the articulation of popular culture through the study of sound. Through an examination of the noises of a baby monitor, it has become possible to analyze changes in domestic space beginning in the 1960s, including a new organization of the parent–child relationship. Looking at sub-bass frequencies as they were produced by the sound systems of 1970s disco culture allows us to explore crucial aspects of a history of listening that is related to a history of nonlistening (Papenburg's "Enhanced Bass"). Through a study of loudness in popular music, we can identify listening practices that are designated as "legitimate" or as "illegitimate" (Binas-Preisendörfer's "Loudness Cultures"). Observations of video game sound design and location-aware technologies facilitate the study of a militarization of the senses and the ambiguity of sensory distance (Bull's "Technological Sensory Training"). An analysis of the sound of popular music productions allows for an elaboration of a conceptualization of recorded sound (Wicke's "The Sonic"; Großmann's "Phonographic Work"; Papenburg's "(Re-)Mastering Sonic Media History"). And, as shown above, the Californian Ideology can be investigated through a study of the sounds of consultants and clouds.

The French economist, author, ex-bank director, and former presidential adviser Jacques Attali developed a similar methodological approach to sound. Even if Attali's historiography and lofty adherence to a romantic notion of the autonomous Subject might appear to some rather dubious and outdated today, his concept of music, and the methodological implications provided the necessary momentum for the development of *sound as popular culture* as a method. For Attali, music "is a way of perceiving the world. A tool of understanding" (1985, 4). Thus, Attali's book is not about music "itself," not about music as an "autonomous" entity, and moreover, not about music's "Other"—noise. Instead, it is about attaining an understanding, an awareness of the power relations in culture and society through music and about the changing boundaries between music and noise.[8]

Using sound as an epistemological "tool" in this way leads to at least two problems or epistemological risks: the problem of either only functionalizing and *instrumentalizing* sound or only of *metaphorizing* sound. On the one hand, an epistemological functionalization of sound can be very productive for cultural analysis. Using sound as a tool enables us to move beyond widespread approaches to sound from any art or fan tradition.[9] In this approach, we do not refer—at least not primarily—to any specific sound because we think it is a great piece of art or because we are affected by it as fans. On the contrary, we refer to a particular sound because it has the potential to assist in the study of a certain aspect or relevant issue of popular culture. In other words, sounds that are not aestheticized as music can also be useful for dealing with these research questions. This point is reflected in this book's contributions on multiple levels. On a very basic quantitative level, it includes a diverse set of contributions that are quite explicit in not dealing with sound as music. On a more specific level, the contributions that do deal with sound as music use it as more of an indirect tool for the study of a particular question concerning popular culture. With that being said, on the other hand, the employment of sound as a mere tool for the study of the organization of popular culture and its power relations has an apparent blind spot: the fact that sound has a constitutive function for popular culture. Sound does not only reflect popular culture; it plays a part in the transformations of popular culture. It is therefore necessary to transform Attali's *tool* of understanding into an integral approach for the study of popular culture as a whole.

We can add here that in sound as popular culture the "as" does not merely articulate a metaphorical relationship, a purely hermeneutic interpretation of sound as something (popular culture) that is not sound in the literal sense. Rather than marking a metaphorical construction based on a similarity, the "as" here works as a metonymy based on contiguity. Thus, the contributions to this book analyze sound as a crucial part of popular culture, and not as a metaphorical "as if" interpretation. Or, to put it differently: popular culture is not silent, but (re)sounding. In contrast to the methods discussed above, this book studies sound as both an integral and constitutive part of popular culture. Therefore, the "as" in *sound as popular culture* contributes more to an integral understanding than to a functional one: how the two concepts, sound and popular culture, are interrelated, interwoven, and how they reciprocally represent and reference each other. In sound as popular culture, the preposition "as" does not mark a definite and systematic functionalizing of the one for the other: If that were the case, the presumably broader concept—popular culture—would overshadow the narrower—sound. In similar situations, the presumably narrower concept never even really appears in that kind of research concept; it is often seen as no more than a used and abused concept, a rather minor distinction. In this volume, however, it is the symmetry between the two concepts that finally makes it possible to explore their actual and specific interrelations. The choice of using "as" instead of "and" or "in" situates this book within a specific methodological discourse. The usage of "as" marks a strict distinction from methodologies that reproduce a dualistic subject–object schema, leading to sound being analyzed as sound "itself," thereby decontextualizing it as such, or to it being examined as an object or text that exists in a dualistic relationship to a subject or a context. We deploy the preposition "as" as a kind of exoteric code word that permits access to an epistemological world that differs from the epistemologies of conjunctions such as "and" or "in."[10] The conjunction "and" tends to reproduce an epistemology in which "sound" as well as "popular culture" are two entities or objects that exist primarily independently from one another and interact with each other secondarily, selectively—rarely, in fact. While "and" tends therefore to reproduce epistemologies that are based on this subject–object schema, the other option "in" tends to reproduce epistemologies that are based on a context–text schema. The epistemological consequences of the usage of "as" are thus quite radical; they do not only imply a collapse of dichotomies of musical texts and contexts; they imply that music theory (in a general sense) *is* cultural theory—and vice versa.[11]

Besides this problematic subject-object dualism, the epistemological value of the study of sound and music as "text" is itself limited. Wicke (2003) points out that the transfer of the concept "text," which came to music analysis from linguistics and literary studies, has the quite unspecific and often misleading function of somehow associating music with "meaning." The concept of text in literary studies is—according to Wicke—a "category which is bound to writing as a linearly constructed code," which would also imply the fundamental metaphor of "reading," which is actually external to many musical and sonic practices. In his anthropology of the senses, David Howes explores precisely this difference in critical differentiation to "language games. Culture as discourse. World as text" (2004, 1). Instead of "reading" culture, Howes argues for the "sensing of cultures"; instead of cultural texts, he analyzes "empires of the senses." Along similar lines, the authors represented in this volume are not analyzing sound as text—instead, they are analyzing sound through all its performative actualizations as a complex, layered, and convoluted sensory and cultural artifact. The historical, contemporary, and future cultures of the popular appear to be accessible through sound and to function through it as well. Sound is not incidental to popular culture: it is fundamental to it.

Notes

1. TED started as a conference forum as early as 1984. Since 1990, there has been an annual TED conference, which took place in California until 2013. Since they were made available online in 2006, the TED Talks have gained significant momentum. In 2001, the journalist and publisher Chris Anderson became the TED Talks curator.

2. See the growing body of manuals on topics such as sound branding, sound design as a part of product design, or—though a bit outdated—ringtones (e.g., Jackson 2003; Schifferstein and Hekkert 2008; Bronner, Hirt, and Ringe 2014).

3. On the cultural analysis of the transformation of mediated listening spaces and mobile forms of music, see also Born 2013; Gopinath and Stanyek 2014.

4. The study of the exploration and economization of the "sound" of popular music as "hit sound" began in the 1970s and 1980s (see Kealy 1979 and Théberge 1989). The "sound" of music is still a crucial and productive subject in the research on the specific status of "produced music," which has been an active research field for a few years now (see Frith and Zagorski-Thomas 2012) and which studies music as situated "in terms of its sensory materiality" (Johnson and Cloonan 2009, 13). Even if the analysis of the "sound" of music is not solely a musicological endeavor (see

Gracyk 1996), it provides the possibility of specific musicological access to popular music studies. Therefore, Shepherd and Wicke identified a specific musicological contribution to the "essentially interdisciplinary undertaking" of popular music studies stemming from the analysis of sound in popular music (2003, 94). Musicological approaches in this research field analyzing "the specific character of sounds recognized as musical within popular music" (2003, 94); see also Garcia Quiñones's ("Sound Studies versus [Popular] Music Studies") as well as Großmann and Hanáček's ("Sound as Musical Material") contributions in this vol. to the well-established and still extremely productive discourse on sound in musicology.

5. For a delineation of this research field, see the pertinent volumes by Bull and Back (2003), Pinch and Bijsterveld (2012), Sterne (2012), and Novak and Sakakeeny (2015).

6. As may become clear, we prefer the concept of popular culture vis-à-vis other adjoining concepts, such as vernacular culture or everyday life culture. There are certainly smooth transitions between these concepts, but each concept establishes a specific focus. While vernacular culture remains a historically broad concept that is not restricted to modernity, we can conceive of popular culture as being inherently connected to modernity. The same is also true for everyday life, which has become a prominent object in academic research (see Lefebvre [1947] 1991; de Certeau [1980] 1988). The prominence of the category everyday life in current sound research is astonishing (see, e.g., DeNora 2000; Bull 2007; Sloboda 2010; Herbert 2011). The category has nearly developed a cultural critical potential in the present, when even the most marginal phenomenon is presented and marketed as an "event." By all means, the study of daily or ordinary routines and their individual appropriations is definitely necessary; nevertheless, popular culture also encompasses distinctly more exceptional spheres, such as recording studios and clubs.

7. In the United States, the study of popular culture gained institutional momentum within academia through the Popular Culture Association (founded in 1969), and the *Journal of Popular Culture* (first issue in 1967), which put the study of subjects such as detective stories, comics, movies, popular music, and TV programs on its agenda (see Browne 2002). For the discussion of popular culture in popular music studies that is critically based in British cultural studies see Wicke 1992 and Kassabian 1999. For an overview of the multiple meanings of the concept "popular culture," see also Hecken's chapter in this vol.

8. We can therefore read Attali's method as an early example of an acoustic epistemology. The knowing of the world through sound became a prominent method for sound studies relying on concepts such as "acoustemology" (Feld 1996). While Steven Feld based the concept of acoustemology—which he defines as "local conditions of acoustic sensation, knowledge, and imagination" (1996, 91) or as "sounding as a condition of and for knowing" (97)—on ethnographic research he conducted outside of Western modernity (in the "exotic" rainforests of Papua New Guinea), the concept has been picked up and developed to analyze sonic modernity as well, most prominently through anthropological sound research (Porcello 2005, 270; Rice 2012; Born 2013), and also beyond its focus on the Western world (see Eisenberg 2013). We can add here that Attali's acoustic epistemology is—contrary to Feld's—not interested in the description of a coherent and consistent "world" but is instead focused on the tensions, contradictions, and power relations. In this regard, it is close to Michel Foucault's analysis of power, and thus includes

categories such as the "Other" or the "Outside." This becomes particularly evident in Attali's analysis of music, in which he positions it in relation to its Other, that is, noise.

9. These traditions are also still very much alive in popular music studies. While the art tradition is interested in proving that selected forms of popular music are "art" (whatever that is exactly), the fan tradition is interested in raising the reputation of the worshiped hero. While the art tradition has lost momentum over recent years, the fan tradition has gained momentum. Both approaches have remained relatively unproductive at the level of cultural theory.

10. Early implementations of this usage of the preposition "as" can be found in ethnomusicology, in relation to what Richard Middleton (2003) called the "cultural turn in ethnomusicology." In the early 1970s, ethnomusicologist Alan Merriam (1977, 204) defined ethnomusicology as "the study of music as culture." Based on that definition, Bruno Nettl (2005, 217–228) distinguished three approaches to, or perspectives on, (ethno)musicological research: (1) "music in its cultural context" (2), "music in culture," and (3) "music *as* culture." While Nettl argued that the first two approaches presuppose a specific concept of music that exists independently of culture, he sees the third approach as one in which music is an integral part of culture: "The study of music as culture would require the integration of music and its concepts and its attendant behaviour and indeed, all musical life, into this kind of a model of culture" (ibid., 217).

11. Especially in the case of musicological research that is grounded more in the fields of cultural studies and anthropology than in a tradition of "art" scholarship, numerous examples for the productivity of the "sound as" or "music as" approach can be found. Robert Fink (2005) studied disco and "minimal music as cultural practice," John Shepherd—still relying on the problematic text analogy—analyzed music "as social" (Shepherd 1991) and "as cultural text" (Shepherd 1992). The study of sound and music not as sound and music in and of itself but as something else— that is, as popular culture in the case of this book—bears some similarities to what Georgina Born (2010) called a "relational musicology" and what Richard Middleton (2003) defined as a "cultural study of music." Whereas Fink used minimal music and disco to analyze an "excess of repetition" in "postindustrial, massmediated consumer society" (Fink 2005, x), Anahid Kassabian (2013) used "muzak" to analyze the formation of the concept of "ubiquity," which later became prominent in concepts such as "ubiquitous computing."

References

Bibliography

Attali, Jacques. 1985. *Noise: The Political Economy of Music*. Trans. Brian Massumi. Minneapolis: University of Minnesota Press.

Barbrook, Richard, and Andy Cameron. 1996. The Californian Ideology. *Science as Culture* 6 (1): 44–72.

Betts, Raymond F., and Lyz Bly. 2013. *A History of Popular Culture: More of Everything, Faster and Brighter*. London: Routledge.

Born, Georgina. 2010. For a relational musicology: Music and interdisciplinarity: Beyond the practice turn. *Journal of the Royal Musical Association* 135 (2): 205–243.

Born, Georgina. 2013. *Music, Sound, and Space: Transformations of Public and Private Experience*. Cambridge: Cambridge University Press.

Bronner, Kai, Rainer Hirt, and Cornelius Ringe. 2014. *Audio Branding Yearbook 2013/2014*. Baden-Baden: Nomos.

Browne, Ray B. 2002. *Mission Underway: The History of the Popular Culture Association/American Culture Association and the Popular Culture Movement, 1967–2001*. Madison, WI: Popular Press.

Bull, Michael. 2007. *Sound Moves: iPod Culture and Urban Experience*. London: Routledge.

Bull, Michael, and Les Back, eds. 2003. *The Auditory Culture Reader*. Oxford: Berg.

de Certeau, Michel. (1980) 1988. *The Practice of Everyday Life*. Trans. Steven Rendall. Berkeley: University of California Press.

DeNora, Tia. 2000. *Music in Everyday Life*. Cambridge: Cambridge University Press.

Eisenberg, Andrew J. 2013. Islam, sound, and space: Acoustemology and Muslim citizenship on the Kenyan coast. In *Music, Sound, and Space: Transformations of Public and Private Experience*, ed. Georgina Born, 186–202. Cambridge: Cambridge University Press.

Feld, Steven. 1996. Waterfalls of song: An acoustemology of place resounding in Bosavi, Papua New Guinea. In *Senses of Place*, ed. Steven Feld and Keith H. Basso, 91–136. Santa Fe, NM: School of American Research Press.

Fink, Robert. 2005. *Repeating Ourselves: American Minimal Music as Cultural Practice*. Berkeley: University of California Press.

Frith, Simon, and Simon Zagorski-Thomas, eds. 2012. *The Art of Record Production: An Introductory Reader for a New Academic Field*. Burlington, VT: Ashgate.

Gopinath, Sumanth, and Jason Stanyek, eds. 2014. *The Oxford Handbook of Mobile Music Studies*, vols. 1 and 2. New York: Oxford University Press.

Gracyk, Theodore. 1996. *Rhythm and Noise: An Aesthetics of Rock*. Durham, NC: Duke University Press.

Hardt, Michael, and Antonio Negri. 2001. *Empire*. Cambridge, MA: Harvard University Press.

Herbert, Ruth. 2011. *Everyday Music Listening: Absorption, Dissociation, and Trancing*. Burlington, VT: Ashgate.

Hosokawa, Shuhei. 1984. The Walkman effect. *Popular Music* 4:165–180.

Howes, David, ed. 2004. *Empire of the Senses: The Sensual Culture Reader*. Oxford: Berg.

Jackson, Daniel. 2003. *Sonic Branding: An Introduction*. New York: Palgrave Macmillan.

Johnson, Bruce, and Martin Cloonan. 2008. *Dark Side of the Tune: Popular Music and Violence*. Burlington, VT: Ashgate.

Kassabian, Anahid. 1999. Popular. In *Key Terms in Popular Music and Culture*, ed. Bruce Horner and Thomas Swiss, 113–123. Malden, MA: Blackwell.

Kassabian, Anahid. 2013. *Ubiquitous Listening: Affect, Attention, and Distributed Subjectivity*. Berkeley, CA: University of California Press.

Kealy, Edward. 1979. From craft to art: The case of sound mixers and popular music. *Work and Occupations* 1:3–29.

Lefebvre, Henri. (1947) 1991. *The Critique of Everyday Life*, vol. 1. Trans. John Moore. London: Verso.

Merriam, Alan P. 1977. Definitions of "comparative musicology" and "ethnomusicology": An historical-theoretical perspective. *Ethnomusicology: Journal of the Society for Ethnomusicology* 21 (2): 189–204.

Middleton, Richard. 2003. Introduction: Music studies and the idea of culture. In *The Cultural Study of Music: A Critical Introduction*, ed. Martin Clayton, Trevor Herbert, and Richard Middleton, 1–15. London: Routledge.

Nettl, Bruno. 2005. *The Study of Ethnomusicology: Thirty-One Issues and Concepts*. Urbana: University of Illinois Press.

Novak, David, and Matt Sakakeeny. 2015. *Keywords in Sound*. Durham, NC: Duke University Press.

Pias, Claus, and Sebastian Vehlken. 2010. Einleitung: Von der "Klein-Hypothese" zur Beratung der Gesellschaft. In *Think Tanks: Der Beratung der Gesellschaft*, ed. Thomas Brandstetter, Claus Pias, and Sebstian Vehlken, 7–15. Zurich: Diaphanes.

Pinch, Trevor, and Karin Bijsterveld, eds. 2012. *The Oxford Handbook of Sound Studies*. New York: Oxford University Press.

Porcello, Thomas. 2005. Afterword. In *Wired for Sound: Engineering and Technologies in Sonic Cultures*, ed. Paul D. Greene and Thomas Porcello, 268–281. Middletown, CT: Wesleyan University Press.

Rice, Tom. 2012. Sounding bodies: Medical students and the acquisition of stethoscopic perspectives. In *The Oxford Handbook of Sound Studies*, ed. Trevor Pinch and Karin Bijsterveld, 298–319. New York: Oxford University Press.

Schafer, R. Murray. 1977. *The Tuning of the World*. New York: Knopf.

Schifferstein, Hendrik N. J., and Paul Hekkert. 2008. *Product Experience*. San Diego, CA: Elsevier.

Schulze, Holger. 2007. Die Audiopietisten: Eine Polemik. *Kultur und Gespenster* 2 (3): 122–129.

Shepherd, John. 1991. *Music as Social Text*. Cambridge: Polity Press.

Shepherd, John. 1992. Music as cultural text. In *Companion to Contemporary Musical Thought*, vol. 1, ed. John Payntor, Tim Howell, Richard Orton, and Peter Seymour, 128–155. London: Routledge.

Shepherd, John, and Peter Wicke. 2003. Musicology. In *Continuum Encyclopedia of Popular Music of the World*, vol. 1: *Media, Industry, and Society*, ed. John Shepherd, David Horn, Dave Laing, Paul Oliver, and Peter Wicke, 90–94. London: Continuum.

Sloboda, John A. 2010. Music in everyday life: The role of emotions. In *Handbook of Music and Emotion: Theory, Research, Applications*, ed. Patrik N. Juslin and John A. Sloboda, 493–514. Oxford: Oxford University Press.

Sterne, Jonathan. 2003. *The Audible Past: Cultural Origins of Sound Reproduction.* Durham, NC: Duke University Press.

Sterne, Jonathan. 2011. The theology of sound: A critique of orality. *Canadian Journal of Communication* 36 (2): 207–225.

Sterne, Jonathan. 2012. *The Sound Studies Reader.* Abingdon: Routledge.

Théberge, Paul. 1989. The "sound" of music: Technology, rationalization, and the production of popular music. *New Formations* 8 (2): 99–111.

Toffler, Alvin. 1980. *The Third Wave: The Classic Study of Tomorrow.* London: Collins.

Wicke, Peter. 1992. "Populäre Musik" als theoretisches Konzept. *PopScriptum* 1 (1): 1–42.

Wicke, Peter. 2003. Popmusik in der Analyse. *Acta Musicologica* 75 (1): 107–126.

Sound and Media

Treasure, Julian. 2009. The 4 ways sound affects us. TED Global. https://www.youtube.com/watch?v=rRepnhXq33s (accessed January 19, 2015).

I   Outlining a Non-Discipline: The Theory of Sound as Popular Culture

Conceptualizing Sound

1   The Sonic: Sound Concepts of Popular Culture

Peter Wicke

Sound is one of the primary media of popular culture—and one that reaches far beyond music. Whether it is the soundscape of the stadium, the sound world of computer games, the sound image of YouTube videos, the auditory aspects embedded within advertisements, the sound design of feature films, or the functional sounds that penetrate an all-around technologized day-to-day—hardly any typical manifestations of popular culture are not accompanied by characteristic sounds, or conveyed through a sound world of their own. As a symbolic medium par excellence that already contains the reference to its source (a sound is always the sound of something), sound constitutes a foundation of the human relationship to the world, and the hearing appropriation of sound is a central form of self-assurance for the Subject. Thus, the production and adoption of sounds is one of the foundations of human activity that can be found as a constitutive practice in every culture. However, within popular culture contexts, the medium of sound occupies a position of particular importance, as it has distanced itself from all acoustic realities because of its technical design, and has become the object of an extremely real form of auditive imagination.

Nowhere is that made clearer than in the case of pop music production. This departure from acoustic principles began as early as the 1950s, with the introduction of Sam Phillips's "slapback" echo, which deviated (rather unintentionally at the time) from the laws governing acoustic reverberation. By means of two tape decks, Phillips generated the time-offset echo effect using the delay time between the recording and playback head. This effect lacked the typical decay of room reverberation and would become one of the sound trademarks of Phillips's label, Sun Records. Although it is physically impossible to have an echo without the characteristic decay, it does result in a powerful sound presence, which gave Elvis Presley's early recordings, for example, their overwhelming character.[1]

What was actually only supposed to compensate for the unfavorable acoustics of Sam Phillips's studio recording space—a transformed former radiator repair shop on Memphis' Union Street—would prove in retrospect to have been an important step toward technologically synthesized sound images, as they have become characteristic for today's popular culture. Bruce Swedien, a five-time Grammy winner (and not just one of the most profiled, but also one of the longest-working present-day sound engineers), has set an undisputed standard with his sweeping, long, and impressive list of recordings, ranging from Count Basie to Mick Jagger to Paul McCartney to Michael Jackson. He has the following to say on the matter:

> When I discovered through my own experimentation that I could successfully record sonic images that previously only existed in my imagination, recording music became extremely exciting and inspiring for me. ... I include my own brand of sonic fantasy in my recordings and mixes and, at the same time, keep whatever sonic reality I think is necessary. (Swedien 2013, 12)

When sound came into contact with audio technology at the end of the nineteenth century, it was technologically imprinted so that it gradually became emancipated from the, so to speak, "natural" attachment to the mode of sound generation (striking, bowing, blowing, etc.), the particularities of sound generators (construction, resonance relations, etc.), and the conditions set by the spatial relations of its dispersal (reverberation). At the beginning of the recording process (especially during the era of mechanical sound recording), that imprinting came up against its own technological limitations. It was initially experienced as an unavoidable "unnaturalness" of the technically mediated sound. So it was at first in the case of the introduction of electrical sound recording and the associated transformation of the sound event into the particular state of matter of the electrical signal in film and radio drama production (Bayley 2009; Chion 1994; Flückiger 2006; Wodianka 2013) and soon after in music production (Schmidt-Horning 2013), that sound design became accessible as its own level of design independent of the conditions of sound generation.

Thus, alongside the organization of sounds in time and space, an additional level of design had emerged—that of the organization of sound in its physical materiality. Although this development acted on the fiction of "unnaturalness" at the start based on the dictum of "high fidelity," by confronting it with the most acoustic realism possible, sound design quickly broke away from the imitation of live sound after the introduction of the magnetic recording tape. Sam Phillips's high-energy sound spaces, the innovative overdubbing experiments of the swing guitarist Les Paul, and Phil Spector's "Wall of Sound" are early evidence of these kinds of imagined sound images that have become standard in pop music.

The Historicity of Sound

Sound's material carrier is acoustic matter, and sound is thus bound to corresponding laws. In a certain way, sound, in contrast to noise, is structured acoustic matter, whose structure manifests itself in a way subject to its generation as a periodical vibration phenomenon in characteristic processes of attack and decay and a range of formative harmonic multiples (partial tones), which emerge as resonance vibrations over the generated fundamental vibration (fundamental tone).[2] From the point of view of physics, sound consists of complex, structured acoustic pressure relations, which have historically obtained prominent cultural status as a medium for making music.

Since the modes of sound generation have changed at a very slow pace over centuries, from body-borne sound (clapping the hands or body) to sound generation using the human voice to the mechanical sound machines that we know as musical instruments, one could have the impression that the relationship between sound and the auditive practices of its usage as music in particular are more or less static and fundamentally given, defined by the sound generators (musical instruments) that are available. However, this impression is a deceptive one, and fails to consider the reality that has resulted from at least a century of audio-technological developments. In the studio, there have been fundamental transformations regarding not only the making of music but also the material through which that is executed—the sonic materiality of the music. Alongside the technical interventions into the sonic materiality of music—changes to the frequency composition using equalizers, modifications of the formant structure by means of the Aural Exciter, or the purely technical synthesis of sounds—in a much more fundamental sense, what has been transformed here is the conceptualization of sound (sound design) that is at the foundation of the recorded music (Wicke 2009).

The way in which the physical parameters of acoustic matter are linked to cultural and aesthetic practices, the way in which acoustic matter is structured in order to function as sound within the framework of a cultural practice such as that of music making, is at the basis of historical transformation. The function of acoustic material as sound is neither given nor fixed. The acoustic material in its physicality as measurable sound pressure does not function in the music simply as sound; it is instead drawn into a culturally defined reference system (tone systems, aesthetic paradigms, etc.) via music making, the process of intentional sound generation. Only then does it become aesthetically relevant and a component of that which is practiced and understood as "music" in different cultures, and in a completely different way in each of those cultures. That is the basis of the main distinction between acoustic matter as the physical-acoustic carrier of sound and sound as the material medium of music. This kind of differentiation between acoustic matter, as that which is physically hearable, and sound, as acoustic matter structured in a particular way, changes nothing about the objective character of the acoustic pressure relations exhibited by the sound. However, as a medium of cultural practices such as music, they are perceived with varying focuses, interpreted differently, and are relevant in different ways depending on the respective historical and cultural context. So, although it is possible to measure differences in the physical manifestations of a sound during a musical performance, for example, that sound is perceived as the same sound at different locations within the space—even though it clearly no longer exhibits the same acoustic pressure relations, or in other words, the same acoustic matter. In the reference system according to which acoustic events are interpreted as sound within the framework of a given culture, differences in acoustic matter appear irrelevant even if they are measurable; and others, such as the fundamental frequency as root, can appear extremely relevant even though this relevance has more to do with the frame of reference (tonal system) than with the degree of measurable difference between that and the existing partial tones.[3] If acoustic matter is generated in order to be heard, it loses its autonomy and ends up in a relationship of interdependency with the different contexts in which it is heard. The identity of a sound, and that of its iteration, is the result of a socially constructed auditive practice. It is thus not only music but certainly also sound that has a history which still needs to be written.[4]

Sound as Medium: The Sonic

All of the particular ways in which sound functions as a medium are based on cultural imprinting that must be received by the acoustic matter in order to be receptive to the ascriptions and inscriptions within the framework of a given culture. This is the level of its "cultural formatting," as it were—an analogy that is in no way purely metaphorical. So just as digital storage media need formatting in order to be recordable, acoustic matter must be culturally formatted in the form of the specific respective sound concepts, in order, for instance, to be able to record, or to "store," that special form of human interaction that we call music. It is here that operators, technologies of articulation, and their discourses, which are bound by their conceptual and processing parameters, come into play (see Shepherd and Wicke 1997, 117–124). The way in which sound is conceptualized, organized as music, and transformed into music is always connected to the principal modes of its generation, and thus to the leading technologies of sound generation. Clapping and vocal forms of sound production, which form the foundation of music making for almost all indigenous peoples, constitute a musical universe quite distinct from that of the mechanical sound machines that we know by the term "musical instruments" and which have made sound—in the form of discrete, individual events (tones) that are clearly distinguishable from one another—into the foundation of music making.

To conceptualize this level of the culturally and discursively constructed concept of sound—and therefore the level between the acoustic and the cultural—that is at the basis of a society's audio culture, we can look to a specific concept—the sonic. This term can be traced back to its Latin form, *sonus*, and was once widely employed in medieval literature on music.[5] Although it points to the materiality of the acoustic in its most general and original meaning, that is, as an appellation for that which is audible, it is nevertheless linked to the hearing Subject through this reference to hearing, and is thereby also connected to a cultural-historical dimension. In contrast to the concept of sound used by acoustics, the hearing Subject is included within the term "sonic." It is sound as not just acoustic matter structured in a particular way (in contrast to noise), but rather structured acoustic matter in relation to the respective conditions of relevance within the framework of a given culture. The sonic is then *culturalized acoustic matter*—or, in other words: the concept of sound that is linked to the respective modes of sound generation and its technology, as well as to the soundscapes (Schafer 1980) of a particular time and society. This is inscribed into the instruments of sound generation just as it is into the modes of music making—although this level goes far beyond music and makes it much more identifiable as an integral component of a society's audio culture (Großmann 2013).

The most accessible is the historicity of sound as a certain formation of the sonic (in contrast to the history of its specific forms of organization within the framework of cultural practices such as music), in the recording studios, even if that includes only the last century. Under the technological imperative, from the outset the transformation of acoustic matter as the material of hearability into sound, as the sensory medium of music making that is conditioned by the intervening technology, must first be painstakingly restored, since the culturally given context could no longer adjust itself by itself, in light of the limits of recording apparatuses. Based on the "high fidelity" principle, this restoration process became a reconstruction process, which eventually became a design process of its own kind as part of its further course of development—sound design, the creation and the concept of material into which the sound gestalts are inscribed.[6] In a talk entitled "The Studio as Compositional Tool" held in New York in 1979—and which ended up being published in print form in the magazine *Down Beat* in 1983—Brian Eno described how this development came to be:

> You're working directly with sound, and there's no transmission loss between you and the sound—you handle it. It puts the composer in the identical position of the painter—he's working directly with a material, working directly onto a substance, and he always retains the options to chop and change, to paint a bit out, add a piece, etc. (Eno 1983, 67)

Alongside the dimension of sound creation—or more specifically the musical dimension—music making had taken on an engineering/sound technology dimension. In the field of popular music, both became dependent on one another, owing to the dominance of sound carriers and the audiovisual mass media, so that an understanding of how the music developed is no longer possible without being able to look at both sides simultaneously. In the twentieth century, the history of music was written by musicians and audio technicians in equal measure. That should not only be understood as meaning that the songs first took on their identity through the studio recording process. On the contrary, they were conceived in the first place based on an awareness of the vastly enhanced possibilities that the medium of sound now had thanks to technical sound design—if they had not already resulted out of an interactive process with audio technology from the very beginning. Mike Stoller and Jerry Leiber, who were one of the most successful author and producer teams in the history of pop music based on their work for Elvis Presley, got right to the heart of it when they retrospectively commented: "We didn't write songs, we wrote records" (as cited in Pareles and Bashe 1983, 322).

Our concepts of sound—the particular format of the sonic that emerges out of the fundamental relationship between technology, sound, and the corresponding strategies of hearing—have not only inscribed themselves into music, but have developed into a signature of present-day auditive culture as a whole, which permeates all of their manifestations, and despite promising beginnings in relation to some of their individual aspects (see, e.g., Doyle 2005; Papenburg 2011; Sterne 2003), they are still awaiting systematic analysis.

Translated by Jessica Ring

Notes

1. Although Sam Phillips made the slapback echo into the signature sound of his label starting in 1954, the guitarist Les Paul (Les Paul and Mary Ford, "How High the Moon," Capitol Records [1951] 2001), as well as Bill Putnam (Little Walter, "Juke," Checker Records [1952] 1997), known as "the father of modern recording," had experimented with it at an earlier point in time.

2. The term "sound" (*Klang*) is not meant here in the narrower musical sense as a tone composed of precise harmonics (*Zusammenklang*) in contrast to (pure) "tone" (*Ton*), but rather in the fundamental meaning of acoustics, where it stands for acoustic matter with a particular structure—the very structure that mechanical sound generators, i.e., all common musical instruments, produce.

3. In Western music, the specific orientation toward that one particular fundamental note is culturally defined, based on a hierarchical system of tones constructed around root notes. This hierarchical concept leads to an analogous form of sound perception causing the fundamental to be heard even when it is not present (because it has been filtered out technologically, for example). In contrast, the Tuva people, who live in the border region between Siberia and Mongolia, have a so-called overtone music—a form of making music in which a person modulates the overtones by changing the resonance space in the mouth and throat while maintaining the same fundamental note. In that case the musical focus is on the overtones, the partials are treated as musically equal and are not linked to any root concept. This method became known in the West through the Siberian singer Sainkho Namtchylak.

4. One noteworthy exception in this regard is the study of the history of the sound of bells conducted by the historian Alain Corbin, who describes the process of the cultural formation of the sound of bells as an instrument for the construction of territorial identity in nineteenth-century

France. That was thus de facto a piece of scholarly work on the sonic and its history, even if Corbin did not use that terminology (see Corbin 1998).

5. In the course of its long history, the term *sonus* has been attributed a variety of meanings. However, its original and fundamental level of meaning designated the material of audibility, the as yet unformed musical tone, and so a level beyond the acoustic matter, but on this side of music. This usage is found in Boethius's *De institutione musica* (1500).

6. The construction of instruments also executes a particular design of the acoustic material in each case, a particular formation of the sonic with wide-ranging effects on music making. However, those effects take place over such long periods of time that they often go unnoticed.

References

Bibliography

Bayley, Amanda, ed. 2009. *Recorded Music: Performance, Culture and Technology*. Cambridge: Cambridge University Press.

Chion, Michel. 1994. *Audio-Vision: Sound on Screen*. Ed. and trans. Claudia Gorbman. New York: Columbia University Press.

Corbin, Alain. 1998. *Village Bells: Sound and Meaning in the 19th-Century French Countryside*. Trans. Martin Thom. New York: Cambridge University Press.

Doyle, Peter. 2005. *Echo and Reverb: Fabricating Space in Popular Music Recording, 1900–1960*. Middletown, CT: Wesleyan University Press.

Eno, Brian. 1983. Pro Session: The studio as compositional tool. *Down Beat* 50:65–67.

Flückiger, Barbara. 2006. *Sound Design: Die virtuelle Klangwelt des Films*. Zurich: Schüren.

Großmann, Rolf. 2013. Die Materialität des Klangs und die Medienpraxis der Musikkultur: Ein verspäteter Gegenstand der Musikwissenschaft? In *Auditive Medienkulturen: Techniken des Hörens und Praktiken der Klanggestaltung*, ed. Axel Volmar, and Jens Schröter, 61–78. Bielefeld: transcript.

Papenburg, Jens Gerrit. 2011. Hörgeräte: Technisierung der Wahrnehmung durch Rock- und Popmusik. PhD diss., Humboldt-Universität zu Berlin.

Pareles, Jon, and Patricia Romanowski Bashe, eds. 1983. *The Rolling Stone Encyclopedia of Rock & Roll*. New York: Rolling Stone Press/Summit Books.

Schafer, R. Murray. 1980. *The Tuning of the World: Toward a Theory of Soundscape Design*. Philadelphia, PA: University of Pennsylvania Press.

Schmidt-Horning, Susan. 2013. *Chasing Sound: Technology, Culture, and the Art of Studio Recording from Edison to the LP*. Baltimore, MD: The Johns Hopkins University Press.

Shepherd, John, and Peter Wicke. 1997. *Music and Cultural Theory*. Cambridge: Polity Press.

Sterne, Jonathan. 2003. *The Audible Past: Cultural Origins of Sound Reproduction*. Durham, NC: Duke University Press.

Swedien, Bruce, with Bill Gibson. 2013. *The Bruce Swedien Recording Method*. Milwaukee, WI: Hal Leonard.

Wicke, Peter. 2009. The art of phonography: Sound, technology, and music. In *The Ashgate Research Companion to Popular Musicology*, ed. Derek. B. Scott, 147–170. Farnham: Ashgate.

Wicke, Peter. 2010. From schizophonia to paraphonia: On the epistemological and cultural matrix of digital generated pop-sounds. In *Approaches to Music Research: Between Practice and Epistemology*, ed. Leon Stefanija, and Nico Schüler, 71–78. Frankfurt am Main: Lang.

Wodianka, Bettina. 2013. Intermediale Spielräume im Hörspiel der Gegenwart: Zwischen Dokumentation und Fiktion, Originalton und Manipulation, akustischer Kunst und Radiophonie, Theater und Installation. In *Auditive Medienkulturen: Techniken des Hörens und Praktiken der Klanggestaltung*, ed. Axel Volmar, and Jens Schröter, 339–358. Bielefeld: transcript.

Sound and Media

Little Walter. 1997. *His Best*. CD, Chess Records, CHD-9384.

Paul, Les, and Mary Ford. 2005. *The Best of Capitol Masters: 90th Birthday Edition*. CD, Capitol Records, 09463-11411-2-6.

2   Sound/Music

Diedrich Diederichsen

To name what would broadly be considered pop music, we need to consider the difference between sound and music: an artistic-musical field focused on the exposition and marking of sound against the background of the traditional (but not only traditional) forms of music through which it is communicated. When Mario Perniola calls "sound, understood precisely in the neutral and inorganic indifference evoked by this word" (1999, 88–95) the essence of the development of rock and pop music, a posthuman hard core, he needs musical scenarios and terminology, as well as musical action, which—as its medium—generates this hard core in the first place; even though he might really want it to emerge out of the music alone. But pop music's inorganic hard core and its sexuality can only be had if it contrasts itself with the expressive, subject-centered classification of a type of music that is to a degree still conventional music. The same sounds in the context of the white cube or in the framework of a public installation are used and contextualized by other frameworks and media.

Asymmetry and Other Things about Music

Initially, however, the difference between music and sound appears asymmetrical. Our intuition at first says to us that there can be sound without music, but there cannot be music without sound. In this view, music would be an instance of sound. All instances of sound that are not music—according to different descriptions and definitions of music of course—would be sound, and thus also those sounds that one would colloquially call noises and which are characterized by the fact that their form was not purposefully designed; they are thus still within the same intuitive model, but just farther from the sound that is not music, but which could indeed potentially be capable of music. A tone struck on the piano belongs to the category of music because both the instrument and the action are correspondingly aligned with a particular conception of music. A wine glass incidentally struck by one's knuckle produces a sound capable of becoming or being used in music, but it is not music because it was not made to be struck based on a particular conception of music. It would, however, be capable of becoming music since its acoustic characteristics are suited to existing music-making practices. The noise that is caused by the same wine glass if it is thrown against the wall is not even capable of becoming music, as there is no conception of music that would integrate this noise.

Wait: of course, that is not true. The world is full of compositions that base a new application of music on precisely those sounds considered not capable of being music. Whether this applies first to the sirens of Edgard Varèse ("Ionisation," 1933), or to the piece by Arseni Avraamov ("Sinfonia Gudkov"/"Symphony of Sirens," 1921–22), or to the ideas in Luigi Russolo's manifesto (*L'Arte Dei Rumori/The Art of Noise*, 1913), or first to the unintentional noises that John Cage successfully introduced into the history of concepts of music (completely independent from the level of personal investment John Cage had in this musical concept), plays only a minor role: Cage suspected that Varèse wanted to express himself through the noises instead of allowing the noises to speak for themselves as articulations or documents of their own kind. A decisive factor for these considerations is that the concept of music permits a characterization of sounds that is either based only on their physical, acoustic qualities or based only on the social sound acts (in terms of speech acts using sound) with which they are linked: as what they are meant to be, or based only on the experiences that a person can have with them, is disrupted by the definition of music since it occupies its part of the differentiation with more than just something objectifiable—a phenomenon that exists alongside other phenomena, but with which it has something in common: being audible—but always with something programmatic and normative. Music is not conceivable without normativity. Even when it permits everything, it permits everything to be music, and that is naturally something different from just permitting everything.

The difference between sound and music is thus always overlapped by two other differences: first, the difference between music and its absolute Other, that which is not music and does not claim to be music but only shares audibility with music; and second, the difference between music and other sound arts (and perhaps third, the sound events that are not, could not be, and do not want to be art). One can surely test and clarify the possible differences between sound and music using the discursive effects of empirical materials.

One would have to observe the most representative discourses possible in order to attempt to do away with the inherent normativity of the concept of music: one can search for the commonalities between phenomena, innovations, and developments that are perceived as sound and also discussed outside of musical debates—in contrast to those that have gone unheeded outside of "musical" thought. Outside of such musical thought could mean: within discourses that are dedicated to other arts (for instance, sculptural, installational) or are located far outside of art-related discourses. The social practices of shivaree, for example, or other forms of intentional, targeted noise that Jacques Attali described as a "simulacrum of violence" (1985, 26) would be described by sociopolitical, sound material-acoustic and urban-sociological terminology—and only by musical terminology in a very limited and selective way. Sound psychology and perception theory are interested in alarms and sirens, but not much more than a Peter Brötzmann album acknowledges police sirens as musical material. For that matter, plenty of hip hop and rock music uses police sirens as montage snippets for specific music.

Alarm as Medium

All of the following would be necessary subtopics and side topics for the differentiation of sound and music using the example of a police siren: Peter Brötzmann's album *Alarm* (1981), certain music and sound-psychological effects of the fourth interval used as a siren by the German police in their patrol cars, the complaints by non-German road users that the fourths of the police are difficult to locate amid the traffic, while an alarm signal that accelerates rhythmically along with the speed of the vehicle would be safer (based on the US-American example), as well as the *musique concrète* sampling strategies and aesthetic-political components in montages and beyond montages, and the lifeworld meaning of police sirens for the different types of audiences interested in hip hop. In my opinion, it would be more useful to assume a terminological relationship in which sound and music are not asymmetrically linked to one another, rather than combing through the phenomena according to how they intersect, categorizing the one or the other based on intuition in order to then force differences again into these overlapping elements and to improve the categories. One could demonstrate how in the case of one arbitrary sound phenomenon there is one side, one level, that is music and one that is sound—not in the sense of an absolute sound ontology, but respective to the specific relations concerning communicative addressees, types of attention and/ or aesthetic strategies.

I suggest setting this relationship parallel to one that Niklas Luhmann described as the relationship between medium and form (1997, 165–213). For Luhmann, a medium is something that makes a form possible in the broadest sense, in that it makes it recognizable—whether that be as a physical carrier, as a basis for contrast, or as a means of contrast. It can do that, however, only if the medium itself is not considered. That does not mean that it remains invisible or concealed, but rather that, in terms of the perception and processing of a stimulus as form, the medium recedes and is no longer considered. Central to Luhmann's proposal of this differentiation is that it does not refer back to fixed characteristics of a material or an object that also allow themselves to be observed outside of the system. Medium and form are only that which they are for one another within the system (ibid., 167).

Thus, for sound and music a relationship emerges that can change owing to external influences and framings, as well as through inner shifts in function; they do not have to allow themselves to be determined based on fixed, external assignments of function, such as those of a legal or physical kind. The relationship can also shift within a sound-musical object or process. Tone pitches can become the media for sound design; sounds can become the media of tone pitches.

Although, more often, the case should be that individual musical genres—especially in the area of pop music—differentiate themselves in that they assign a sound (such as technologically innovative sounds, novel sounds, signal sounds, sounds known from certain external world contexts) with the quality of attraction and need a medium for it, which can be music; while other musical genres fundamentally make aesthetic decisions so that tone pitches and the usage of instruments, human virtuosity with instruments, and other musical parameters belong to the set of phenomena marked as attractions. By genres, I am referring to a generic term for practices that incorporate a certain amount of precedents (whose scope can have something to do with their social, popular, or industrial constitution, but does not have to). Independent of specific compositions or interpretations, something in them is already determined that is no longer left to the musician or composer in the individual case. The question of to what extent music or sound are medium and form for one another pertains for the most part to those precedents that constitute the genre.

Material/Art

What is interesting about the overlap between aesthetic and other more classificatory questions, however, is that before the historical avant-garde of the twentieth century, the differentiation of sound from music was set parallel to the differentiation of material from art, rather than to that of form and medium (as defined by Luhmann). Naturally, this also affects contemporary ways of speaking. The form is aligned with art, the medium with the material. When the material—the noise, the sound "in and of itself"—emancipates itself, as a rule, the first to emerge are models of noise art or sound art, which distance themselves from music and orient themselves toward other oppositions or deviations from music: spatial relations instead of temporal relations, sculpturality instead of linearity, public spaces instead of closed concert halls, immobility of the installation instead of the mobility of the replicated sound carriers. Sound as art that wants to emancipate itself from music repeatedly sought out one of those paths. That is true of the various attempts to found a sound art that would establish itself completely outside of the music scene and also beyond the fine arts, as has been the case at least since the works of Max Neuhaus (such as his "Public Supply," 1966; see Licht 2007, 8–71).

The differentiation between nonintentional noise (or at least noise not intended as art) and the invariably intended, mostly artistically intended music does not follow the same path as the differentiation between sound and music. Intention is only in question when an artistic and/or technical reason is given that presupposes or requires a directionality linked to human-subjective action. This is not only aesthetically questionable since Cage; it also can no longer be found technologically in sounds generated by software and algorithms. The scale of intention and directionality in sound events inside and outside of music is no longer a generic or technical prerequisite; it is subjected to free will. Intention, therefore, has also become capable of both form and medium.

The Extremism of Pop Music

Nevertheless, it can be asserted that neither the historical avant-garde nor the neo-avant-garde in the tradition of John Cage worked with this differentiation when they integrated noises or nonintentionally created sounds into music. Two closing questions thus arise: Is there a practice that actively works with the medium–form differentiation, thereby according more to this approach than just the legitimation of terminological elegance? And: are there sounds that are not integratable into music, and that thus permit the music–sound differentiation to be subsequently accorded a transcendental basis originating beyond the relationship that the two sides of this differentiation have to one another?

First: the practice that actually works with this differentiation—while unspoken for the most part—is, of course, pop music. Musical components (rhythm, tone pitches, acoustic colors) are often repetitive and largely determined by genre conventions. Sound effects, however, are of central importance to pop music, and not just for their appeal and effect, but also for the social cohesion of subcultural groups who unite around one individual sonic signal (for instance) that they have defined as specific to them. Even the equiprimordiality of radio advertisements, DJ moderation, rock 'n' roll, and other classic forms of pop music have worked on the shift from marking to demarcating certain sound and music aspects: a song that has actually been identified as a hit based on its musical elements becomes the rhythmic background for the moderator or emcee's performance—the music becomes medium and the radio announcer's verbose act becomes form. The emergence of toasting in reggae during the late 1960s and '70s and that of US-American rap are further developments. Inversely, the electronic sound design machines by Moog and others, which were originally celebrated as liberating musical sound, are often ostentatiously melodically employed in pop music: only through the medium of completely conventional and familiar music (*Switched-On Bach*) could the new sound be recognizable as form.

Second: sounds are always integratable into music to such an extent that one could use the music concept parallel to the nominalist usage of the *white cube* concept in the fine arts; as a result, everything that is symbolically framed as music, is music; other stronger music concepts are not readily available in the same way. This construction is nevertheless tautologically contaminated: we call music that which we call music. The (old avant-garde) gesture of inclusion is not possible according to it. It is therefore not very surprising that its high relevance is already declining and that it is being increasingly renounced; and that the extramusical sound genre could gain stability and establish itself as its own separate field with its own terminology, as many have demonstrated, including Brandon LaBelle (2006, xv).

Sound as musical extremism requires the framework of music as a medium—that is one of the fundamental concepts of pop music.

Translated by Jessica Ring

References

Bibliography

Attali, Jacques. 1985. *Noise: The Political Economy of Music*. Trans. Brian Massumi. Minneapolis: University of Minnesota Press.

LaBelle, Bruce. 2006. *Background Noise: Perspectives on Sound Art*. New York: Continuum.

Licht, Alan. 2007. *Sound Art*. New York: Rizzoli.

Luhmann, Niklas. 1997. *Die Kunst der Gesellschaft*. Frankfurt am Main: Suhrkamp.

Perniola, Mario. 1999. *Der Sex-Appeal des Anorganischen*. Vienna: Turia + Kant.

Russolo, Luigi. 1913. *L'Arte Dei Rumori*. English translation, 1967. *The Art of Noise*. Trans. Robert Filliou. New York: Something Else Press.

Sound and Media

Brötzmann, Peter. 1981. *Alarm*. LP, FMP, 030.

3 Popular Culture

Thomas Hecken

Popular Culture: A Disputed Term

Over the course of recent years, I have collected around a thousand statements that claim to explain what should be understood as "popular culture." For our purposes, two contemporary explanations will serve as examples. The first is from Maase, who describes popular culture in the following way: popular culture objects are characterized by their "broad popularity throughout all classes" (1997, 23). The second originates from Burke, who states that popular culture is the current, mass media "common culture," which also incorporates the "elites" but from which the intellectuals exclude themselves (1984, 12).

However, what is the point of this pedantic approach, this seemingly quite boring act of collecting definitions (analyzed in Hecken 2006, 2007, 2009)? Why should we not just dedicate ourselves to the popular culture phenomena, at the very least to the (more or less) interesting and exciting popular sounds? The answer to that question, simply stated, is that the popular cannot be understood in isolation. In nominalist or constructivist terminology: "terms or names like 'popular culture'" are "designations, categorizations, differentiations, definitions," which "could conceivably have gone by other names, definitions, metaphors," because, no matter how familiar the term may sound, "there is not really a 'field' or a 'place' out there with a sign above the gate saying 'popular culture'" (Bowman 2008, 198).

Coming from that perspective, the same would of course be true for all terms. The term "popular culture," however, is of particular importance as it holds a central position in political and aesthetic coordinate systems. Exceedingly often, an unmistakable valuation occurs along with the act of describing or designating something as being a part of popular culture. This valuation is frequently accompanied by a negative sign; popular culture is considered unaesthetic, bad art, or not art at all; its creators are deemed morally ambiguous, and its consumers are presumed to be uneducated or easily manipulated, politically irresponsible subjects who should be reformed through education or kept under police surveillance. Not uncommonly, and particularly in the last three decades, one also encounters opposing values; even the current popular culture (not just some past class culture, *Volkskultur*, or community culture, which conservatives often invoke and which nationalists want to promote) is vehemently defended politically and/or put in a positive light aesthetically or artistically against the pretensions of the elite, the educated, the dominant class, and so on. On the other hand, liberals favor a definition of popular culture that suggests that popular culture is not the result of a collective substratum, but rather is in fact the result of the combined sum of individual voting decisions.

A Collection of Circulating Definitions

Looking back at the years since the Second World War, quite a series of meaningful differences has been observed in the usage of the term "popular culture" in academia alone. *Is popular culture the culture of the powerless?* This concept has a variety of important manifestations: popular culture has been defined as the culture of the people, meaning the people of the lower class, or as the culture of the ruled classes in early modern Europe, that is, the tradespeople and farmers, among others (Muchembled 1978; Burke 1996; Clark 1983), or different historical formations of the working class (Hoggart 1957; Thompson 1974; Yeo and Yeo 1981); as such, popular culture has been placed in opposition to the culture of the elite (Reay 1985), as that which is excluded from the legitimate culture (Bourdieu 1993) as opposed to high culture, as all "taste subcultures" that are different from "high culture" (Gans 1966, 551); popular culture would then be a specific form of uneducated, nonacademic reception and/or productive appropriation (Chartier 1984). But could it not also be a set of very skillfully combined and utilized consumption practices (Certeau 1980), an exciting oppositional appropriation of mass media products (Fiske 1992)? Finally, it would then be understood as a culture that exhibits similarities (such as straightforwardness and authenticity) to the "folk culture," in contrast to mass culture (Handlin 1961; Hall and Whannel 1964; Kellner 1995); and it would then stand in radical opposition to autonomous, complex "'high' art" (but also to "folk art") (Lowenthal 1950). Historically, popular culture would be an open-ended battleground, an important place where those who have been powerless up until now can break through the hegemony of the ruling classes, a place where the powerless might even welcome this dominance, because it still appears prudent to them to do so. Popular culture could be one of the decisive "sites where this struggle for and against the culture of the powerful is engaged; it is also the stake to be won or lost in that struggle" (Hall 1981, 239). So, *does popular culture serve in the abolition of the traditional opposition of higher and lower culture?* It has been seen as mere postmodern culture, or as a no longer strictly differentiable part of the postmodern hybrid culture (Chambers 1988;

Collins 1989; During 2005); and as such it would constitute a modern overcoming of the separation of higher and lower culture, as the dominant "middle of the road forms," as the "new consensus" (Nowell-Smith 1987, 83). In contrast: *does popular culture represent the culture of a qualitatively undetermined, quantitatively measurable and variable mass?* Is it really a collection of all "aspects of culture, whether ideological, social, or material, which are widely spread and believed in and/or consumed by significant numbers of people" (Hinds [1988] 2006, 363; Levine 1992)? However, is popular culture not also a strand of culture that, by these means, actually reaches a large public "that cannot be simply described by a single social variable, such as class or gender or age" (Grossberg, Wartelle, and Whitney 1998, 37)? Does it consist merely of an ensemble of products that are directed toward average tastes (van den Haag 1957)? Consequently: *is popular culture just schema culture?* Popular culture would then be nothing more than a collection of degenerate, conventionalized, modern representational techniques that provoke simple recognition responses (Greenberg 1939); it could then be understood as a selection of works that are characterized mainly by formulas and patterns (Nord 1980). Is it a form of culture whose ideals and substance are "conformity and conventionalism" (Adorno [1954] 1957, 478)? Or *is popular culture a culture of stimulus and pleasure?* Popular culture has been conceptualized as a culture that "often inscribes its effects directly upon the body: tears, laughter, hair-tingling, screams, spine-chilling, eye-closing, erections" (Grossberg 1992, 79)—and as such it is experienced as "pleasurable" by many (Jenkins, McPherson, and Shattuc 2002, 26). Finally: *is popular culture identical with entertainment culture* (Shusterman 2003; Hügel 2003)? Currently, this is probably the most widely accepted and generally distributed definition of what popular culture could be. And is this not simply obvious when we listen to music and sound?

Popular Music and Sound

With the scope of different evaluations and determinations regarding popular culture, one ends up with results that are far-reaching and quite ideologically and/or institutionally ingrained; popular music as: an organically developed folk song? The authentic expression of oppressed but vital social classes? Vulgar, common? What is on the charts, sells well, satisfies widespread demand, or serves false needs brought about by manipulation? Merely pleasant or intense entertainment? The music of any youth subculture, or of subcultures with an anti-elite, recoding, and oppositional approach? Formulaic, highly conventionalized music? Functional music? The most important component of postmodern bricolage and retro sample combination art?

Moreover, since "sound" can in no way be equated with "music"—in fact, it is sometimes even contrasted with "music"—further differentiations are also possible. That can be observed in the case of positions that would like to strictly separate popular culture and "unculture" from one another, and use sounds to do so. There are four classifications of particular historical importance that use the standard of sounds to determine whether something is popular culture or unculture: first, there is Johann Gottfried Herder's position. Herder accords the folk song with the highest cultural order, and thus opposes elite conceptions of art. As much as he stands for the light, the simple, and the natural, however, he draws both an aesthetic and social distinction just as vigorously in his separation of poetry and song from particular sounds: "The riffraff on the street never sing or recite, they but scream and garble" (Herder [1779] 1990, 239). Many similar statements directed against the chaotic noise generated by rebellious crowds were made in the centuries following; for example, when they surged through the streets and town squares, or protested against the "primitive" music of supposedly inferior groups (such as in the case of the conservative and National Socialist condemnation and persecution of so-called "nigger music").

Second, to reconcile itself with the sounds of the assembled masses, the relevant authoritarian version consists of completely aligning the people with the leader. The authoritarian argument goes like this: the modern mass democracy undermines the immediacy and authenticity of the acclamations owing to the isolation of the voters. Only when the people agree with the leader can democracy prevail. This does not happen after collective, discursive proceedings; even at smaller events—such as at a marketplace—the people can express themselves only through statements of agreement or disagreement (Schmitt 1927, 33–34). From this authoritarian perspective, with a right-wing leader and a homogeneous public body, it can naturally only be about an acclamation of agreement. The kinds of fantasies discussed above, and the political, often racist, and violent measures toward the creation of a projected homogeneous *Volk* regularly associated with such fantasies, are linked to animosity against music considered "foreign to the people" (*volksfremd*), as well as to its persecution. It is the sound of music discredited in this way that often draws the attention of authoritarian nationalists and friends of *Volk* culture—its supposedly shrill, cacophonous sounds that are not cleanly separated from one another.

Third, given the international distribution of pop and rock music, that criticism loses its powerful effect. But the result is the same: for pop music, sounds carry particular importance. In *Variety*, for instance, one author stated that "R&B is strictly a sound phenomenon" (Anonymous 1955). In the following decades, however, this kind of "sound" music began to receive a high value judgment (including often in academic categorical analysis; see, e.g., Wicke 2003). The enthusiasm for electrified, unique, exciting, artistic, or expressive sounds constitutes an important *topos* for the reception of pop music. As such, so-called popular music is an important part of popular culture in many countries (often cross-nationally), nationalist and racist views of a pure, homogeneous body of people are more difficult to enforce. And as for the fourth position, a part of the liberal view of sound is that the recognition of sounds that were earlier separated from positively perceived music and culture should not reach into the political and employment domains. Liberal conceptions support the liberated individual—however, they often also fear that these "atomized" people, released from their corporative communities, could easily allow themselves to be united by leaders into an illiberal majority. The tone of speech used by such reactionary leaders—an aggressive, passionate sound that often comes close to screaming—thwarts liberals with a political sound ideal consisting of the controlled, expressionless voice of experts and administrators. Politicians are indeed permitted to adopt populist tones during the election campaign; the victor is then required, however, to moderate it down to a tone of calm and distanced authority. Sounds—which have now come to belong to the scope of legitimate music and culture in liberal societies—are in fact not considered as such in the current hegemonic political sphere.

An Assignment for Future Sound Research

At this point, my assignment for future sound research will not come as a surprise: when analyses of popular culture coming from the perspective of sound examine the meaning of certain sounds within popular culture, they must account precisely for the definition of "popular culture" taken as a basis. Considering the multitude of competing language usages, just speaking of "popular sounds" without additional comment contributes substantially to the confusion.

The second consequence is more profound and extends beyond the scope of the academic. I am unable to provide a scientific basis for this, but I am able to bring it to the political forefront: other than for historical research into their linguistic usage, the terms "popular," "popular culture," "popular sounds," and so on should no longer be utilized. Too much misery has already been perpetrated in the name of "the people," (especially in the name of the German *Volk*), in the name of the supposed common good. The rhetoric of "the people" should be replaced with argumentations in the name of particular interests and specific groups. Even with its small scope, sound research could play its part in this effort. It should not be difficult to specify who produces and receives which sounds without the use of the "P"-word.

Translated by Jessica Ring

References

Adorno, Theodor W. (1954) 1957. Television and the patterns of mass culture. In *Mass Culture: The Popular Arts in America*, ed. Bernard Rosenberg and David M. White, 474–488. Glencoe: Free Press.

Anonymous. 1955. Music biz now R&B punchy: Even hillbillys are doing it. *Variety*, February 9, 51, 54.

Bourdieu, Pierre. 1993. Sagten Sie "populär"? In *Praxis und Ästhetik: Neue Perspektiven im Denken Pierre Bourdieus*, ed. Gunter Gebauer and Christoph Wulf, 72–92. Frankfurt am Main: Suhrkamp.

Bowman, Paul. 2008. *Deconstructing popular culture*. Basingstoke: Palgrave Macmillan.

Burke, Peter. 1984. Popular culture between history and ethnology. *Ethnologia Europaea* 14:5–13.

Burke, Peter. (1978) 1996. *Popular Culture in Early Modern Europe*. Aldershot: Scolar Press.

Certeau, Michel de. 1980. *L'invention de quotidien: 1. L'art de faire*. Paris: Union Générale d'Éd.

Chambers, Iain. (1986) 1988. *Popular Culture: The Metropolitan Experience*. London: Methuen.

Chartier, Roger. 1984. Culture as appropriation: Popular culture uses in early modern France. In *Understanding Popular Culture: Europe from the Middle Ages to the Nineteenth Century*, ed. Steven L. Kaplan, 229–253. Berlin: Mouton.

Clark, Stuart. 1983. French historians and early modern popular culture. *Past and Present* 199:62–99.

Collins, Jim. 1989. *Uncommon Cultures: Popular Culture and Post-Modernism*. New York: Routledge.

During, Simon. 2005. *Cultural Studies: A Critical Introduction*. London: Routledge.

Fiske, John. (1989) 1992. *Understanding Popular Culture*. London: Routledge.

Gans, Herbert J. 1966. Popular culture in America: Social problem in a mass society or social asset in a pluralist society? In *Social Problems: A Modern Approach*, ed. Howard S. Becker, 549–620. New York: Wiley.

Greenberg, Clement. 1939. Avant-garde and kitsch. *Partisan Review* 6:34–49.

Grossberg, Lawrence. 1992. *We Gotta Get Out of This Place: Popular Conservatism and Postmodern Culture*. New York: Routledge.

Grossberg, Lawrence, Ellen Wartella, and D. Charles Whitney. 1998. *Media Making: Mass Media in a Popular Culture*. Thousand Oaks, CA: Sage.

Hall, Stuart. 1981. Notes on deconstructing "the popular." In *People's History and Socialist Theory*, ed. Raphael Samuel, 227–240. London: Routledge & Kegan Paul.

Hall, Stuart, and Paddy Whannel. 1964. *The Popular Arts*. London: Hutchinson Educational.

Handlin, Oscar. 1961. Comments on mass and popular culture. In *Culture for the Millions? Mass Media in Modern Society*, ed. Norman Jacobs, 63–70. Boston: Beacon Press.

Hecken, Thomas. 2006. *Populäre Kultur: Mit einem Anhang "Girl und Popkultur."* Bochum: Posth.

Hecken, Thomas. 2007. *Theorien der Populärkultur: 30 Positionen von Schiller bis zu den Cultural Studies*. Bielefeld: transcript.

Hecken, Thomas. 2009. *Pop: Geschichte eines Konzepts, 1955–2009*. Bielefeld: transcript.

Herder, Johann Gottfried. (1779) 1990. *Volkslieder: Nebst untermischten andern Stücken, Zweiter Teil*, vol. 3 of *Werke: Volkslieder, Übertragungen, Dichtungen*, ed. Ulrich Gaier, 229–430. Frankfurt am Main: Deutscher Klassiker Verlag.

Hinds, Jr. Herold E. (1988) 2006. Popularity: The sine qua non of popular culture. In *Popular Culture Theory and Methodology: A Basic Introduction*, ed. Herold E. Hinds Jr., Marilyn F. Motz, and Angela M. S. Nelson, 359–370. Madison: University of Wisconsin Press.

Hoggart, Richard. 1957. *The Uses of Literacy: Aspects of Working-Class Life with Special Reference to Publications and Entertainments*. London: Chatto & Windus.

Hügel, Hans-Otto, ed. 2003. Einführung. In *Handbuch Populäre Kultur: Begriffe, Theorien und Diskussionen*, 1–22. Stuttgart: Metzler.

Jenkins, Henry, Tara McPherson, and Jane Shattuc. 2002. Defining popular culture. In *Hop on Pop: The Politics and Pleasures of Popular Culture*, ed. Henry Jenkins, Tara McPherson, and Jane Shattuc, 26–42. Durham: Duke

Kellner, Douglas. 1995. *Media Culture: Cultural Studies, Identity, and Politics between the Modern and the Postmodern*. London: Routledge.

Levine, Lawrence W. 1992. The folklore of industrial society: Popular culture and its audiences. *American Historical Review* 97:1369–1399.

Lowenthal, Leo. 1950. Historical perspectives on popular culture. *American Journal of Sociology* 55:323–332.

Maase, Kaspar. 1997. *Grenzenloses Vergnügen: Der Aufstieg der Massenkultur, 1850–1970*. Frankfurt am Main: Fischer.

Muchembled, Robert. 1978. *Culture populaire et culture des èlites dans la France modern, XVe–XVIIIe siècles*. Paris: Flammarion.

Nord, David P. 1980. An economic perspective on formula in popular culture. *Journal of American Culture* 3:20–29.

Nowell-Smith, Geoffrey. 1987. Popular culture. *New Formations* 2:79–90.

Reay, Barry, ed. 1985. Introduction: Popular culture in early modern England. In *Popular Culture in Seventeenth-Century England*, ed. Barry Reay, 1–30. London: Croom Helm.

Schmitt, Carl. 1927. *Volksentscheid und Volksbegehren: Ein Beitrag zur Auslegung der Weimarer Verfassung und zur Lehre von der unmittelbaren Demokratie*. Berlin: de Gruyter.

Shusterman, Richard. 2003. Entertainment: A question for aesthetics. *British Journal of Aesthetics* 43:289–307.

Thompson, Edward P. (1963) 1974. *The Making of the English Working Class*. Harmondsworth: Penguin.

van den Haag, Ernst. 1957. Of happiness and of despair we have no measure. In *Mass Culture: The Popular Arts in America*, ed. Bernard Rosenberg and David M. White, 504–536. Glencoe: Free Press.

Wicke, Peter. 2003. Popmusik in der Analyse. *Acta Musicologica* 75 (1): 107–126.

Yeo, Eileen, and Stephen Yeo. 1981. *Popular Culture and Class Conflict, 1590–1914: Explorations in the History of Labour and Leisure*. Brighton: Harvester Press.

4 Sound Practices

Carla J. Maier

Fluteboxing

When Nathan Lee walks on stage, the audience falls silent. He starts playing the flute in a lucid, melodic fashion, featuring Indian classical scales. Standing alone in the middle of the huge stage, he appears introverted and almost fragile; his eyes are lowered, and his body makes only small movements, no more than a shy nod of the head, creating little vibrations in his shoulders. The sound is crisp and fills the entire concert space; the sound waves are processed with reverb, enhancing Queen Elisabeth Hall's capacious architecture. Then, after a breath of a pause, Lee starts playing the flute in a completely different way: using a lot more breath, and pressing the air through his half-closed lips into the mouthpiece of the flute, he produces thudding, syncopated beats that make his whole body vibrate. The vibration is transported into the floorboards and out into the room, the thick sonic texture becoming the substance that puts the audience under Lee's spell. Just when this energetic rhythmic explosion reaches the audience and makes them burst into cheers, Nathan drops the beat and returns to the melodic, caressing tune, smoothing down the frenzied sound waves he had created (for more detailed analysis, see Müller-Schulzke 2012).

This essay's main objective is to propose the usage of the concept of sound practices for an analysis of sound in contemporary popular culture. Following an outline of the concept's theoretical implications, the fluteboxing technique of London-based flute player and beatboxer Nathan Lee will be explored as a concrete sound practice. Fluteboxing is an extended playing technique in which beatboxing (making beats with the mouth) is applied to a concert flute. This chapter will inquire into how this particular sound practice becomes relevant not only within musical discourse, but more generally with regard to how sound is perceived, used, and interacted with in everyday life. After the detailed analysis of this particular and music-specific case, the chapter will propose how the concept of sound practices can be transferred to other realms of popular culture as well: from recorded music to radio and film, and from the sounds of electronic devices such as mobile phones and microwaves to traffic signals and the mechanical sound of escalators.

The Concept of Sound Practices

The sounds we are surrounded with in everyday life—may they be musical, natural, mechanical, electronic—accompany us everywhere we go and in everything we do. However, the sounds of music coming from shopping mall loudspeakers, the sounds of the mobile devices we carry with us, the car horns and the train signals have become so integral to our activities that we tend to take them for granted, either being very quick in detecting their meaning, or paying hardly any attention to them at all. How can we investigate sounds as an activity, as something we create, use, and transform—as a sound practice? As a concept, sound practice means putting sound on the agenda of the field of cultural studies and asking how sound becomes meaningful as an artistic and intellectual activity and a discursive practice in and of itself.

The term "sound practice," coined by Altman (1992) in reference to film sound, is proposed here as part of interdisciplinary terminology that functions across the boundaries of a single discipline. The aim is not, however, to provide a universal defi- nition of sound, but rather to conceptualize sound in relation to a particular practice. For clarification, Nathan Lee's fluteboxing technique can serve as an example. Inves- tigating his fluteboxing performance requires listening beyond conventional musico- logical categories that are based on the musical score and classical musical praxis, and paying attention to the technically and technologically processed attributes of the spe- cific fluteboxing sound. As will be elaborated below, the fluteboxing sound depends on a number of factors: not only how the flute is played, but also how the sound of the fluteboxing technique is processed by a reverb that highlights this performance as a technologically enhanced event, characterized by the use of microphones, preampli- fiers, amplifiers, and speakers. Another facet is that of the acoustic properties of the concert space, as well as the resonance of the audience, which also forms part of this auditory event. Sound is here something that is molded and invested with meaning by a whole conglomeration of corporeal, physical, technological, and architectonical attributes. These attributes do not carry any inherent meaning, but they are part of a sonic and cultural practice that challenges musical conventions and rigid concepts of sound.

In particular, with regard to an analysis of sound practices that is informed by a cultural studies approach, as it is proposed here, it is important to ask how sound practices imply an engagement with the ways in which sound is perceived, produced, and performed as a form of identification and socialization. In regard to Nathan Lee's fluteboxing technique, his performance also challenges and transgresses ethnocen- tric perspectives on Indian classical music as belonging to a homogeneous traditional

Indian culture. Fluteboxing can thus be explored as a transcultural sound practice that is firmly situated in contemporary club music culture. Conceiving of sound practices as a transcultural practice therefore means thinking about how sound is related to and intervenes in discourses of race, class, and gender and how specific sound practices generate and transform these discourses (and thus become discursive practices in their own right).

The questions of how we experience sound, how we use sound and interact with sound in everyday life have become crucial within recent sound studies approaches, and the contributions in this companion reader are certainly testimony to that. To name only a few important works in this regard: Bull and Back's (2003) concept of auditory culture, which emphasizes a shift of the study of sound from sound as an objectified representation to an everyday cultural practice; Kassabian's (2013) theory of ubiquitous listening, which analyzes how the ubiquity of sound in everyday life has changed the ways we listen; and Goodman's (2010) notion of "unsound," which elaborates the materiality and politics of contemporary bass music and the ways in which sub-bass and ultrasound have been used as sonic weapons in the wars of the twenty-first century.

What these approaches share, although they explore different sounds from different sources and different theoretical angles, is that they construe sounds not as a side product of visual culture, or as a mere illustration of the "real product," but as an activity, as something that is perceived, used, manipulated, and thus as something that becomes part of a specific sound practice.

Analyzing Sound Practices: Nathan Lee's Fluteboxing Technique

"Fluteboxing," a term Nathan Lee coined for this particular technique, implies a combination of flute playing and beatboxing: Lee creates rhythmic patterns with his mouth, while at the same time sending his breath and the guttural sounds he makes through the transversal hole of the flute. The fluteboxing technique thus extends the sonic and technical repertoire of the concert flute. In contrast to the sustained, airy, high-frequency flute sounds that feature in the melodic parts of his performance, when using the fluteboxing technique, the streams of compressed air that hit the flute's transversal hole produce short, thudding, and low-frequency flute sounds. In this sound practice, sound cannot be grasped in terms of the classical musical repertoire; the practice therefore challenges the listening conventions of the classical flute sound. Listening to the way in which Lee adds bass to the flute, while creating multirhythmic and polyphonic beat patterns, sound is here primarily a sonic fabric in the form of the materiality of the bass. Extending the frequencies and timbres usually associated with the classical flute, the fluteboxing technique reinvents the instrument and questions the boundaries of musical styles: the application of the beatboxing technique transgresses the confines of classical music, thereby linking this extended performance technique and the specific sound that is produced to the realm of hip hop and urban dance music. This particular sound practice situates Lee's performance at the intersection of different musical styles, technical conventions, and sound qualities. Another facet of Lee's fluteboxing technique is its connection to sound production technology that is characteristic to hip hop, jungle, and related electronic dance music traditions. Lee's fluteboxing structures his performance more in the form of an electronic dance music track than as a piece of classical Indian music. Technology is regarded here in a broad sense, acknowledging that "sound technologies are vital elements of the musical text rather than supplementary to its unfolding" as Weheliye (2005, 2) notes in *Phonographies*. This statement implies that technologies such as the microphone, turntables, or the sampler are not merely transmitters of a musical idea, but that the use of the technology within and beyond its mechanical or electronic limits plays an integral part in the creation of music. This means that the technological tools are conceived not just as operational tools that reproduce the "right" or "appropriate" sound, but as hybrid instruments that are "worked" in a way that constantly manipulates and enhances their given function. Dysfunctionality is an appropriate term for the description of this process, in which the boundaries between noise and sound, or noise and music, are renegotiated. Therefore, the melodic, sonic, and rhythmic attributes of a musical track always interact and resonate with the technology that is used. From this perspective, the extracting and rearrangement of sound snippets in sampling practices, the scratching of records in turntablism to create noisy rhythms, or—in the case of Nathan Lee's fluteboxing technique—the distorted noises that are created with the mouth in beatboxing have been part of a particular Afro-Caribbean and diasporic production culture. These sonic disturbances have been the driving force in forging new sounds and generating new musical styles, from reggae to dub, to garage, to drum 'n' bass. Thus, conceiving of fluteboxing as a sonic technology leads to being able to analyze it alongside other sound practices in the production culture of urban club music such as sampling or turntablism.

Finally, the fluteboxing technique can be conceived of as a discursive practice: the ways in which Indian classical scales are traversed by the syncopated, bass-oriented beat patterns deconstruct the binary oppositions often established in musical discourses between "traditional" Indian music and urban music styles such as hip hop or electronic dance music. One of the main problems with this framework is that Indian sounds are expected to be authentic reflections of the traditional cultural heritage of the musicians who use these sounds, while technology is primarily defined as a Western practice. However, Afro-diasporic music styles in particular have used digital sound reproduction and manipulation techniques since the 1970s, together with all kinds of performance techniques such as sampling, scratching, filtering, and so on.

Sound practices mark key moments within a sonic performance in which a specific idea is created, a shift of auditory focus implied, or a new sonic relationship established through a specific interaction of a sound and a rhythmic pattern, a sound and a specific effect, or between different layers of sounds and sonic enactment.

Cultural Analysis of Sound Practices in Everyday Life

The concept of sound practices proposed with regard to the music-specific example of Nathan Lee's fluteboxing technique indicates a move away from conceptualizations of sound that are primarily bound to a musicological perspective of scores and conventional playing techniques, toward an understanding of sound as an activity that engages with sound as a fabric that can be molded according to influences and techniques that stem from more diverse cultural and artistic knowledge and skills. For example, Nathan Lee's musical skills stem from his activity as a jungle MC and beatboxer, and therefore his sound practice is genuinely influenced by electronic dance music culture. Thus, when he started learning to play the flute much later in his musical career, being taught Indian and Western classical music, he developed a unique style, his fluteboxing technique, that functions within the logic of his affinity for beatboxing and electronic music, and the flute becomes just another tool to further diversify his artistry. The understanding of Lee's fluteboxing technique as a sound practice requires a broader perspective than that permitted by traditional musicological terminology.

Transferring the concept of sound practices to other aspects of popular culture in everyday life, the functional sounds of car horns, traffic signals, or mobile devices can serve as a starting point for a cultural analysis of the sonic and semiotic particularities of these sounds (see Schulze, Maier [née Müller-Schulzke], and Schneider 2012–2014). Rather than defining these sounds as merely bound to one function or inherent meaning, they need to be analyzed as: malleable "forms of sonic knowledge" (Schulze 2007) that are first generated during the process of listening to, producing, performing, and mediating sound—a process that is situationally, culturally, and spatially specific (Maier [née Müller-Schulzke] 2012, 117).

Analyzed in that way, everyday functional sounds become multifaceted articulations of popular culture. The example of a cell phone's ringtone can be introduced here in order to expand on how a specific sound practice that involves everyday sounds in the urban space can be analyzed (cf. Müller-Schulzke 2012). On a very basic level, a cell phone's ringtone can be described as a signal indicating that someone is calling the owner of the cell phone, which also includes the implied request to answer the phone. On a more complex level, the meaning of this functional sound depends on the temporary, spatial, and social components of a specific situation in which the sound is heard: if the sound of a cell phone's ringtone is heard on a crowded train platform on a weekday morning, the sound will intervene in that particular everyday situation. Depending on the loudness, duration, and handling of this sonic event, the sound will either just merge into the general sonic rhythm of other background sounds—the train, the escalator, the nearby building site's jackhammer, the movements and footsteps of people or random conversations—or it will cause an irritation. The sonic rhythm is coded through seemingly nondescript parameters that guide the sonic, cognitive, and corporeal perception of people who are used to certain sounds occurring in a particular place in a certain interval and duration. This rhythmic code is not universal, but it might be shared among people in the context of similar daily activities. If the ringtone does not stop for a longer period of time, bystanders might be drawn to notice the call's receiver and the way he or she handles the situation, or to the sonic signature of the phone (e.g., the characteristic marimba sound of the iPhone), or they may drift off and develop their own associations with regard to the ringtone, for instance being reminded of the appointment with the hairdresser they still need to cancel. The relational meanings of the ringtone allude to the drifting quality of everyday functional sounds as either "unheard" or irritating, identity-establishing, insignificant, disruptive, or transformative (cf. Müller-Schulzke 2012).

As the above examples clearly demonstrate, there is not one set of methods that can be easily applied to any and all musical or everyday functional sounds. Rather, the analysis of sound practices, as has been elaborated in this essay, requires scholars to use close analysis, and, depending on the specific research question, to critically engage with established concepts of sound and attend to the relational, transactional, and multisensorial facets of specific sound practices. The concept of sound practices that was proposed here was thus developed as an extension and critique of rigid frameworks of musical analysis by putting the focus on a flexible conceptualization of sound and sound practices rather than on conventional notions of musical practice (a term that is largely restricted to conventional musical works, musical scores, and playing techniques). This new conceptualization of sound practices extends beyond musical sound to the various facets of sound in everyday life and popular culture at large. In fact—and this was also emphasized in regard to the example of Nathan Lee's fluteboxing technique—sound practices do not dismiss the particularities of music for a broadened concept of (extramusical) sound—they also function as musical discourse. In many instances, there is not a clearly definable boundary between musical and nonmusical sounds anyway; examples include the sounds of car horns and engines and how they are invested with new meanings when used in the context of a wedding celebration in Berlin, or the work of sound designers who create sounds not on the basis of preconceptualized meanings but rather in the in-between spaces of communication, jargon, and musical and pop cultural influences based on the knowledge of specific films, musical pieces, and computer games (see Schulze, Maier [née Müller-Schulzke], and Schneider 2012–2014).

To conclude, this chapter is suggesting that sound not be treated as something that is already invested with a given meaning, but as something that is part of a sound practice generated in and through a particular constellation of sonic, cultural, technical, and social activities.

References

Bibliography

Altman, Rick, ed. 1992. *Sound Theory, Sound Practice*. New York: Routledge.

Bull, Michael, and Les Back, eds. 2003. *The Auditory Culture Reader*. Oxford: Berg.

Müller-Schulzke, Carla J. 2012. Transcultural sound practices: South Asian sounds and urban dance music in the UK. PhD diss., Goethe-Universität Frankfurt.

Müller-Schulzke, Carla J. 2012. Driftende Klangzeichen: Zur semiotischen Klanganalyse in den Sound Studies. *Zeitschrift fur Semiotik* 34 (1–2): 109–123.

Schulze, Holger. 2007. Wissensformen des Klangs: Zum Erfahrungswissen in einer historischen Anthropologie des Klangs. *Musiktheorie* 22 (4): 347–355.

Schulze, Holger, Carla J. Maier [née Müller-Schulzke], and Max Schneider. 2012–2014. Sound Studies Lab: Funktionale Klänge. Kultur- und Gestaltungstheorie non-verbaler Klangzeichen. DFG-Projekt Humboldt-Universität Berlin.

Weheliye, Alexander. 2005. *Phonographies: Grooves in Sonic Afro-Modernity*. Durham, NC: Duke University Press.

Sound and Media

Lee, Nathan "Flutebox." 2008. Performance at Queen Elisabeth Hall London, October 18. https://www.youtube.com/watch?v=wC1wKtgvWRg (accessed January 19, 2015).

5   Sound as Musical Material: Three Approaches to a Material Perspective on Sound and Music

Rolf Großmann and Maria Hanáček

Regarding sound as a determining force in today's music, especially in pop music, is not really new. As Chris Cutler, Theodore Gracyk, Peter Wicke, Paul Théberge, and many others have pointed out, cultural industries, mechanical reproduction, and studio work are essential not only for the distribution but also for the musical process of popular music itself. Also, a new perspective on classical music in relation to the media-materialized phonographic representation of "the work" is developing (Ashby 2010). Studio producers and engineers are claiming their part in the creative process, and remixes are making new stars out of people who have never touched a musical instrument. They are using technical devices to play and remix already played music and working with audio workstations with built-in musical knowledge such as graphic interfaces and virtual instruments. In the tradition of *musique concrète* or DJ cultures like techno and hip hop, sound has turned into a concrete (media-) materiality that can be mixed and transformed at a haptic level, as in the case of tape splicing and disc scratching.

However, what is the significance of this materialization of sound, and what does it imply for the understanding of today's musical production and its forms and styles? Focusing on the general cultural impact of "stockpiling music" (Attali 1985) may be helpful, but it might lead to speculative results concerning musical structure itself. We argue that it is not enough to address this issue in terms and categories of cultural change, of enhancing and modifying "real music" or of the novelties of sound effects. Aesthetic strategies such as sampling, recombination, and remix, and forms like tracks instead of songs, clearly show that working with phonographic sound material has become an established practice of making and composing "music" itself. Even sound effects—for example, filtering or reverb—are essential means used in popular music production and live performance—driven by dub, DJ culture, techno, and live electronics.

To take an almost randomly selected example from the charts: the intro of David Guetta's number one hit "Titanium" (2011)[1] starts with a totally dry hook line, the second part of which receives a specially designed reverberation. This virtual space is cut precisely on the one beat of the hook repetition—and then again for a second time when the intro is over—and then the song starts. Certainly, it is a simple reverb effect, but it is calculated to have a maximum impact on the listener. There is no claim to any real or authentic reference outside of the music itself. The aesthetic strategy can be interpreted in several ways, perhaps as playing with connotations of space or as an extended gated reverb, such as what Steve Lillywhite as producer and Hugh Padgham as engineer originally invented to "fatten" Phil Collins's snare, and which is now applied to an entire phrase in order to create a special rhythmic effect.

For our purposes, the preferred interpretation is not important. Rather, what is remarkable is the method itself—the analytically based interpretation of sound composition. In contrast to its individual usage in a special case, it is possible at a theoretical level that derives from a historical perspective on widely established practices and the systematic description of a technical and corresponding aesthetic strategy in general.

In this case, the historical perspective can be traced back to the Ultraphon (invented by H. J. Küchenmeister; see figure 5.1), an early gramophone (1924–25) with a second pickup shifted by approximately 8 centimeters, creating a constant delay of approximately 100 milliseconds and hence a broader or more spatial sound; or to a theatrical bathroom reverberation effect used by the Harmonicats' hit "Peg O' My Heart" (1947) (cf. Großmann 2013), to the slapback echo used by Sun Records (Elvis Presley, "Baby, Let's Play House," 1955), to gated reverb, multitap delay, and

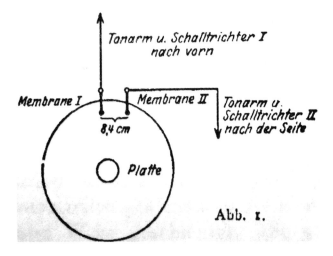

Figure 5.1
How the Ultraphon works (membrane I, membrane II, pickup arm, and horn I facing forward, pickup arm and horn II facing to the side, record) (diagram taken from Kappelmayer 1925, 1104).

many more popular delay FX. Those kinds of effects have been researched extensively, especially regarding their early usage (see, e.g., Doyle 2005) and in relation to recordings of the human voice.

However, this effect leads us to a more fundamental question. Since the human ear normally derives localization, spatial information, and spatial associations from the interaural difference in time, and the delayed reflection ratio in the signal (which is created through sound waves being cast back at the boundaries of a space), reverb and delay effects are usually discussed under the category of space. Correspondingly, in the traditional aesthetics of music, musical spaces are assigned to the respective performance, to the resounding of the musical structure. At the level of an elementary physical-acoustic materiality, space is required for the diffusion of sound waves, but as a rule, its specific form remains accidental as far as the substance of music compositions is concerned.

So, before we come to questions of the (media) material used for designing popular (and of course many types of contemporary) music, we have to take a step back to look at the debate concerning the term "material" and its relation to basic musical substance. Is music made of elaborated melodic and harmonic structures that must be retraced through conscious listening? Or is music an initiated affective process derived from the forces of (acoustic) vibrations? How we answer these questions about the "primary material of music" is essential for the role that media technology can play in musical contexts. Even if we adopt the Solomonic approach of "on the one hand but on the other," we will usually also make some form of judgment that is informed by musical aesthetic traditions of thought.

Material One: The Sonic

This brings us to an initial concept of material that is closest to sound, the circumstances of how it is generated and distributed and—on the aesthetic level—its corporeal impact. It is based on physical and physiological contexts resulting in the recipient being affected at the level of materiality. This materiality of sound in the sense of "sonic materialism" (Cox 2011), "vibrational forces" (Goodman 2010), or "the sonic" (see Wicke's contribution to this vol.) is presented in an almost demonstrative manner in popular genres such as rock, techno, or dub, and determines their aesthetic form and impact.

This concept of material is in an ongoing conflict with the tradition of music aesthetics for several reasons. Already with Immanuel Kant, as "Materie des Wohlgefallens [the matter of the liking]" ([1892] 1987, §13, 223), music was seen as a vehicle of pure sensual stimulus, the subjectivity of which is at odds with the form of aesthetic beauty. To Kant, because of its immediate proximity to the sensual stimuli, music itself "has less value than any other of the fine arts,"[2] and in his aesthetics it earns only a subordinate status. The tendency toward placing the formal structure of music at the center as aesthetic substance (which leads to form, idea, or work) is evident in a wide variety of aesthetic theories, from Hanslick to Hegel to Adorno. One possible explanation of the animosity toward sensuality in Western art music is given by Kurt Blaukopf with recourse to Max Weber. Blaukopf (1996) argues that Christian reformist rationalization is associated with a displacement of the body from religious and advanced civilization rituals. In his view, what counts, both in life and in art, is the subjection of "immediate" emotions to a rational order.

This notion of associating societal order with an economy of sensual stimuli leads directly to Theodor W. Adorno's critique of the aesthetic enhancement of sonic material. In the popular music of the mass media, the "blind and irrational emotions" (Adorno [1938] 2002, 295) of being affected are formed to match market demand. Sonic material (in its "certain richness and roundness of sound" and its "rhythmical patterns" that are constructed to fulfill the recipient's "desire to obey" [Adorno 1941]) becomes a standardized product of cultural industry, to which the means of production that shape it belong, in a classical capitalist sense.

This can easily be explained using the example of the Harmonicats' hit referenced above. The desire to transcend an infinite space is "objectified" by the Harmonicats' artificial "supernatural" reverberation effect and becomes sellable as a commodity. At the same time, a modified relationship develops between physicality, sonic material, and traditional musical structure. Physicality—and the same argument can be applied to the digital code of virtual instruments (see Hansen 2006)—always also becomes the object and tool of societal power relations via technical design and embodiment. Especially in this area of affect control, the technical producibility of sound causes a deep mistrust of any nonrational practice that contrasts with the declared level of reflection about musical structures in art music. Adorno argues that regressing to a level below that of critical-discursive reflection, a level that has already been attained, would result in the forfeiture of self-determined, advanced "structural listening," and hence in externally controlled immaturity.

Despite continuing to be of influence today, both positions—on the one hand the reduction of the aesthetic to the *Kunstschöne* of the form, and the culture industry commodity form of sound on the other—tend to have more of a historical character. The strict language of forms in Western art music has long been called into question in the transcultural process of intermingling global music cultures, and its structure-centric score is losing its dominance owing to hybrid forms of controlling and editing audio-material at digital audio workstations. So the notion of a central role of abstract and rationally reflected music structure requires at a minimum an extension and a discussion of the altered historical and technocultural situation. In contrast, regarding Adorno's line of argument, the detailed critique of cultural studies and their successors applies. However, it has hardly resulted in the development of a theory in its own right in the field of music aesthetics. This is particularly apparent in popular music studies; musical analysis in this field has been repeatedly problematized: owing to a supposed lack of "musical substance," music aesthetics has been deemed unsuitable for examination, and the focus is instead placed on the song lyrics (Tagg 1982).

Further criticism of the "sonic" or "sonic materialism" (Cox 2011) is leveled at its conceptual construction. While the interconnection of physical acoustics with human physicality, perception, and cultural practice allows a hitherto neglected focus on precisely those links, it comes under the suspicion of constituting a new essentialism and an imagery possessing the physical power to explain phenomena (ibid., 157). Concepts such as resonance and synchronization target a description of the "immediate" impacts of vibration and repetitively organized time at a physical level. The sonic is in close proximity to a tradition reducing popular music to its "vibrating matter" and thus bears the risk of physicalizing cultural practice and its resulting aesthetic forms.[3] Affective listening, which was developed most influentially in meditative forms created by Buddhist cultures and which became integrated into a global musical culture in the 1950s and '60s through various influences such as John Cage and the Beatles, is a corresponding mode of listening. "Affective listening is listening to sounds neither *as* sounds nor as music. There is no *as* involved, because conscious perception has not yet entered. One feels into the variations of the intensity of sounds, the movement of sonic molecules" (Wang 2012).

Thus an "ontology of vibrational forces" becomes a walk on a tightrope, upon which physical and musical terminology are parallelized and interpreted. Instead of being material in a physical sense, the sonic constitutes a conceptual construction that is difficult to grasp and is a crucial element in the development of research on the perception of sounds that Helmholtz ([1863] 1875) had in mind, but also goes beyond that with regard to a speculative discourse on "nature and matter themselves" (Cox 2011, 157).

Material Two: Notation (the Score, Phonography, and the Digital Code)

The second concept of material refers to the media of memory, archives, communication and distribution, and productivity. Composing, producing, and performing music is carried out in close interaction with its written codification. The original understanding of art music (as *Tonkunst*) and the work as an aesthetic form is deeply linked to the score as a medial representation of tonal structure. However, after the gramophone won the competition against the musical automata in the 1920s, a new notation was established: *phonography*. Just like the score, this medium effects a transformation of sound structures: temporal events are projected onto lasting arrangements of material. Thus, just like note material, they become the material of memory, reflections, and artistic work. Now, however, it is the vibration of the air itself, that is, the sonic material described above, that can be recorded and technically reproduced.

Similar to the score which leads to the *"geistfähigem Material"*[4] of the notes and their structure discussed by Eduard Hanslick ([1854] 1891),[5] this new form of musical writing constitutes a further, more comprehensive material basis for the cultural practice of music. As "secondary orality" (Ong 1982), it simultaneously contains both the aspect of oral transmission and that of written codification, interconnected in a technical process of writing and reading.

Paradoxically, because of this codification of the material into written form, the components of music that could only be transmitted orally in the era of the score are also attributed qualities resembling those of a work of art: "phonographic orality" (Toynbee 2006) is joined by a "phonographic artifact." What results is not a work of art in the nineteenth-century sense (see Goehr 1992), but an object that endures independently of individual and historical contexts that, similar to written music's "opus perfectum et absolutum" of the sixteenth century (see Loesch 1998, 342), can form the starting point for a very diverse range of work concepts. In an "iterative mode" (Toynbee 2006), it can make dynamic changes available as individual artifacts (as in practices such as "versioning," see Hebdige 1987), and it can also record the final products of artistic work that claim to be "works." A wide spectrum of new methods and aesthetic concepts opens up with the cultural practice of music captured in phonographic material. Moreover, as a result of cultural work, the technical process of reading and writing itself can be dealt with in artistic work and developed as an object of aesthetic action.

Since the 1950s, groundbreaking innovations in music have resulted from the productive use of this new media material with its indivisible link between *orality, literacy,* and *technology.* Whereas the last innovation in the paper-based notation of Western art music was the "emancipation" (Schönberg [1946] 1975, 258) of dissonance, noise, and chance, experiments to emancipate phonographic media material from its role of simple reproduction and transmission had already begun in the 1920s. The first crooners were very successful in using electronically amplified microphones, while Ernst Toch, Paul Hindemith, and later multitrack artists such as Les Paul worked with time-based manipulations and the layering of phonographic material, and Pierre Schaeffer initiated the aesthetic concept of *musique concrète* with an experimental exploration of the *objets sonores.* In parallel with these more technologically based innovations, blues and jazz made use of phonographic orality to achieve an extension of sound in terms of musical parameters and a dynamization of the concept of a musical work (Großmann 2013).

However, the most momentous innovations have proven to be the result of a transcultural (media) practice of popular music in which reproduction media were reinterpreted as media of production and improvisation in performances: this took place in the context of early forms of phonographic aesthetic strategies in Jamaican dub and New York disco, which are relevant to contemporary music as a whole. Sampling, remixing, and the nonsimulative usage of sound effects that go beyond the traditional written form and the canon of values in Western art music are the result of this particular popular culture (see Rose 1989).

Generative work involving media material has been accelerated by MIDI and digital audio. The operation of music automata and phonographic material has been given a new, standardized, digital form of writing as control data and digital phonography value lists. Here, the materiality of sound undergoes a fundamental process of abstraction. Sound and musical structure can now be written down and handled both as a generative process and as a phonographic image. As audio data, a program code, or a user interface, this writing becomes the object of an extended technocultural practice of music. In this new written form, the traditions of analog synthesizer control and phonographic design are continued in part—for example, as algorithmic computer music or sampling in hip hop—but also broken down into hybrid concepts—for example, in live sequencing. Composition and performance techniques are elaborated and sedimented, for example, in synthesizer presets, in sequencing GUIs, or in virtual studio (VST) instruments and effects.

Material Three: Musical Material

The two material concepts of sound discussed above refer to its physical existence and its technological literacy. The third approach comes from a musical poetics perspective and focuses on the material with which the composer and the performer work.

One historical concept from the era of nineteenth-century classical work has already been mentioned: Hanslick's music aesthetics' *geistfähiges Material*. In contrast, any conceivable concept of musical material today is simultaneously extended and biased by Adorno's music philosophy. "Material … is what artists work with: It is the sum of all that is available to them, including words, colors, sounds, associations of every sort and every technique ever developed. To this extent, forms too can become material; it is everything that artists encounter about which they must make a decision" (Adorno [1970] 1997, 148). While this very broad definition clearly appears to be suited to an integration of sound, Adorno's theoretical background entails a number of fundamental problems. His proposition of the material's autonomy and immanent tendency, which is to be recognized and implemented by the composer, was the final point of structural thought in western European artistic music, and had already reached its limits in the mid-twentieth century.

As Adorno ([1954] 2002) himself noted, there are several reasons for this concept's failure when applied to twentieth-century music, or to popular music. Here, material development is reduced to an immanent logic of tonal structure that can be applied neither to extended concepts of musical composition, as with John Cage, nor to the aesthetic strategies for the processing of already recorded music in DJ culture. Even though an awareness of the perhaps possible "gramophonic montage" in the context of film montage technologies can be attributed to him (Levin 1990, 46), for Adorno, phonography was purely a sound-capturing medium and was not in any way conceived as a means of aesthetic production.

The culture industry's usurpation of technical reproduction media and their material described above is a further problem. Adorno generally understands them as tools of an economically oriented mass culture. Correspondingly, while popular music is discussed at the sound level under the heading "musical material," as a standardized mass commodity (Adorno 1941), it simply remains a means of affect control. If the concept of musical material is to play a meaningful role in today's discourse, new perspectives will be required for these problematic issues.

Nevertheless, the proposition of progress in musical material generally makes sense, as it allows for the identification of aesthetic methods and a description of their historical development. However, the assumption of any uninterrupted linearity or single historically valid status of material is misleading. Instead, a complex combination of lines of development and breaks in continuity can be observed that are indeed possible to accurately describe, for example as forms of "phonographic work."[6] Those kinds of lines of development, the courses of which are becoming increasingly transcultural in a globalized media culture, include both the development of sonic material in the sense of the first concept of material introduced here and the new written form of phonographic material.

An appropriate view of popular sound culture focuses on the dynamic aspects of material development in addition to its standardization. While it principally remains in the societal field of tension between power, hierarchy, and identity, it can also be conceived of in a positive sense. If affecting is understood no longer as a passive process but as an active and dynamic one, it can be discussed as "affectivity," as the "capacity of the body … to deploy its sensorimotor power to create the unpredictable, the experimental, the new" (Hansen 2004, 6). Here, the issue is not that of subjecting affects to the hierarchies of instrumental rationality in societal power, referred to by Adorno as the dark side of enlightenment. On the contrary, the technical rationality sedimented in the machines can become the object of sensory experiments and aesthetic acquisition. Thus, very much in the sense of sonic materialism, instead of representing musical symbols or an assigned meaning, the cultural practice of handling, its forms of production and adoption become musical knowledge themselves. Sonic effects lose their significance as manipulators of a real sound and turn into the means of designing phonographic work, synthesizer presets are tried out under the extreme conditions of the "out of control" status, and loops become experimental arrangements of a breakbeat science of sonic experience beyond symbolic reference structures (see Schloss 2004).

However, the concept of sonic materialism also reduces the character of the material to a preconscious practice separated from cultural reflection. Its dynamic, which Cox (2011, 157) rightly stresses as an argument against the accusation of essentialism, is by no means random, but is formed within lines of traditionalism that can definitely be referenced and named. In the micro range, the periodic ordering of time in music comprises the *vibrations*, and in the macro range the *rhythms* (in corresponding metaphors of *effect*, *resonance*, and *synchronization*—see above). Both aspects are culturally preformed and are highly differentiated. Rhythms are a codification of movements, of performed or stylized dances, or forces of motor synchronization. They contain both physical knowledge and the knowledge of rituals and conditions of social significance. The new treatment of temporal circumstances in program control data and digital phonography that can be edited to the millisecond is developing rhythmic orders out of these traditions (see Großmann 2014). This happens in technocultural configurations that are themselves capable of reflection, and also in their methods.

Returning to the example with which we began, if delay effects are understood as sedimented material in creating music, they attain a certain degree of independence (if not autonomy) from the context of simulating the real spaces from which they have evolved. Reverb and delay have "emancipated" themselves as musical material. The issue is no longer that of musical imaging's credibility, authenticity, or high fidelity, but of a meaningful aesthetic application. In this field, popular music can be composed in a highly differentiated, artful, and experimental manner. In Squarepusher's electronically produced "My Red Hot Car," for example, there is no continuous "natural" space. Passages that are completely "dry" in their electronic creation alternate with very precisely dosed reverb and delay passages. Spatial perception and association become independent aesthetic means. Here, the artificial design of musical (not real) space assumes a status similar to that of tonal design. Just like notes and chords, as melody and harmony became the *geistfähiges Material* of the nineteenth century, the forms of effects sedimented in apparatus, software, and application strategies are becoming the musical material of the sound culture of the twentieth and twenty-first centuries. It is thus necessary to rethink the concept of musical material in popular and contemporary sound cultures: as sedimented and established forms of culturally formed sonic material, as material of phonographic notation and program codes, and as material accessible in hybrid technocultural configurations.

Translated by Mike Gardner

Notes

1. Produced in 2011, Guetta's "Titanium" was a top hit on the UK charts in 2012.

2. Kant ([1892] 1987). Original: "weniger Werth als jede andere der schönen Künste," §53, 328.10–11; cf. Hanáček 2008.

3. "Vibrating matter is the condition to which certain forms of music (rock, most frequently) have been reduced, in descriptions that seek to account for the transformative impact of such musical forms on human bodies"(Straw 2012, 230).

4. "*Geistfähiges Material*" is translated by Gustav Cohen as "material capable of receiving the forms which the mind intends to give" (Hanslick [1854] 1891, 72).

5. "The crude material which the composer has to fashion … is the entire scale of musical notes and their inherent adaptability to an endless variety of melodies, harmonies, and rhythms" (Hanslick [1854] 1891, 66–67).

6. See Großmann's "Phonographic Work," this vol.

References

Bibliography

Adorno, Theodor W. (1938) 2002. On the fetish-character in music and the regression of listening. In *Essays on Music*, ed. Richard Leppert, trans. Susan H. Gillespie et al., 288–317. Berkeley, CA: University of California Press.

Adorno, Theodor W. (with George Simpson). 1941. On popular music. *Studies in Philosophy and Social Science* 9:17–48.

Adorno, Theodor W. (1954) 2002. The aging of the new music. In *Essays on Music*, ed. Richard Leppert, trans. Susan H. Gillespie et al., 181–202. Berkeley, CA: University of California Press.

Adorno, Theodor W. (1970) 1997. *Aesthetic Theory*. Ed. Gretel Adorno, and Rolf Tiedemann, trans. Robert Hullot-Kentor. Minneapolis: University of Minnesota Press.

Ashby, Arved. 2010. *Absolute Music, Mechanical Reproduction*. Berkeley, CA: University of California Press.

Attali, Jacques. 1985. *Noise: The Political Economy of Music*. Trans. Brian Massumi. Minneapolis: University of Minnesota Press.

Blaukopf, Kurt. 1996. *Musik im Wandel der Gesellschaft: Grundzüge der Musiksoziologie*, 2nd ed. Darmstadt: Wissenschaftliche Buchgesellschaft.

Cox, Christoph. 2011. Beyond representation and signification: Toward a sonic materialism. *Journal of Visual Culture* 10 (2): 145–161.

Cutler, Chris. 1985. *File Under Popular*. London: November Books.

Doyle, Peter. 2005. *Echo and Reverb: Fabricating Space in Popular Music Recording, 1900–1960*. Middletown, CT: Wesleyan University Press.

Goehr, Lydia. 1992. *The Imaginary Museum of Musical Works: An Essay in the Philosophy of Music.* Oxford: Clarendon Press.

Goodman, Steve. 2010. *Sonic Warfare: Sound, Affect, and the Ecology of Fear.* Cambridge, MA: MIT Press.

Gracyk, Theodore. 1996. *Rhythm and Noise: An Aesthetics of Rock.* Durham, NC: Duke University Press.

Großmann, Rolf. 2013. Die Materialität des Klangs und die Medienpraxis der Musikkultur: Ein verspäteter Gegenstand der Musikwissenschaft? In *Auditive Medienkulturen: Techniken des Hörens und Praktiken der Klanggestaltung,* ed. Jens Schröter and Axel Volmar, 61–78. Bielefeld: transcript.

Großmann, Rolf. 2014. Sensory engineering: Affects and the mechanics of musical time. In *Timing of Affect: Epistemologies, Aesthetics, Politics,* ed. Marie-Luise Angerer, Bernd Bösel, and Michaela Ott, 191–205. Zürich: Diaphanes.

Hanáček, Maria. 2008. Das Sonische als Gegenstand der Ästhetik. *PopScriptum* 16 (10). http://www2.hu-berlin.de/fpm/popscrip/themen/pst10/pst10_hanacek.htm.

Hansen, Mark B. N. 2004. *New Philosophy for New Media.* Cambridge, MA: MIT Press.

Hansen, Mark B. N. 2006. *Bodies in Code: Interfaces with New Media.* New York: Routledge.

Hanslick, Eduard. (1854) 1891. *The Beautiful in Music.* Trans. Gustav Cohen. London: Novello.

Hebdige, Dick. 1987. *Cut 'n' mix: Cultural Identity and Caribbean Music.* London: Methuen.

Helmholtz, Hermann von. (1863) 1875. *Die Lehre von den Tonempfindungen als physiologische Grundlage für die Theorie der Musik.* Brunswick. Translated as *On the Sensations of Tone,* 3rd ed., trans. Alexander J. Ellis. London: Longmans, Green.

Kant, Immanuel. (1892) 1987. *Kritik der Urteilskraft. Gesammelte Schriften, Bd. 5 Preussische Akademie der Wissenschaften.* Translated as *Critique of Judgement,* 2nd rev. ed., ed. and trans. Werner S. Pluhar. London: Macmillan.

Kappelmayer, Otto. 1925. Das Ultraphon. *Der Radio-Amateur* III:47.

Levin, Thomas Y. 1990. For the record: Adorno on music in the age of its technological reproducibility. *October* 55 (winter): 23–47.

Loesch, Heinz von. 1998. "Musica" und "opus musicum": Zur Frühgeschichte des musikalischen Werkbegriffs. In *Musikwissenschaft zwischen Kunst, Ästhetik und Experiment,* ed. Reinhard Kopiez, 337–342. Würzburg: Königshausen & Neumann.

Ong, Walter. 1982. *Orality and Technology: The Technologizing of the Word.* London: Methuen.

Rose, Tricia. 1989. Orality and technology: Rap music and Afro-American cultural resistance. *Popular Music and Society* 13 (spring): 35–44.

Schloss, Joseph G. 2004. *Making Beats: The Art of Sample-Based Hip-Hop.* Middletown, CT: Wesleyan University Press.

Schönberg, Arnold. (1946) 1975. *Style and Idea: Selected Writings of Arnold Schoenberg*. Ed. Leonard Stein, trans. Leo Black. New York: St. Martin's Press.

Straw, Will. 2012. Music and material culture. In *The Cultural Study of Music: A Critical Introduction*, 2nd ed., ed. Martin Clayton, Trevor Herbert, and Richard Middleton, 227–236. London: Routledge.

Tagg, Philip. 1982. Analysing popular music: Theory, method, and practice. *Popular Music* 2:37–67.

Théberge, Paul. 1997. *Any Sound You Can Imagine: Making Music/Consuming Technology*. Hanover, CT: Wesleyan University Press.

Toynbee, Jason. 2006. Copyright, the work, and phonographic orality in music. *Social and Legal Studies* 15 (1): 77–99.

Wang, Jing. 2012. Affective listening: China's experimental music and sound art practice. *Journal of Sonic Studies* 2 (1). http://journal.sonicstudies.org/vol02/nr01/a11.

Wicke, Peter. 1990. *Rock Music: Culture-Aesthetic-Sociology*. Cambridge: Cambridge University Press.

Sound and Media

Guetta, David, feat. Sia. 2011. "Titanium." Twelve-inch single, no label (David Guetta), SIA2.

Harmonicats. 1947. "Peg O' My Heart." Ten-inch Shellac, Vitacoustic.

Presley, Elvis. 1955. "Baby Let's Play House." Seven-inch single, Sun Record Company.

Squarepusher. 2001. "My Red Hot Car." On *Go Plastic*. LP, Warp Records, WARPLP85.

Questioning Disciplines

6   Sound Studies versus (Popular) Music Studies

Marta García Quiñones

Popular Music: A Deaf Spot in Sound Studies?

An interest in the use of recorded music (primarily popular music) as a sonic background or foreground to all kinds of everyday activities—in short, what Anahid Kassabian has called "ubiquitous listening" (Kassabian 2013; García Quiñones, Kassabian, and Boschi 2013)—was what drove me to look into popular music studies and what is now known as sound studies at the beginning of this century, in a refreshing detour from my original background in philosophy. Coming from that orientation, I find it difficult to think of popular music studies and sound studies as two different disciplines that are developing in parallel, and intersecting only occasionally. Yet, as attested by some of the most recent additions to this thriving field (Birdsall and Enns 2008; Sterne 2012; Bull 2013), if sound studies scholars even deal with music at all, they demonstrate a preference for either aesthetic practices that blur the frontier between sound and music—such as specific forms of avant-garde music, sonic art, soundscapes—or for usages of music through other media (on this, see also Coates 2008), or for the technological aspects of popular music, that is, studio recording, electronic music production, and the like. In other words, scholars who see their research as belonging to the field of sound studies often avoid dealing with musical practices and repertoires that are normally associated with music studies. In particular, the relationship between sound studies and popular music studies is hardly ever addressed.

In that sense, Devon Powers's (2013) recent contribution to the virtual panel "Sonic Borders,"[1] titled "Popular Music Studies: An Audible Discipline?," is a rare exception: it not only deals with the matter openly, but also elicited a couple of interesting comments from popular music scholars Barry Shank and Steve Waksman. In a nutshell, while acknowledging the importance of sound studies, Powers made the case for the preservation of popular music studies by arguing for the specificity of music as an intentional, expressive embodiment of human interaction with sound. Shank's remark pushed the issue further by asking whether sound studies may offer a particular approach to popular music, that is, "if there is a sound studies way of discussing popular music that would not fit" the category of popular music studies. Finally, Waksman's comment called attention to the, in his opinion, scarce presence of music in sound studies, affirming that "sound studies seems progressively to be defining itself as the study of non-musical sound." While I would object to Powers's definition of music on the basis of intention and expression,[2] both her arguments and the questions posed by Shank and Waksman indicate a "deaf spot" in the field of sound studies—one that, precisely at this vibrant moment for the new field, I think it necessary to confront.

For that purpose, the first section of this essay will elucidate the relationship of popular music studies to music studies by drawing on the precedent of the field of visual studies and its critical relationship to art history, which in recent decades has raised a series of questions that may also be relevant to the disciplinary positioning of sound studies. Then I will attempt to apply the same notions to the link between sound studies and popular music studies, and finally, I will make the case for a consideration of (popular) music studies within the context of sound studies.

Visual Studies versus Art History

One could certainly object to the convenience of discussing disciplinary issues related to sound research by referring to the traditional "villain" of its history, namely the visual. However, it is difficult to avoid the fact that the formation of sound studies has drawn legitimacy from the emergence of visual studies as an academic field, which took place around the late 1980s (Smith 2008, 4–8),[3] preceding sound studies by a couple of decades (Sterne 2012, 3). Thereafter, visual scholars engaged in a series of public debates on disciplinary questions—beginning with the question of whether visual studies should be considered a proper discipline, or rather an interdisciplinary field, or even, as some have preferred to put it (Mitchell 1995, 542), "an indiscipline"—from which sound studies could perhaps learn some interesting lessons.

First of all, a few considerations about the very name of the field, "visual studies," are apropos, as it continues to elicit controversy even today (Smith 2008, 8–11). While scholars like W. J. T. Mitchell distinguish between "visual culture" as "the object or target of study" and "visual studies" as "the field of study" (Mitchell 2005, 337), other scholars like Nicholas Mirzoeff prefer to name it "visual culture"[4] to stress the political dimension of studying visuality, as well as its challenge to established disciplines (Mirzoeff 2012, 6), and a third group of scholars including Marquard Smith (Smith 2008) favors the tag "visual culture studies." Disagreements over the denomination of the field have sometimes been interpreted (Moxey 2009) as a productive tension between those representatives of the "iconic turn" who seem to be fascinated by the mere presentational power of images (for example Mitchell, or Georges Didi-Huberman), and those others (such as Mirzoeff) who focus mainly on images as representations or incarnations of human intentions. In contrast, a wide consensus has grown recently around the name "sound studies" (see its increasing use in monographs, edited collections, and handbooks, e.g., Schulze 2008; Pinch and Bijsterveld 2012; Sterne 2012; Bull 2013), which seems to have successfully displaced other previously used names, such as "auditory culture" (Bull and Back 2003) or "aural cultures" (Drobnick 2004). This displacement, which involves the replacement of an adjective ("auditory") referring to human perception by the name of the perceived object ("sound") (Sterne 2012, 7),[5] as well as the substitution of "culture" with "studies," may have far-reaching consequences and certainly deserves attention. Yet, until recently it has not been considered particularly controversial.[6]

Visual studies, visual culture, or visual culture studies are currently meant to deal with such different objects as theories of visuality (that is, particular conceptions of visual perception, and the technologies and practices associated with them), the pedagogy of sight, the study of different (artistic and nonartistic) image libraries, and the way images are configured and interpreted in different historical, cultural, and social contexts. It has even been argued that the field has no object as such, but is constantly contributing to the emergence of its object (Smith 2008, 12)—which could probably also be said of almost any other field. In spite of the origins of the denomination "visual culture" in art history, visual studies' commitment to study artistic and nonartistic images—in other words, their blurring of the border that traditionally separated artistic images from images of other kinds—has elicited debates about the legitimacy of their approach in contrast to art history.[7]

While the arguments exchanged in those debates are too complex to be analyzed here, W. J. T. Mitchell has provided, in my opinion, one the most useful explanations of the relationship between visual studies (or rather, visual culture, as he prefers to call it) and art history in observing that visual culture is to art history "primarily an 'inside-out' phenomenon." On the one hand, as an "'outside' to art history," it opens up the traditional field of visual art to "the larger field of vernacular images, media, and everyday visual practices," some of which (the ones generated by the mass media) are considered to belong to other disciplinary traditions (cinema and media studies), while others, for instance vernacular and scientific images,[8] appear as new focuses of interest. On the other hand, it operates as "a deep 'inside'" to art history insofar as it reveals that history must necessarily be "more than a history of works of art," and so it must inevitably include those models of spectatorship and intersubjective structures that allow for the very existence of art as an institution (Mitchell 1995, 542). As the late José Luis Brea pointed out (Brea 2006), recognizing this last possibility involves being ready to enter a new territory that cannot be legitimately claimed by any single disciplinary tradition, but whose exploration requires a concerted transdisciplinary effort.

The Popular in Music Studies, and the Music in Sound Studies

To a great extent, the challenge posed by visual studies to art history, in the terms defined by Mitchell, is similar to the one that popular music studies posed to the field of music studies when it appeared back in the late 1970s. The field of popular music studies was originally conceived as a response to the perceived need, notably among those already involved in pedagogical tasks, to study rock, pop, and other mass-appealing genres of the 1960s and 1970s—a musical repertory that was enormously influential at the time, particularly among the young, but which was totally ignored by traditional musicologists (Fabbri 2010, 77–78). Yet, in dealing with these musics, popular music scholars soon recognized the urgency to understand the music technologies that made them possible, and to also elaborate a musicology of the media through which they were produced and played in everyday life. In other words, popular music studies presented itself, in the first place, as an "'outside' to music history," in that it expanded the domain of music studies beyond the limits of the classical tradition. This movement of expansion continues even today, in more than one direction. For instance, popular music scholars increasingly admit that, in spite of the many ambiguities and contradictions implied by the concepts of "popular culture" (see Hecken, "Popular Culture," this vol.) and "popular music" (Fabbri 2010, 85–90), any serious attempt to think about popular music must date it back to at least the nineteenth century, when a "third type" of musical culture was born (Scott 2009).[9] It has also become clear that the field of popular music studies must certainly include many other musics, besides the popular musics of Europe and North America that still constitute its main focus.

In the second place, just as in the case of visual studies versus art history, popular music studies has revealed its potential to be considered (recalling again Mitchell's term) a "deep 'inside'" to conventional music studies. Since the very emergence of the field, popular music scholars have accepted that they cannot just employ the tools of the musicological trade—in other words, they cannot rely on the same text-based methods and analytical vocabulary—to deal with their object of study, and have incorporated notions from the fields of sociology, politics, semiotics, and communication studies, among others, often behaving more like a transdisciplinary field than a proper discipline. By stressing popular music's *sound event* aspect (*Klanggeschehen*, Wicke 2003)—an aspect that is ultimately shared by all musics—popular music studies has revealed music's performative dimension. It has created the theoretical tools for the understanding of the social, economic, and media structures that allow for its production, distribution, and appreciation as a sound event, including those that have historically sustained the purported "autonomy" of classical music (Tagg and Clarida 2003). In that sense, Philip Tagg's statement (Fabbri 2010, 78), at the time of the foundation of the International Association of Popular Music, that "what was really needed wasn't an association for the study of popular music, but an association for the popular study of music," far from being just a witticism, reveals the profound implications of popular music studies as an invitation to rethink the entire field of music studies.

What about sound studies? What is its contribution and how does it relate to the popular study of music? On the one hand, sound studies—like popular music studies—may legitimately be considered as a supplement or "outside" to the whole field of music studies, as the field has brought scholarly attention to new objects: not only nonmusical sounds—for example, natural sounds (Schafer 1994; Pinch and Bijsterveld 2012; among others), everyday sounds and noises of the past and the present (e.g., Corbin 1998; Picker 2003; Smith 2004; Bijsterveld 2008), all kinds of media soundtracks (e.g., Kahn and Whitehead 1992; Altman 1992; Chion 1994; Douglas 1999; just to name some of the pioneering titles), or the various scientific uses of sound (Spehr 2009; Schoon and Volmar 2012)—but also the very question of audition as culturally and historically constructed (e.g., Sterne 2003; Thompson 2004), including a whole range of nonmusical auditory practices (e.g., Pinch and Bijsterveld 2012). In my opinion, the addition of these new subjects and the possibilities that they open up for the drawing of new associations is sound studies' most valuable contribution—its "deep 'inside'" to music and popular music studies—since these new subjects involve a consideration of sounds of any kind within a larger context, for example, the histories of technology, science, urbanization, or cultural performance.

In Michele Hilmes's words, the field of sound studies holds (like visual studies) the promise and the potential to "draw parallels and establish continuities between disciplines separated more by historic accident than logical coherence" (Hilmes 2008, 115). As Jonathan Sterne has also stated, "sound studies' challenge is to think across sounds, to consider sonic phenomena in relationship to one another—as types of sonic phenomena rather than as things-in-themselves—whether they be music, voices, listening, media, buildings, performances, or another path into sonic life" (Sterne 2012, 3). This challenge—I would like to add—should not exclude any kind of music, much less popular music. Even the question of auditory practices, which is one of the main topics in the field of sound studies, may benefit from the reference to specifically musical sounds and contexts: to give just one example, the tuning practices of rock guitar players may be placed and studied alongside other expert auditory skills (Pinch and Bijsterveld 2012), such as stethoscopy (Sterne 2003; Rice 2010) or the techniques developed by telegraphers at the end of the nineteenth century (Sterne 2003). In sum, sonic epistemologies should not leave music out, since they will ultimately reveal their potential in elucidating how music can be a form of knowledge through sound, in dialogue with other sounds.

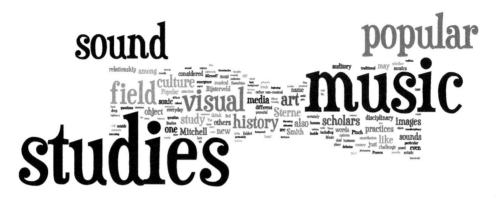

Figure 6.1
Tag cloud of the essay made with wordle.net (diagram courtesy of wordle.net).

Notes

1. The virtual panel "Sonic Borders" (available at http://iaspm-us.net/submissions/sonic-borders/) was hosted by the sound studies blog *Sounding Out!* and the US branch of the International Association for the Study of Popular Music (IASPM-US) during the months of January and February 2013 in preparation for the 2013 IASPM-US conference, which took place in Austin, Texas, February 28 to March 3. The panel featured contributions by ten sound and popular music scholars: Liana Silva, Regina Bradley, Marcus Boon, Barry Shank, Tavia Nyong'o, Theo Cateforis, Tara Betts, Shana L. Redmond, Airek Beauchamp, and Devon Powers. *Sounding Out!* (available at http://soundstudiesblog.com/) is a pioneering sound studies blog founded by Liana Silva (managing editor), Jennifer Stoever-Ackerman (editor in chief), and Aaron Trammell (multimedia editor).

2. Shank's contribution to the same panel, "On Popular Music Studies" (Shank 2013), suggested instead that it may be explained on the basis of listening, as "what we hear when we are engaged in musical listening."

3. Smith (2008) specifically mentions two events as main turning points toward the formation of the field. The first was the publication of the collected volume *Vision and Visuality*, edited by Hal Foster (1988). The second one was the establishment, in 1989, of the first US-based graduate program in visual and cultural studies, at the University of Rochester.

4. The first occurrences of the term "visual culture" are normally attributed to two influential art historians, Michael Baxandall and Svetlana Alpers (Baxandall 1974; Alpers 1983), who studied the relationship between the forms of visuality embedded in the artistic production of particular epochs—in Baxandall's terms, the "period eye" of *Quattrocento* painters and of seventeenth-century Dutch art, respectively—and other (nonartistic) contemporaneous visual practices.

5. Sterne discusses the difference between "auditory culture" and "sound studies" as a matter of acknowledging (or not) the centrality of the human ear to the world of sound (defined then as a specifically human way of perceiving a particular range of mechanical vibrations). However, his own preference for the name "sound studies" is explicitly based on euphony (Sterne 2012, 13).

6. However, see Schulze, "Sonic Epistemology," this vol., on the difference between "auditory" and "sonic" epistemologies.

7. The hostility that the field of visual studies has generated among some art historians is well represented by the tone of the anonymous 1996 "Visual Culture Questionnaire," which was sent to a selection of scholars by the art journal *October*, and which implicitly charged visual studies with considering images as ahistorical and disembodied, and ultimately with producing "subjects for the next stage of globalized capital" (Alpers, Apter et al. 1996, 25).

8. On the study of scientific images, see Daston and Galison 2007 and Elkins 2007.

9. Some other scholars maintain that popular music emerged even earlier, in the sixteenth century, when the first system of music printing was invented (Grosch 2012; Meine and Noske 2011, 23).

References

Alpers, Svetlana. 1983. *The Art of Describing: Dutch Art in the Seventeenth Century*. Chicago: University of Chicago Press.

Alpers, Svetlana, Emily Apter, et al. 1996. Visual culture questionnaire. *October* 77:25–70.

Altman, Rick, ed. 1992. *Sound Theory, Sound Practice*. New York: Routledge.

Baxandall, Michael. 1974. *Painting and Experience in Fifteenth Century Italy*. Oxford: Oxford University Press.

Bijsterveld, Karin. 2008. *Mechanical Sound: Technology, Culture, and Public Problems of Noise in the Twentieth Century*. Cambridge, MA: MIT Press.

Birdsall, Carolyn, and Anthony Enns, eds. 2008. *Sonic Mediations: Body, Sound, Technology*. Newcastle upon Tyne: Cambridge Scholars Publishing.

Brea, José Luis. 2006. Estética, Historia del Arte, Estudios Visuales. *Estudios visuales* 3:8–25.

Bull, Michael, ed. 2013. *Sound Studies*. 4 vols. London: Routledge.

Bull, Michael, and Les Back, eds. 2003. *The Auditory Culture Reader*. Oxford: Berg.

Chion, Michel. 1994. *Audio-Vision: Sound on Screen*. Ed. and trans. Claudia Gorbman. New York: Columbia University Press.

Coates, Norma. 2008. Sound studies: Missing the (popular) music for the screens? *Cinema Journal* 48 (1): 123–130.

Corbin, Alain. 1998. *Village Bells: Sound and Meaning in the 19th-Century French Countryside*. Trans. Martin Thom. New York: Columbia University Press.

Daston, Lorraine, and Peter Galison. 2007. *Objectivity*. New York: Zone Books.

Douglas, Susan J. 1999. *Listening In: Radio and the American Imagination*. New York: Times Books.

Drobnick, Jim, ed. 2004. *Aural Cultures*. Toronto: YYZ Books.

Elkins, James. 2007. *Visual Practices across the University*. Munich: Fink.

Fabbri, Franco. 2010. What is popular music? And what isn't? An assessment, after 30 years of popular music studies. *Musiikki* 2:72–92.

Foster, Hal, ed. 1988. *Vision and Visuality*. Seattle: Bay Press.

García Quiñones, Marta, Anahid Kassabian, and Elena Boschi, eds. 2013. *Ubiquitous Musics: The Everyday Sounds That We Don't Always Notice*. Aldershot: Ashgate.

Grosch, Nils. 2012. ¿Música popular en la Galaxia Gutenberg? Un intento de una reinterpretación desde la perspectiva de la historia de los medios. Actas del VI Congreso IASPM-América Latina, Buenos Aires, August 23–27, 2005. http://www.iaspmal.net/wp-content/uploads/2012/01/nilsgrosch1.pdf (accessed May 31, 2014).

Hilmes, Michele. 2008. Foregrounding sound: New (and old) directions in sound studies. *Cinema Journal* 48 (1): 115–117.

Kahn, Douglas, and Gregory Whitehead, eds. 1992. *Wireless Imagination: Sound, Radio, and the Avant-Garde*. Cambridge, MA: MIT Press.

Kassabian, Anahid. 2013. *Ubiquitous Listening: Affect, Attention, and Distributed Subjectivity*. Berkeley: University of California Press.

Meine, Sabine, and Nina Noske eds. 2011. Musik und Popularität: Einführende Überlegungen. In *Musik und Popularität: Aspekte einer Kulturgeschichte zwischen 1500 und heute*, 7–24. Münster: Waxmann.

Mirzoeff, Nicholas, ed. 2012. *Visual Culture Reader*. London: Routledge.

Mitchell, William J. T. 1995. Interdisciplinarity and visual culture. *Art Bulletin* 77 (4): 540–544.

Mitchell, William J. T. 2005. Showing seeing. In *What Do Pictures Want? The Lives and Loves of Images*, 336–356. Chicago: University of Chicago Press.

Moxey, Keith. 2009. Los estudios visuales y el giro icónico. *Estudios visuales* 6:8–27.

Picker, John M. 2003. *Victorian Soundscapes*. New York: Oxford University Press.

Pinch, Trevor, and Karin Bijsterveld, eds. 2012. *The Oxford Handbook of Sound Studies*. Oxford: Oxford University Press.

Powers, Devon. 2013. Popular music studies: An audible discipline? Paper given at Sonic Borders: A Collaborative Virtual Panel for the IASPM-US 2013 Conference. http://iaspm-us.net/

sonic-borders-virtual-panel-devon-powers-popular-music-studies-an-audible-discipline/ (accessed May 31, 2014).

Rice, Tom. 2010. Learning to listen: Auscultation and the transmission of auditory knowledge. Special issue, *Journal of the Royal Anthropological Institute*, 41–61.

Schafer, R. Murray. 1994. *The Soundscape: Our Sonic Environment and the Tuning of the World*. Rochester, VT: Destiny Books.

Schoon, Andi, and Axel Volmar, eds. 2012. *Das geschulte Ohr: Eine Kulturgeschichte der Sonifikation*. Bielefeld: transcript.

Schulze, Holger, ed. 2008. *Sound Studies: Traditionen, Methoden, Desiderate; Eine Einführung*. Bielefeld: transcript.

Scott, Derek B. 2009. The popular music revolution in the nineteenth century: A third type of music arises. In *De-Canonizing Music History*, ed. Vesa Kurkela and Lauri Väkevä, 3–20. Newcastle upon Tyne: Cambridge Scholars Publishing.

Shank, Barry. 2013. On popular music studies. Paper given at Sonic Borders: A Collaborative Virtual Panel for the IASPM-US 2013 Conference. http://iaspm-us.net/sonic-borders-virtual -panel-barry-shank-on-popular-music-studies/ (accessed May 31, 2014).

Smith, Mark M., ed. 2004. *Hearing History: A Reader*. Athens, GA: University of Georgia Press.

Smith, Marquard, ed. 2008. Introduction. In *Visual Culture Studies*, ed. Marquard Smith, 1–16. London, Thousand Oaks, CA: Sage.

Spehr, Georg, ed. 2009. *Funktionale Klänge: Hörbare Daten, klingende Geräte und gestaltete Hörerfahrungen*. Bielefeld: transcript.

Sterne, Jonathan. 2003. *The Audible Past: Cultural Origins of Sound Reproduction*. Durham, NC: Duke University Press.

Sterne, Jonathan. 2012. Sonic imaginations. In *The Sound Studies Reader*, ed. Jonathan Sterne, 1–17. London: Routledge.

Tagg, Philip, and Bob Clarida. 2003. *Ten Little Title Tunes: Towards a Musicology of the Mass Media*. New York: The Mass Media Musicologists' Press.

Thompson, Emily. 2004. *The Soundscape of Modernity: Architectural Acoustics and the Culture of Listening in America, 1900–1933*. Cambridge, MA: MIT Press.

Wicke, Peter. 2003. Popmusik in der Analyse. *Acta Musicologica* 75 (1): 107–126.

7   Sound and Racial Politics: Aural Formations of Race in a Color-Deaf Society

Carlo Nardi

In 2008, the news spread that the French film industry had a racist policy regarding the casting of dubbing actors. More precisely, it was reported that casting directors deemed White dubbing actors as having "universal voices" suitable for dubbing any kind of actor, and Black and Asian dubbing actors as having a distinct tone of voice suitable only for dubbing Black and Asian actors, respectively (Modestine 2008). The unfairness of such a policy is twofold: first, because it excludes particular groups of individuals on the basis of their racial background, and second because it rests upon a false premise—just as there is no scientific evidence for dividing up humanity on the basis of presumed racial differences,[1] there is likewise no necessary biological relationship between phenotypic traits such as skin color and the quality of the voice.

Though its consequences are very real, race is a slippery concept in virtue of its being a social construct. It is unstable not only historically and geographically, but also semantically, as it shifts between the labeling of different qualities, including phenotype, culture, personality, ethnicity, and nationality. In fact, it can be better understood as a process: Omi and Winant use the term "racialization" to signify "the extension of racial meaning to a previously racially unclassified relationship, social practice or group" (2015, 111) and the term "racial formation" to indicate "the process by which social, economic and political forces determine the content and importance of racial categories, and by which they are in turn shaped by racial meanings" (Omi and Winant 1986, 61). Bonilla-Silva (2014, 9) distinguishes between two levels of racialization: a racial structure "as the *totality of the social relations and practices that reinforce white privilege*," and a racial ideology as "*the racially based frameworks used by actors to explain and justify* (dominant race) *or challenge* (subordinate race or races) *the racial status quo.*" There is reciprocity between these two levels: racial formations legitimize forms of discrimination, which, in turn, naturalize racial formations.

It is worth noting that not only is racial ideology made of stereotypes about how individual races are supposed to be, it also has a cognitive dimension concerning how race is dealt with in everyday life. Drawing on the "sensual turn" in scholarship, which "treats cultures as ways of sensing the world" (Howes 2003, 29), I shall argue that this cognitive dimension of racialization involves particular ways of sensing race, that is, ways of understanding and signifying race through our senses. If we concur with Classen's (1993, 136) claim that "the way a society senses is the way it understands," then we should assume that the investigation of aural/auditory formations of race is a crucial step in the dismantling of racial ideology.

Race and the Senses

It is generally accepted that the modern concept of race was developed in the early eighteenth century when it became a means to reconcile capitalist expansion with ideas about natural law and justice.[2] In this sense, the belief in the coexistence of different stages of human evolution was used to justify the civilizing mission of imperial countries, to maintain a system of exploitation of the colonized and enslaved peoples, and, more recently, to legitimize restrictive immigration policies and racial profiling in the context of crime control. Intellectual resources have been variously mobilized in the service of this ideological task, even though it meant the betrayal of the universalistic and egalitarian spirit of the Enlightenment, and tampering with the scientific method. Notwithstanding all their flaws and lack of evidence, theories about racial difference established themselves in many areas of popular culture as well by intersecting nationalist ambitions and rationalizing feelings of ethnic identity.[3]

Racism endures in contemporary democratic societies, although its rationale has shifted for the most part from a biological to a cultural explanation that justifies discrimination as the consequence of intrinsic and unchangeable cultural differences. Nonetheless, Bonilla-Silva (2014) sees a contradiction between the professed color blindness of most Whites and the persistence of color-coded inequality. However, the resolution not to look at race, far from signaling that racism has been defeated, prevents us from effectively tackling race issues. Among these, aural/auditory racial formations are possibly the hardest to identify, being scarcely conceptualized; with the consequence that racial thinking often perpetuates itself without even being noticed. Color blindness, hence, is complemented by a widespread "color deafness," intended as a lack of awareness of and competence about ongoing racialization processes.

Apparently, race and visual culture are strictly related, both in the present and in the past: as racial distinction is embodied in phenotypic traits, visible boundaries segregate groups and populations at various levels. Representations of subordinate races have been laden with primarily visually demeaning imagery. Moreover, the increasing reliance of modern science on optical methods and technological devices[4] prompted scientists to retrieve visual evidence of racial distinction through the measurement of skulls and bones and aided law enforcers and bureaucrats in their administrative tasks by tracking criminals through fingerprints. The justification of the subordination of

Black people during slavery even included the pursuit of textual evidence from the Bible (Hayes 2002).

This emphasis on visuality in modernity has been variously interpreted as an emanation of the ethos of the Enlightenment (Leppert 2004; Classen 1993) and of the role of print in producing a culture based on literacy in place of one based on orality (Lowe 1982; Ong 1982; McLuhan 1964, 1962). On the other hand, Sterne (2003, 3) contends that modernity is defined by "audile techniques," as much as by ways of seeing: such as the adoption of focused ways of listening, which permitted a more efficient operation of the telegraph, or the use of the stethoscope as a means of social distancing. This helps to explain why racial formations also characterize the aural/ auditory domain. The entanglement of linguistics and colonialism, aimed at supporting cultural superiority through a hierarchization of linguistic varieties (Hutton 2002; Pennycook 1998), provides an example of the pervasiveness of racial ideology, regardless of the sense considered. Another example of the salience of auditory racial cues is provided by linguistic profiling, which Baugh (2000) defines as the conflation of racial profiling (based on visual cues) and profiling based on auditory cues of overheard voices in court cases.[5]

This perspective suggests that voice and sound are subject to forms of racialization in a way that is analogous to visible phenotypic traits.[6] This should not be a surprise, given the centrality of the voice in producing and mediating culture and social relations: according to Dolar (2006, 14), "we are social beings by the voice and through the voice." As a specific form of objectification, staged voices in media—voices are recorded, edited, processed, and synchronized, hence technologically and aesthetically mediated—leave further maneuvering room for processes of racial formation, as we shall see in the next section.

Racialized Voices

Altman (1980) argues that images and sound in cinema have a complementary relationship: sometimes sound completes and reinforces the image, and at other times it "uses the image to mask its own action" (79), like a ventriloquist (67). The effect of voice dubbing provides an exemplification of what Altman means by that. Whether the voices are recorded before or after the image, they need to reproduce the illusion of issuing from the same source. Since the basis for this audiovisual illusion is the result of sedimented discourses and practices, the criteria that guide choices in dubbing can reveal how media professionals conceive of race. In particular, the metaphor of the ventriloquist suggests that racialization is the result of the concurrent action of audio and video, which, by drawing on an ingrained albeit not necessarily conscious symbolism, "gives voice" to visual representations of race (and vice versa), thus naturalizing them.

I shall now examine two examples of dubbing, starting with the animated series *The Simpsons*. Here, as unrealistic as it may be, the yellow complexion of its main characters, as well as of the majority of side characters, is intended as an expression of White normativity. On the other hand, certain characters have features that mark their racial difference: not only are they colors other than yellow, but their voices are also more or less racialized, accentuating their "ethnic otherness" (see Denzin 2002, 21–22). For instance, Dr. Hibbert is dark brown and has a speech tune culturally associated with Black people in the media; interestingly, although he and his wife and children are a parody of *The Cosby Show*, their voices and "jive talk" are more racialized than those of the Huxtables in the famous sitcom.[7] Realtor Cookie Kwan is less yellow than the "White" characters, has epicanthic folds, and a heavy Chinese accent. Bumblebee Man is light brown and speaks a "mock Spanish" that includes catch phrases designed for a generic American audience.[8] The nerdy and obnoxious Jewish character Artie Ziff has a prominent nose, curly hair, and the recognizable voice of Jewish comedian Jon Lovitz. Shopkeeper Apu, who is light brown and incarnates a range of stereotypes about Indian immigrants, including a degree in computer science, is dubbed by voice actor Hank Azaria with a thick South Asian accent that Davé (2013, 2005) labels a "brown voice."

Commenting on Channel 4's Black-oriented programs, Gilroy (1983, 131) objects that proponents of multiculturalism "view racial differences as relatively immobile cultural attributes of the communities involved—a position which, it has been pointed out, precisely parallels the contours of contemporary racist thought." In the example discussed above, visual and aural elements convey a stereotypical representation where ethnicity *and* race overlap at various levels: in particular, skin color and somatic traits are complemented by a voice that is "appropriate" for the ethnicized *and* racialized character.[9] In an animated series in which images and situations are largely implausible and often nonsensical, the choice to provide such a consistent and, to quote Gilroy, immobile image of race is striking. The effect is that, even for a product of fiction, stereotypes are reinforced through the mobilization of specific sets of correspondences, namely race–ethnicity on the one hand and image–sound on the other, ultimately reinforcing White normativity.[10]

The following second example will demonstrate how decisions regarding the features of a voice in translated versions relate to racial representations. There is an additional complication in this case, owing to the fact that the cultural background of the target audience may be considerably different from that of the audience of the original version. As a consequence, speech and linguistic features that are racialized in the country of origin may not have an equivalent in the target culture. Translations, Ferrari (2010, 102) maintains, "are bridges between different cultures, not only between different languages."

The Italian adaptation of *Gone with the Wind* was first released in 1948 with subtitles and in 1950 with dubbed Italian language voices. A distinctive feature of this version is the voice of the Mammy figure typical of Blackface. In the original film, a nostalgic and romanticized version of the pre-abolition South, Mammy (Hattie McDaniel) does not speak "proper" English, expresses her ideas frankly and loudly and has a strong ethnic *and* racial identity marked by "Black" intonation and the usage of pidgin. In the 1950 Italian version, in an attempt to translate the character, Mammy was given a guttural and melodic prosody and ill-formed language (all verbs are in the infinitive mood), which reproduced a different racist stereotype that draws on racist representations of Africans that were popular during the colonial expansion of the Fascist regime and, to a certain extent, have remained popular until today.[11] In 1977, Roberto De Leonardis made a new adaptation removing all racialized features in the dubbing. This version, however, did not meet with the favor of the audience and, since then, only the 1950 version has been shown on television and made available on VHS and DVD. With few exceptions, such explicitly racist dubbing practices were abandoned soon afterward based on a "homogenizing convention" that tended to remove sociolinguistic markers (Chiaro 2009, 158). Nonetheless, the lasting success of the original dubbing of *Gone with the Wind* suggests that the exposure to racist stereotypes, even if they seem outdated, creates a "racial grammar" (Bonilla-Silva 2012, 180) that can exert a continuing hold on the collective imagination.[12]

Conclusion

A common misunderstanding has characterized, and to a certain extent continues to characterize, media studies that mainly emphasize verbal and visual communication while paying considerably less attention to sound. Nevertheless, audiovisual media confront viewers with messages that are composites, and both emotionally and cognitively engage their entire sensory capacity. Media content requires consistency between voice and character: the unease resulting from a perceived incongruity between voice and body reveals something about these expectations and how deeply they are ingrained in our culture (Simon 2004). Just as sensory perception forms interconnected patterns that are experienced as natural and inevitable, similarly, ideas about race are entrenched in our ways of understanding the world, linking concepts to sensory abilities (and their shortcomings). In addition to this, sensing involves an emotional component, as the failure of an expectation can generate the discomforting feeling that our system of thought is wrong. Moreover, the realization, for viewers who do not consider themselves racist, that they have a racial bias can be even more discomforting, and threatening to their self-image. In this way, the naturalizing effect that sound has on media images can be very powerful in reproducing racial thinking even without our awareness.

It must be noted that racial identity, in certain historical moments, has been also used for emancipatory purposes. The choice to embrace a particular racial identity, rather than leaving it for the oppressor to define, has characterized racial consciousness movements around the world. As a consequence, certain cultural traits, including ways of speaking, have been deliberately adopted to express racial identity in positive terms.[13] The reproduction of those traits under the ideological umbrella of a dominant White discourse, however, would likely neutralize their emancipatory potential.[14] Therefore, racial formations in the media ultimately serve the interest of the normative and dominant group at the expense of racial minorities, as the initial example of this chapter illustrates. Moreover, and to conclude, the subtlety of certain racialization processes that establish arbitrary associations and make them appear natural almost inevitably perpetuates not only racial stereotypes but also "ways of sensing" racial difference, thus providing the cognitive means for the reproduction of racial ideology.

Notes

1. The argument that race is an inaccurate description of biological variation within the human species is widely accepted; for a comprehensive overview of this issue, see Mukhopadhyay, Henze, and Moses 2007.

2. Ferro (1997) stresses the complementary relationship between colonialism and the modern concept of race as a hierarchical system of human differentiation. On race and capitalism, see Brodkin 2000; Balibar 1999; Wilson 1996.

3. On how racism cannot be simply considered a variety of heterophobia, see Bauman 1989, 62.

4. Poole (1997) argues that the idea of race was shaped by a particular "visual economy" that included the use of photographic archives. About the relationship between biometric methods of surveillance and the state, see also Breckenridge 2014.

5. On linguistic profiling, see also Baugh 2003; Purnell et al. 1999.

6. Stoever-Ackerman (2010, 2011) uses the term "sonic color-line" to indicate the mutually constitutive relationship between sound, listening, and race. For a multisensory construction of race from a historical perspective, see also Back 2011; Smith 2006.

7. The basso profondo Thurl Ravenscroft dubs the character's singing voice and Harry Shearer dubs his speaking voice. The latter dubs several other characters in the series, and the ethnic *and* racial characterization of Hibbert can be considered as a mark of distinction. This strategy, however, is still laden with ideological undertones. Regarding the Huxtables, Rickford and Rickford (2000, 6) recall Bill Cosby's hostility toward the official recognition of Black American English, also known as African American Vernacular English (AAVE).

8. "Mock Spanish, with its relentlessly anglicized and even hyperanglicized and boldly mispronounced phonology and pidgin grammar, assigns native Spanish fluency to the realm of the 'un-American'" (Hill 2008, 148).

9. It is worth noting that, in *The Simpsons*, the animation is made after the dialogue is recorded (Rhodes 1990).

10. For a discussion on how racializing forms in everyday language elevate "White America" as they disparage minority languages and their speakers, see Hill 2008.

11. In 1937, the Italian Ministry of Popular Culture issued a propaganda campaign against so-called racial hybridity, encouraging morally and physically degrading representations of "colored races" (Randazzo 2008, 265). On the stereotypical African in Fascist cinema, see Boggio 2003.

12. The continuing accessibility of racist stereotypes in Italian media is reflected by the revival of similar characterizations between the late 1970s and the early 1980s, such as American actress and singer Edith Peters's interpretation of a character called Mamie, which conjures up all of the same stereotypes linked to the homonymous character from *Gone with the Wind* (*Il Bisbetico Domato*, starring famous Italian singer, actor, and entertainer Adriano Celentano), and Nigerian-Italian actor Isaac George's interpretation of a stereotyped African immigrant speaking a racialized "pidgin Italian" in several films, television shows, and ads. Going back to the former example, Ferrari (2010, 104) shows how the Italian translation of *The Simpsons* reproduces the ethnic and racial multiplicity of the American version by reterritorializing the original accents "across Italian geographical and stereotypical lines." Interestingly, Gracie Films, coproducer of the series with FOX, was directly involved in the translation process for the Italian version (ibid., 101–102).

13. This is especially apparent in music and its articulation of racial identity (see, e.g., Rose 1994). On "Black" verbal styles, see Lanehart 2009; Smitherman 2006; Mitchell-Kernan 1999; and Smitherman 1985. On Black consciousness, see Mngxitama et al. 2008; Guillory and Green 1998; Lewis and Bryan 1988; and Stuckey 1987.

14. Cashmore (1997) argues that Blacks people's success in the entertainment industry is contingent on their conforming to Whites' images of Blacks. The case of Hattie McDaniel's performance in *Gone with the Wind* is symptomatic of the consequences of this coercion, in that she was the first Black person to win an Academy Award, but, for the same reason, she was sometimes subjected to harsh criticism within the Black community (Banks 2008, 69).

References

Bibliography

Altman, Rick. 1980. Moving lips: Cinema as ventriloquism. *Yale French Studies* 60 (Cinema/Sound): 67–79.

Back, Les. 2011. Trust your senses? War, memory, and the racist nervous system. *Senses and Society* 6 (3): 306–324.

Balibar, Étienne. 1999. Class racism. In *Race, Identity, and Citizenship: A Reader*, ed. Rodolfo D. Torres, Louis F. Mirón, and Jonathan Xavier Inda, 322–333. Oxford: Blackwell.

Banks, Ingrid. 2008. Women in film. In *African Americans and Popular Culture*, ed. Todd Boyd, 67–87. Westport, CT: Praeger.

Baugh, John. 2000. Racial identification by speech. *American Speech* 75 (4): 362–364.

Baugh, John. 2003. Linguistic profiling. In *Black Linguistics: Language, Society, and Politics in Africa and the Americas*, ed. Sinfree Makoni, Geneva Smitherman, Arnetha F. Ball, and Arthur K. Spears, 155–168. London: Routledge.

Bauman, Zygmunt. 1989. *Modernity and the Holocaust*. Cambridge, MA: Polity Press.

Boggio, Cecilia. 2003. Black shirts/Black skins: Fascist Italy's colonial anxieties and "Lo Squadrone Bianco." In *A Place in the Sun: Africa in Italian Colonial Culture from Post-Unification to the Present*, ed. Patrizia Palumbo, 279–298. Berkeley, CA: University of California Press.

Bonilla-Silva, Eduardo. 2012. The invisible weight of Whiteness: The racial grammar of everyday life in contemporary America. *Ethnic and Racial Studies* 35 (2): 173–194.

Bonilla-Silva, Eduardo. 2014. *Racism without Racists: Color-Blind Racism and the Persistence of Racial Inequality in America*, 4th ed. Lanham, MA: Rowman & Littlefield.

Breckenridge, Keith. 2014. *Biometric State: The Global Politics of Identification and Surveillance in South Africa, 1850 to the Present*. Cambridge: Cambridge University Press.

Brodkin, Karen. 2000. Global capitalism: What's race got to do with it? *American Ethnologist* 27 (2): 237–256.

Cashmore, Ellis. 1997. *The Black Culture Industry*. London: Routledge.

Chiaro, Delia. 2009. Issues in audiovisual translation. In *The Routledge Companion to Translation Studies*, rev. ed., ed. Jeremy Munday, 141–165. Abingdon: Routledge.

Classen, Constance. 1993. *Worlds of Sense: Exploring the Senses in History and across Cultures*. London: Routledge.

Davé, Shilpa. 2013. *Indian Accents: Brown Voice and Racial Performance in American Television and Film*. Champaign: University of Illinois Press.

Davé, Shilpa. 2005. Apu's brown voice: Cultural inflection and South Asian accents. In *East Main Street: Asian American Popular Culture*, ed. Shilpa Davé, LeiLani Nishime, and Tasha Oren, 313–336. New York: NYU Press.

Denzin, Norman K. 2002. *Reading Race: Hollywood and the Cinema of Racial Violence*. Thousand Oaks, CA: Sage.

Dolar, Mladen. 2006. *A Voice and Nothing More*. Cambridge, MA: MIT Press.

Ferrari, Chiara. 2010. Dubbing *The Simpsons*: Or How Groundskeeper Willie lost his kilt in Sardinia. In *Beyond Monopoly: Globalization and Contemporary Italian Media*, ed. Michela Ardizzoni and Chiara Ferrari, 101–128. Lanham, MD: Lexington Books.

Ferro, Marc. 1997. *Colonization: A Global History*. London: Routledge.

Gilroy, Paul. 1983. C4—Bridgehead or Bantustan? *Screen* 24 (4–5): 130–136.

Guillory, Monique, and Richard Green, eds. 1998. *Soul: Black Power, Politics, and Pleasure*. New York: NYU Press.

Hayes, Stephen R. 2002. *Noah's Curse: The Biblical Justification of American Slavery*. Oxford: Oxford University Press.

Hill, Jane H. 2008. *The Everyday Language of White Racism*. Malden, MA: Wiley-Blackwell.

Howes, David. 2003. *Sensual Relations: Engaging the Senses in Culture and Social Theory*. Ann Arbor: University of Michigan Press.

Hutton, Christopher. 2002. The language myth and the race myth: Evil twins of modern identity politics? In *The Language Myth in Western Culture*, ed. Roy Harris, 118–138. Richmond: Curzon Press.

Lanehart, Sonja L., ed. 2009. *African American Women's Language: Discourse, Education and Identity*. Newcastle upon Tyne: Cambridge Scholars Publishing.

Leppert, Richard. 2004. The social discipline of listening. In *Aural Cultures*, ed. Jim Drobnick, 19–35. Toronto: YYZ Books.

Lewis, Rupert, and Patrick Bryan, eds. 1988. *Garvey: His Work and Impact*. Mona, Jamaica: University of the West Indies.

Lowe, Donald M. 1982. *History of Bourgeois Perception*. Chicago: University of Chicago Press.

McLuhan, Marshall. 1962. *The Gutenberg Galaxy: The Making of Typographic Man*. Toronto: University of Toronto Press.

McLuhan, Marshall. 1964. *Understanding Media: The Extension of Man*. New York: McGraw-Hill.

Mitchell-Kernan, Claudia. (1972) 1999. Signifying, loud-talking, and marking. In *Signifyin(g), Sanctifyin', and Slam Dunking: A Reader in African American Expressive Culture*, ed. Gena Dagel Caponi, 309–330. Amherst: University of Massachusetts Press.

Mngxitama, Andile, Aamanda Alexander, and Nigel C. Gibson, eds. 2008. *Biko Lives! Contesting the Legacies of Steve Biko*. New York: Palgrave Macmillan.

Modestine, Yasmine. 2008. Cinéma: "Le Métier du Doublage a un Problème avec la Couleur." *Rue89*, April 5. http://rue89.nouvelobs.com/2008/04/05/cinema-le-metier-du-doublage-a-un-probleme-avec-la-couleur.

Mukhopadhyay, Carol, Rosemary C. Henze, and Yolanda T. Moses. 2007. *How Real Is Race? A Sourcebook on Race, Culture, and Biology*. Lanham, MD: Rowman & Littlefield Education.

Omi, Michael, and Howard Winant. 1986. *Racial Formation in the United States: From the 1960s to the 1980s*. New York: Routledge & Kegan Paul.

Omi, Michael, and Howard Winant. 2015. *Racial Formation in the United States*, 3rd ed. New York: Routledge.

Ong, Walter. 1982. *Orality and Literacy: The Technologizing of the Word*. New York: Routledge.

Pennycook, Alastair. 1998. *English and the Discourse of Colonialism*. New York: Routledge.

Poole, Deborah. 1997. *Vision, Race, and Modernity: A Visual Economy of the Andean Image World*. Princeton, NJ: Princeton University Press.

Purnell, Thomas, William Idsardi, and John Baugh. 1999. Perceptual and phonetic experiments on American English dialect identification. *Journal of Language and Social Psychology* 18 (1): 10–30.

Randazzo, Antonella. 2008. *L'Africa del Duce: I Crimini Fascisti in Africa*. Varese: Arterigere.

Rhodes, Joe. 1990. The making of "The Simpsons": Behind the scenes of America's funniest new animated family. *Entertainment Weekly*, May 18. http://www.ew.com/article/1990/05/18/making-simpsons.

Rickford, John Russell, and Russell John Rickford. 2000. *Spoken Soul: The Story of Black English*. New York: Wiley.

Rose, Tricia. 1994. *Black Noise: Rap Music and Black Culture in Contemporary America*. Middletown, CT: Wesleyan University Press.

Simon, Sherry. 2004. Accidental voices: The return of the countertenor. In *Aural Cultures*, ed. Jim Drobnick, 110–119. Toronto: YYZ Books.

Smith, Mark M. 2006. *How Race Is Made: Slavery, Segregation, and the Senses*. Chapel Hill: University of North Carolina Press.

Smitherman, Geneva. (1977) 1985. *Talkin' and Testifyin': The Language of Black America*, rev. ed. Detroit, MI: Wayne State University Press.

Smitherman, Geneva. 2006. *Word from the Mother: Language and African Americans*. New York: Routledge.

Sterne, Jonathan. 2003. *The Audible Past: Cultural Origins of Sound Reproduction*. Durham, NC: Duke University Press.

Stoever-Ackerman, Jennifer. 2010. Splicing the sonic color-line: Tony Schwartz remixes Nueva York. *Social Text* 102:59–86.

Stoever-Ackerman, Jennifer. 2011. The word and the sound: The sonic color-line in Frederick Douglass's 1845 narrative. *Sound Effects* 1 (1): 20–36.

Stuckey, Sterling. 1987. *Slave Culture: Nationalist Theory and the Foundations of Black America*. New York: Oxford University Press.

Wilson, Carter A. 1996. *Racism: From Slavery to Advanced Capitalism*. Thousand Oaks, CA: Sage.

Sound and Media

Gone with the Wind. 1939. Directors: Victor Fleming, George Cukor, and Sam Wood. Feature film, Warner Bros.

Il Bisbetico Domato. 1980. Directors: Franco Castellano and Giuseppe Moccia. Feature Film, Capital Film.

The Cosby Show. 1984–1992. Created by Bill Cosby, Michael Leeson, and E. D. Weinberger. Television series, Bill Cosby, Carsey-Werner Company, and NBC.

The Simpsons. 1989–present. Created by Matt Groening. Animated television series, Gracie Films and 20th Century Fox Television.

8 Sound Studies across Continents: A Multidisciplinary Research Approach

Thomas Burkhalter

Intro

Many musicians and sound artists in Africa, Asia, and Latin America work with similar musical material, aesthetic approaches, and techniques as the musicians and sound artists in Europe and the United States. They would like recognition for their catchy indie tunes and death metal, their electronic beats, or their refined noise and sound collages—and not for being from "far away" or even "exotic" places (World Music 2.0). However, in my fieldwork in Beirut, in my reportages in Belgrade, Bamako, Cairo, and Istanbul, and in my work with international arts councils, I keep hearing musicians and sound artists complain that Euro-American platforms still focus on musical and cultural differences when dealing with music from Africa, Asia, and Latin America. They argue that world music labels, agents, and networks promote musicians who work mainly with "traditional" music, often adapted to the tastes of cultured middle class listeners. Rappers in Africa, Asia, and Latin America also complain. They enter our Euro-American platforms for the most part through nongovernmental organizations (NGOs). They have the most success when they focus on social and political themes (e.g., war, violence, gender, women's rights) and when they represent "voices for change." Sound artists from fields such as free improvised music or computer music get invited by art institutions, with funds from international arts councils and embassies. They also need to concentrate on social or political issues in order to attract interest—and to not be criticized for being "Westernized" (a criticism they receive from both Euro-American and local perspectives). Lebanese composer and pianist Joelle Khoury calls this the "heaviest psychological load," with which she struggles:

A true composition is the fruit of a unique idea, developed into a concept, based in what the composer has heard, aiming at expressing an individual point of the view, a certain personality. Therefore it is neither "western" nor "oriental," taking into account the fact that we have all heard numerous styles of music. This issue is the heaviest psychological load the modern Arab artist has to face. Do we succumb to voyeurism and accept to become a "vitrine" to the West? Do we play it safe and stagnate into repetitions of the same by sticking to the old? The problem is we need the West since they are the main funders of our works. It is very tricky to remain true to oneself and yet please. (Khoury 2011)

In Conflict: Academic Theory and Daily Practice

While terms like "modernity," "locality," and "authenticity" are debated and deconstructed in academia, they are still in use in cultural networks and markets. This discrepancy between academic theory and daily practice is one of the tensions that are important in my work as an ethnomusicologist, music journalist, and cultural producer. I am interested in how the new "democratic" means of production and distribution manage to change power relations between musicians and markets. Do an increased number of musical positions reach listeners in more places than ever? Do new "successful" positions help reconfigure how we understand music—and even how we understand the world? Discussing these possible changes with musicians from the Middle East, Africa, Asia, and Latin America gives my approach an activist twist. In line with Stuart Hall and Lawrence Grossberg—although certainly on a much less experienced level—I would thus position myself as a "political intellectual" (Grossberg 2010). One aim of my approach is to open up debates on how musicians from the "non-Western" world interact with music and cultural markets worldwide. Do they—as they argue above—become confronted with stereotypical ideas, unrealistic demands, and old "North–South" power relations—especially when trying to perform in Euro-American markets?

An Experimental Research Approach

To achieve a close reading of discrepancies and interrelations between musical production, musicians' motives, and reception on local and on Euro-American platforms, one of my goals is to design a multidisciplinary research approach that combines theories from ethnomusicology, sound studies, popular music and media studies, cultural studies, and social anthropology. To stand up to the test, this approach has to be close to the musicians and the music, and it should not ignore daily praxis and power balances in cultural markets.

Step One: Sonic Surroundings, Past and Present

In this approach, sound studies serve as a door opener—sound comes first. I learn about "place" and "time" through listening. While conducting fieldwork, I first focus on the sonic environment with which musicians are surrounded. In addition, I observe "mediascapes," listen to politicians, ads, jingles, and pop music, and—although I do not completely disregard it—I do not place primary focus on traditional music, as many ethnomusicologists have done.

In my field research in Beirut (Burkhalter 2013), I worked with precisely this approach. I observed the present and the past of Beirut through listening first—and through digging through archives and discussing sonic memories with musicians from both younger and older generations. However, analyzing the interrelations between "sonic surroundings" and tracks is only the first step. It can lead to superficial, oversimplified, and romanticized readings such as: musicians are inspired for the most part by the sounds and noises of their environment. There is a necessary second step: interviewing musicians and actors in the field and analyzing the possibilities and challenges in musical and cultural markets. In Beirut, through listening, interviews, and discussions, I compiled interrelations from six major past developments: first, the urbanization of Bedouin and village music and the establishment of *one* Lebanese Music for concert halls and festivals—to which today's key musicians from Beirut are opposed; second, the Europeanization of *maqam*-based music—to which many of today's musicians are opposed; third, noises and propaganda musics from the Lebanese Civil War that inspire many musicians—and which some claim evoke nostalgic sonic memories for them that they then try to imitate or use for parody and other modes of criticism; fourth, leftist protest music—which many musicians today consider outdated (they argue that beautiful poems should not be sung about the horrors of war); fifth, the (psychedelic) rock music of the 1960s and '70s—which most of the musicians do not know, although it paved the way for "alternative" music in Beirut; and sixth, the commercialized pan-Arabic pop music—an omnipresent parallel universe the musicians try to ignore, while at the same time struggling for media representation themselves.

Step Two: Three Research Perspectives

Step two approaches musicians and music from various perspectives:

• The perspective "music making as a practice" involves: the processes of music making and music production (writing, composing, producing, recording, mixing); hardware and software and their inherent laws; the impact of musical influences and references; and trends within the local music circles and transnational niche genres.

• The perspective "music as a media product" involves reception and further processing: it looks at music, culture, and the arts market(s) (including their possibilities); secondary markets (video clips, film music, game soundtracks), and reception ideologies of funding organizations, media, and fans. This category offers various options for action: promotion and networking strategies, representation ideologies, and performance strategies.

• The perspective "musician as an actor" involves all spheres that affect the musician as a human being and artist: for example, the geographical position in which he or she lives; mobility; financial possibilities; position of the musician in his or her country; and knowledge (through socialization and education). All of this leads to the musicians' motives—why he or she makes music. It shows how musicians position themselves in the world.

These three main perspectives overlap, and they are only working categories. However, perspective one (music making as a practice) requires an analysis of music and sound; perspective two (music as a media product) a broad analysis of the reception platforms, including their networks and power structures; and perspective three (musician as an actor) an empirical cultural studies approach.

Reception Test (Music Making as a Process)

Perspective one requires the closest possible analysis of key musicians' tracks, compositions, and songs. I conduct multisited research and use the emic approach and ethical criteria to do so. For my research in Beirut (2013), and for an article on Palestinian singer Kamilya Jubran and Swiss computer musician Werner Hasler (see Burkhalter, Jacke, and Passaro 2012), I carried out reception tests. Inspired by reception tests by Tagg (2001, 9–14) and Steinholt (2005), I sent key tracks to expert listeners (e.g., ethnomusicologists, pop music scholars, musicians, producers) of specific music genres in Europe, the United States, and the Arab world. The goal was to find as many readings of these specific tracks and songs as possible, and to then later discuss these findings with the musicians. Through applying this approach, I hoped to reach deeper levels of discussion.

Reception Platforms (Music as a Media Product)

I use the term *platform* whenever I speak of where these tracks, songs, or video clips (I call them media products) appear: on a local stage in Beirut, on an international stage in London, on a media platform like SoundCloud or YouTube, or in a computer game. With their media products, musicians perform on several of these platforms simultaneously. They might trim their media products to fit on these platforms—or they challenge platform conventions. Similar to concert performances, where musicians take time beforehand to reflect on set lists that will work well, they perform strategically (and knowledgeably, to a greater or lesser degree) on various local and transnational platforms. It is the musicians' strategic use of these platforms and the demands on these platforms that are at the center of interest here. I highlight the interrelations between the production of media products and reception on these possible platforms. In this environment, musicians act strategically within a complex network of organizers, agencies, cultural sponsors, and media.

Multisited Interviews with Musicians and Network Partners (Musicians as Actors)

Using the third perspective, I interview musicians. Through these interviews, I learn more about their education and knowledge, and their financial, cultural, and political position within local and transnational societies and communities, and I understand more about their "cultural capital" to perform and act within local and transnational music markets. Interviews with network partners help to construct a clearer view of these musicians' networks, situating them in the broader artistic, cultural, and civil society milieus of global and globalizing societies. Musicians in Beirut work with actors from Europe and the United States: NGOs, concert organizers, arts curators, multipliers as bloggers and journalists, and embassies. Kiwan and Meinhof (2011) use the term "hubs" to analyze human, spatial, institutional, and accidental bodies that support and interact with musicians:

- Human hubs and their social networks "cross over and link very different geographic spaces" (ibid., 4).
- Spatial hubs refer to the important role of capital and metropolitan cities in the Northern Hemisphere (such as Paris and London) as key nodes for migration flows and migration cultures. According to Kiwan, the cities in the South play similar roles for both the translocal movements of artists within their nations and the transnational multidirectional movements between North and South (ibid., 5).
- Institutional hubs include particular key institutions and organizations that help organize, or are themselves integrated into, artists' networks. They link human and spatial hubs (ibid., 6). Here we find the cultural institutions and the NGOs of the North.
- Accidental hubs involve, for example, the researcher, as Kiwan and Meinhof state: "We were building up the very network structures that we are researching. ... In working with professional or aspiring artists, the chance of our turning into accidental hubs is arguably even stronger than in the anthropology of everyday practices" (ibid., 7).

Discussing Place in Music

This setting is challenging. The ideal is to conduct a multisited ethnography and to achieve results that demonstrate the complexities of musicians' actions in today's globalized and digitalized world. This highly abstract yet flexible theoretical and methodological framework is required for us to be able to approach, interpret, and provide "thick descriptions" (in the sense of Geertz 1987) of very specific examples in music practice. In this manner, I wish to fulfill the fundamental requirement of popular music scholar Binas-Preisendörfer. She argues that "a scientific exploration of musical phenomena in a modern globalized and mediated world demands both reflexive theoretical concepts as well as very specific, small-scale studies" (2010, 103). The approach should be close to the musicians, the music, and the market. Moreover, it follows Veit Erlmann's suggestion to analyze the ways in which histories of cultural, social, or political contexts are inscribed into music (Erlmann 1995, 10).

The following are some resulting research questions:

• Which musical and nonmusical spheres of influence affect the music making of musicians?
• How do these spheres of influence affect them: are they binding, inspiring, or do they offer positioning options or playing opportunities?
• How do these interactions between these various spheres of influence become inscribed in their media products?

Outro

In the best-case scenario, answering these questions and working with the suggested theoretical and methodological approaches would help to change old and simplified conceptions of "cultures" within the creative industries. Like Grossberg, I believe that what we say and do as intellectuals matters, "because bad stories make bad politics" (Grossberg 2010, 290).

In this process, sound studies "across continents" can serve as a door opener. We start our research with open ears, focusing on the daily instead of the nondaily. However, what we hear needs to be challenged by empirical research on and analysis of specific reception platforms. Through sound studies—step one—I gain access to the past, while through step two I gain knowledge of the complexities of actions in the present. Step two's three research perspectives further lead to a "thick description"—and mark the distinction from simple observation.

References

Binas-Preisendörfer, Susanne. 2010. *Klänge im Zeitalter ihrer medialen Verfügbarkeit: Ein Beitrag zu Fragen von Popmusik und Globalisierung.* Bielefeld: transcript.

Burkhalter, Thomas. 2013. *Local Music Scenes and Globalization: Transnational Platforms in Beirut.* New York: Routledge.

Burkhalter, Thomas, Christoph Jacke, and Sandra Passaro. 2012. Das Stück "Wanabni" der Palästinenserin Kamilya Jubran und des Schweizers Werner Hasler im multilokalen Hörtest: Eine multiperspektivische Analyse. In *Black Box Pop: Analysen populärer Musik,* ed. Dietrich Helms and Thomas Phelps, 227–256. Bielefeld: transcript.

Erlmann, Veit. 1995. Ideologie der Differenz: Zur Ästhetik der World Music. *PopScriptum* 3 (3): 6–29.

Geertz, Clifford, ed. 1987. Dichte Beschreibung: Bemerkungen zu einer deutenden Theorie der Kultur. In *Dichte Beschreibung: Beiträge zum Verstehen kultureller Systeme*, 7–43. Frankfurt am Main: Suhrkamp.

Grossberg, Lawrence. 2010. *Cultural Studies in the Future Tense*. Durham, NC: Duke University Press.

Khoury, Joelle. 2011. Music in the Arab World today. Paper presented at roundtable "The Shape of Sounds to Come" at fifth donor's meetings of Arab Fund for Arts and Culture, Beirut, December 17.

Kiwan, Nadia, and Ulrike Hanna Meinhof. 2011. *Cultural Globalization and Music: African Artists in Transnational Networks*. Hampshire: Palgrave Macmillan.

Steinholt, Yngvar Bordewich. 2005. *Rock in the Reservation: Songs from the Leningrad Rock Club, 1981–1986*. New York: Mass Media Music Scholars Press.

Tagg, Philipp. 2001. Music analysis for "non-musos." Paper presented at Popular Music Analysis conference, University of Cardiff, November 17.

Establishing New Methodologies

9 Ethnography and Archival Research in Studying Cultures of Sound

Karin Bijsterveld

Driving the Gray Grooves

In the spring of 2011, a famous Dutch columnist wrote that we should pity people living nearby eight-lane highways. They were, he noted, tragically exposed to the brutal and unbearable noise of traffic, day and night. "That's why the authorities have raised noise-reducing partitions. For miles and miles. In former days, you could watch the meadows, the pollarded willows, the cows, the farmhouses. Whether you enjoyed it or not, it was definitely rustic. Now you drive through a gray groove. But at least you can turn on the radio" (Montag 2011, 37). The columnist, a man in his eighties, was not very positive about what contemporary radio had to offer him, as he admitted in a grumpy-old-man style in the same newspaper item. Yet he made something explicit that had not been articulated often before: a connection between driving the gray grooves of the modern-day highway with its high-rising noise barriers and listening to the radio—which often broadcasts the gray grooves of recorded music.

The remark was like an archaeological snippet of an underobserved culture. It referred to the recent history of noise abatement, to the everyday sensory experiences of driving and suggested a heroic role for the car radio: helping the motorist survive a viewless drive with the auditory imagination offered by radio. Such are the archaeological snippets that scholars studying cultures of sound should have an eye and ear for. In this case, the observation made me wonder whether there might be wider historical connections between the loss of control over driving—caused by highway noise barriers, traffic jams, and electronic matrix panels displaying speed limits—and the rise of in-car audio entertainment such as radio with music and traffic information, as well as audio books and MP3 players. I knew that Michael Bull had already shown how important mobile listening is in today's culture, on the street and in the car, and how audio users create their own "audiotopia" while being on the move (Bull 2000, 2001, 2003, 2007). But I now wanted to know how we had ended up in this situation, historically speaking.

Studying cultures of sound implies an interest in the often taken-for-granted ways in which people give meaning to the sounds they are surrounded with, in how they routinely act upon and use those sounds, and in how that has changed over time. But how can we get access to what is taken for granted in past and contemporary cultures—a task that is even more challenging if it concerns sensory experiences and skills that are hard to grasp? This essay will explain how we can do this by focusing on cultural practices and using theoretical approaches from history and anthropology, as well as their methodological tools: archival research, participant observation, qualitative interviewing, and "experimental" ethnography. It will draw its examples from research on everyday tape recorder usage and the shifting cultural meanings of car radio—examples highlighting the role of sound in popular culture.

History and Anthropology: Unraveling the Mysteries of Tape Recorder Use

It has become a modern-day academic truism to say that a particular topic requires an interdisciplinary approach. How many of us have not said this about sound or sound studies? There is no intrinsic need to approach anything in an interdisciplinary way, however. It does not automatically lead to innovative insights, as science and technology studies (STS) scholars have shown (Barry et al. 2008). More often than not, "interdisciplinary" work actually looks like "disciplinary Halloween," as Jonathan Sterne (2009) has called it. Scholars go to a party, hire a theory from a neighboring workplace to dress up, and go on with our own life the next day. In contrast, we should teach our students that it is not their research topic but their research question that determines whether or not they need more than one discipline, and which aspect of that discipline: theory, concept, method, or style.

One hard-to-explain aspect of doing research, however, is how to find interesting questions. Students are used to going to the library, reading the literature about their topic of interest, and searching for the niche they could jump into and contribute to. They hope to locate the gaps in our knowledge. Yet the best-kept secret about interdisciplinary research is not that it helps to answer new questions, even if it does so at times, but that it generates new questions. It may open your mind to seeing what is uncommon in the common, which is what cultural scholars love to do.

Let me give an example. It was my background in history and in history of technology that made me think, many years ago, that the postwar reel-to-reel recorder, or tape recorder, might be an interesting topic for research. Yet it was my combined openness for ethnography and archival research that generated my main question. Among the first things I did was contact the Philips Company Archives to acquire access to documents about the introduction and marketing of the tape recorder, and check out the websites of a group of consumers highly involved in using tape recorders: the "sound hunters," or amateur sound recordists. I started visiting them and talking to them—at their homes, at interview locations, and later at one of their international meetings. Well, one might say, that was still perfectly within the range of an historian's job. Was that not oral history? Yes, but it was slightly or even significantly more than that.

While meeting with the sound hunters, I made the kind of detailed field notes I had learned from anthropologists. Most of the sound hunters were relatively wealthy men in their forties, fifties, and sixties. Their home archives were extensive and very well-ordered, both in terms of paper documents and recordings. In their meetings, they celebrated the international links between sound hunters by putting small national flags on their tables. In their written and oral evaluations of sound recordings, they highlighted the importance of departing from conventional radio formats, and of recording things that were hard to record, like fast-moving objects. It was thus mainly a male hobby, modeled on hunting, and international in orientation, but it clearly did not attract any youngsters anymore. The sound hunters constantly expressed the conviction—in opening speeches, table talk, and while driving me back to railway stations after our meetings—that photo, video, and the visual orientation of society at large had put them out of the competition. Yet the archives showed that Philips (and other manufacturers such as Grundig, as I would soon find out) had done everything to create a connection between visual practices like taking photos and the auditory practice of making recordings: the visual had been framed as a shining example to follow rather than as the enemy to defeat. Why, then, had amateur recording not reached the status and popularity of taking pictures?

The manufacturers had expected a lot from the reel-to-reel recorder as a family sound album, stressing the similarities between taking pictures and making recordings of important family events. They also mentioned activities such as creating sound letters (for families overseas), making amateur radio plays, or playing recorded music, but the tape recorder was not sold as widely as they hoped for, and the manufacturers almost desperately drew up ever longer lists of what could be done with the tape recorder. However, the sound hunters—the most enthusiastic adopters of the tape recorder—were more interested in hunting for hard-to-record sounds than in making family albums, and had a somewhat ambiguous relationship with radio, taking advantage of the opportunity to broadcast their productions but also departing from studio recording. Moreover, for most tape recorder users, playing music became much more important than making a family sound album. Many of the reasons for the demise of both the tape recorder and sound hunting were rooted in the historical context of the 1950s and 1960s. The analogies between sound recording and photo taking did not work out as the manufacturers expected. While making and collecting photos had a model to follow in painting (framing, putting up, and exhibiting photos in the same way as paintings), which was relatively easy for amateurs to emulate at home, such a model practice—beyond radio—did not exist for sound recording. "Nor was it possible to pass sounds, like photos, from hand to hand at parties or to hang them on the wall" (Bijsterveld 2004, 629).

When a colleague and I finally decided to study "regular" tape recorder users as well—users outside the sound hunters' societies—the relevance of the remark about parties became even more evident. Philips and other tape recorder manufacturers had clearly focused on the similarities between *making* family photo and sound albums, but had not taken into account the practices of *retrieving* and *collectively listening* to recordings. This was clarified in more detail when we learned from anthropologists' work once again. Roger Silverstone and Eric Hirsch (1992) have analyzed how households communicate their status to the outside world with the help of devices they buy for their homes, and have distinguished four phases in this process of domestication: appropriation (buying an object), objectification (expressing norms and social status through the displaying of a device), incorporation (of the object in everyday practices), and conversion (giving the device a role in the relationship between owners and persons outside the owners' households).

From the interviews with tape recorder users, the information they provided about objectification and conversion was most revealing. Initially, most of them had the tape recorder displayed in the living room, but usually out of sight. Moreover, it often changed places from the living room to the bedroom, study, and other rooms. One reason for this was that the women of the house did not want the living room to be a place for tinkering and the messiness of wires and cables. Thus the tape recorder did not become part of the household furniture as happened with the television. Even more problematic, however, was the tape recorder's role in communicating with the outside world. Just playing hours of music was fine, but finding a particular moment on a family recording by winding through the tape could be very time-consuming. Neither notebooks with information about the content of the tapes nor the archival and retrieval gadgets offered by manufacturers of tapes and tape recorders happened to be of long-term assistance to most consumers—unless they displayed the precision in archiving that the sound hunters had developed. And even when everyone eventually listened to the correct tape section, new problems popped up. People felt ashamed of their senseless, recorded talk, or did not recall what they were hearing—they would not recognize the voices, or the recording had simply lost its meaning over time (Bijsterveld and Jacobs 2009).

Those examples demonstrate, therefore, how combining archival research with ethnographic observations, qualitative interviews, and anthropological theory helped us to unpack the auditory cultures of amateur sound recording, sound retrieving, and collective listening.

Archaeology of Anthropology: A Car Radio Manufacturer's Take on Culture

Yet we, as scholars and students, are not the only ones doing anthropological or ethnographic work. When manufacturers prepare for the development and introduction of new technologies, they may also behave like semi-anthropologists in order to make sure that a novel technology "lands" well in a particular society and culture. And when we dive into the history of such introductions, we may thus be involved in an *archaeology of* a manufacturer's *anthropology*.

One brief example comes, once again, from the Philips Archives. In 1929, a Philips correspondent in the United States reported on the first built-in car radio. Apart from noting some problems with the car radio's reception and batteries, he also sent the Philips Company in Eindhoven press clippings about a social controversy in the United States. There were deep concerns about how radio might dangerously distract motorists and cause accidents, notably in urban areas with high traffic density. One of the events that generated these concerns was an accident in which a taxi driver who had his car radio turned on had hit an elderly man. It was assumed that the taxi driver had been distracted by his car radio. This assumption resulted in proposals for a ban on car radios as well as for the restriction of playing the car radio in parked cars only.

When Philips considered introducing car radios in the Netherlands and the rest of Europe, it was not yet clear how this controversy would play out—eventually, radio lobbyists effectively prevented a ban. Prior to that moment, however, Philips had to think about how to anticipate such debates. The company did so by organizing car radio test drives in the areas around their factories and by thinking about how to approach car dealers. Its marketers found a way out of the problem by constructing stories in which they crafted a particular niche for car radio use. They underlined that while car radio might be less fit for driving in urban settings, it was highly recommendable in the countryside, where it would enhance safety by making motorists drive more slowly and keeping them attentive during long and potentially boring stretches. Moreover, drivers would feel emotionally safer with the sonic accompaniment of radio. An additional advantage of this strategy was that one of the problems that had popped up during the test drives—poor reception in an environment with high buildings and narrow streets—would remain less evident. By carefully taking into account the cultural reception of car radio in the United States, the character of Dutch public space, the technicalities of car radio and gender issues, Philips started paving the way for car radio in Europe (Bijsterveld 2010).

When I submitted this analysis to *The Senses and Society*, editor and anthropologist David Howes rightly remarked that what Philips had done was not so much a proper anthropology or ethnography, but rather expressed a particular take on culture, or culturology. I have gratefully adopted that notion. Nevertheless, approaching the history

Figure 9.1
Female driver tuning into the car radio, advertisement, 1950 (photo courtesy of Philips Company Archives).

of technologies that were once new, such as the car radio, by following manufacturers and other relevant actors, such as users, in terms of how they make sense of the culture when producing, introducing, and appropriating new technologies, is a mindset that has proved to be very successful in STS (see, e.g., Oudshoorn and Pinch 2003). In my case, it clarified how Philips articulated and codefined a new auditory culture of listening to the car radio, as listening to music during long drives became one of the most widely appreciated affordances of car radio.

Anthropology as Archaeology: The Car Radio and Coping with the Modern Highway

The snippet of cultural evidence with which I started this essay, and the way it helped to articulate my questions about the car radio, is not so much an example of an archaeology *of* other people's anthropology as discussed in the previous section, but of anthropology *as* archaeology. It is about taking all available cultural evidence into account, about searching for and finding the fragment that does not fit into the picture and that drives the researcher toward an analysis that *is* able to bring all observations under the umbrella of a consistent set of interpretations. Was the link, suggested by the Dutch columnist Montag, between driving highways with high-rising barriers and listening to car radio just an incidental remark, or did it fit into wider sociocultural trends?

It was not that we—the team of researchers working on the history of sound in and of the car—thought that radio had not been associated with escaping the negative aspects of driving (Bijsterveld et al. 2014). We already knew the early history of the car radio as discussed in the previous paragraph, and had found evidence for the gradual transformation of the car radio from being a companion on dull drives to a mood regulator on overcrowded roads between the 1950s and 1970s (Bijsterveld 2010).[1] Even as early as 1937, the BBC television program *Woman's Page* associated the car radio with providing relief during long drives *and* traffic jams. Presenter and British novelist Ursula Bloom indicated that the radio was her "most precious" car possession, "which can while away long waits in traffic jams or to relieve the monotony of a long drive" (*Woman's Page* 1937). Yet the link between the car radio and traffic jams was still rarely established. One reason may simply have been that only a small percentage of British cars had radio installed.

It was first in the 1960s and 1970s that the idea of the car radio as the driver's companion in coping with traffic jams became predominant. In 1971, the BBC television series *Tomorrow's World* had an item about the use of "solid state radio" in the car. The anchorman heralded this type of car radio for its automatic tuning: it would make driving with the car radio safer since it made drivers keep their hands on the wheel while driving. But he also lauded the superb sound quality it provided and the future option—already established abroad—of radio channels that would automatically feed the latest information on local traffic situations into radio broadcasts. These new radio options would transform the car into a "magic carpet" that guaranteed a "smooth drive" on the road, the ultimate dream of every driver (*Tomorrow's World* 1971).

Indeed, providing traffic information for drivers and catering to the commuter with special programs became an increasingly important function of radio. Germany introduced its "ARI" traffic information system in the first half of the 1970s. In the UK, traffic information was gathered with help of "the flying eye" in the 1980s, only to be replaced by networked cameras and computerized control centers in the twenty-first century—as the excellent 2007 BBC television documentary *The Secret Life of the Motorway* shows. At the end of the 1980s, 90 percent of a representative sample of Dutch motorists said that they listened to radio traffic information—44 percent even a few times a day. Getting details about the length of traffic jams and alternative routes were among these drivers' goals (Akerboom 1988, 46, 36–38).

Clearly, the car radio and the desire to be "in control" while driving had become Siamese twins. Today, as British radio disk jockey Tony Blackburn explains in *The Secret Life of the Motorway*, there are "two places where you have a really intimate relation with the listener. First of all there is that relationship with the listener who is in the car, because there is a *captive* audience. The second one is anybody in prison!" (*The Secret Life* 2007). The radio man flashes a joking smile at the interviewer after this remark. Yet what he says is fully in line with the idea of the car radio as mood regulator, as magic carpet or as a musical groove added to the highway groove. Car radio and audio sets literally and metaphorically provide alternatives to being "caught" on the road: caught by the road rage of fellow drivers, caught in traffic jams or speed control, and caught between the oppressing walls of noise barriers (Bijsterveld et al. 2014).

Montag's remark thus signified more than just one original observation. Only a few drivers will experience a life without traffic jams, noise barriers, and matrix panels. Moreover, the car radio's important place in today's auditory culture is closely connected to the characteristics of our mobility culture. An anthropology of contemporary Western culture may indeed start with archaeology, and proceed with cultural history, to arrive at the current situation.

New Anthropology: A Plea for Experimental Ethnography

Yet contemporary anthropology offers more than classical ethnography. One of the newer options is experimental ethnography. Let us return to the tape recorder once more. In *Sonic Souvenirs: Exploring the Paradoxes of Recorded Sound for Family Remembering*, anthropologist Lina Dib and information studies scholars Daniela Petrelli and Steve Whittaker conducted what one might call an ethnographic experiment. They explored "the role of *sound* as medium for social reminiscing" by asking ten families to make sound recordings of their holidays, and by discussing these collections with the family members on return, comparing the sound recording activities with photo taking activities and reminiscing. Even though there was no control group of families without recordings, and the project focused on an in situ situation rather than a laboratory, it was close to being an experiment.

Interestingly, the findings in *Sonic Souvenirs* confirmed some of those from our historical, oral history, and ethnographic work on sound recording amateurs and their "Sound Souvenirs." Dib and her colleagues stressed that both "sounds and pictures triggered active collaborative reminiscing," but that the use of sound had been "more varied, familial and creative" than the use of photos (Dib, Petrelli, and Whittaker 2010, 391). What they meant is that the recordings represented a wide variety of genres, such as ambient sound recordings, mock interviews, family conversations, giggles, pseudo radio shows, commentary about family activities, verbal diaries, abstract reflections, and inside jokes. They explained the creative and exploratory nature of these recordings as being based on the relative absence of a "pre-existing 'cultural' norm" for how to collect sounds compared to taking pictures (ibid., 398). In contrast to the postwar tape recorder manufacturers I studied earlier (Bijsterveld 2004), the researchers had not made the "mistake" of advising families to make a family sound album similar to a family photo album while overlooking the actual differences in both practices. On the other hand, the families had clearly been inspired by several radio conventions.

In addition, the sound recordings "expressed the negative or mundane" more often than the pictures did. Some of the children, for instance, had recorded less flattering family quarrels, whereas the picture taking had focused on creating favorable, performed, and airbrushed images. Moreover, the sound recordings happened to be "harder to interpret" than the photos (Dib, Petrelli, and Whittaker 2010, 391). This last issue is very similar to our earlier finding that sound recording amateurs had a hard time storing and annotating their tapes in such a way that they would remember and understand what they had recorded many years later (Bijsterveld and Jacobs 2009). Also comparable to our work were the respondents' remarks about the sound recordings being more "demanding" than taking photos as the sounds required more of a time investment, both when making the recordings and when relistening to them, even though the reminiscences were rich. The family members had felt less control over the sounds: in a photo of a family playing volleyball, some explained, the photo would have focused on the ball, yet the microphone had picked up more of the people taking part in the game as well as their environment. Moreover, if a particular family member had not been engaged in actually making the recording, the sounds did not work as well as memory cues for that person as they did for the other family members and thus required more decoding. Finally, none of the families "envisaged sharing their sounds outside the immediate family." The researchers thought this might be the result of "the practical details of editing and manipulating sounds" or "the lack of pre-existing social practices associated with sharing sounds" (Dib, Petrelli, and Whittaker 2010, 398). That was, indeed, also what we suggested as an explanation for why the tape recorder did not become the family sound album the manufacturers had hoped it would become (Bijsterveld 2004; Bijsterveld and Jacobs 2009), although sharing sounds in terms of music became a widespread activity.

Concluding Remarks

It is with this kind of work that I see interesting new options for future research—and it might even be helpful to do such experiments with historical audio technologies, not as a way to reenact history, but to elicit past skills and reflections on cultural conventions to which we have no other type of access. Andreas Fickers et al. (2012), for instance, intend to study how families watch and listen to the home movies once created by them with film and video techniques that are now obsolete.

Such new forms of anthropology may be added to the existing tools of combining history *and* anthropology in order to ask fresh questions about the history of (sound) technologies and their appropriation in (auditory) culture; of doing archaeology *of* anthropology to unravel past actors' attempts to understand the culture in which they want to intervene with new (sonic) technologies; and of approaching anthropology *as* archaeology to open up mundane practices for scholarly research. In this way, we do not merely replay a once recorded history, but may create and drive new and exciting grooves through history.

Note

1. The remainder of this section was adapted from *Sound and Safe: A History of Listening Behind the Wheel*, by Karin Bijsterveld, Eefje Cleophas, Stefan Krebs, and Gijs Mom, copyright 2014, and has been reproduced by permission of Oxford University Press.

References

Bibliography

Akerboom, Simone P. 1988. *Het gebruik en effect van (radio)verkeersinformatie: Een schriftelijk vragen-lijstonderzoek*. Leiden: Rijksuniversiteit Leiden.

Barry, Andrew, Georgina Born, and Gisa Weszkalnys. 2008. Logics of interdisciplinarity. *Economy and Society* 37 (1): 20–49.

Bijsterveld, Karin. 2004. "What do i do with my tape recorder...?" Sound hunting and the sounds of everyday Dutch life in the 1950s and 1960s. *Historical Journal of Film, Radio, and Television* 24 (4): 614–634.

Bijsterveld, Karin. 2010. Acoustic cocooning: How the car became a place to unwind. *Senses & Society* 5 (2): 189–211.

Bijsterveld, Karin, Eefje Cleophas, Stefan Krebs, and Gijs Mom. 2014. *Sound and Safe: How We Found Sonic Relief in the Car*. Oxford: Oxford University Press.

Bijsterveld, Karin, and Annelies Jacobs. 2009. Storing sound souvenirs: The multi-sited domestication of the tape recorder. In *Sound Souvenirs: Audio Technologies, Memory and Cultural Practices*, ed. Karin Bijsterveld and José van Dijck, 25–42. Amsterdam: Amsterdam University Press.

Bijsterveld, Karin, and José van Dijck, eds. 2009. *Sound Souvenirs: Audio Technologies, Memory and Cultural Practices*. Amsterdam: Amsterdam University Press.

Bull, Michael. 2000. *Sounding out the City: Personal Stereos and the Management of Everyday Life*. Oxford: Berg.

Bull, Michael. 2001. Soundscapes of the car: A critical ethnography of automobile habitation. In *Car Cultures*, ed. Daniel Miller, 185–202. Oxford: Berg.

Bull, Michael. 2003. Soundscapes of the car: A critical study of automobile habitation. In *The Auditory Culture Reader*, ed. Michael Bull and Les Back, 357–374. Oxford: Berg.

Bull, Michael. 2007. *Sound Moves: iPod Culture and Urban Experience*. London: Routledge.

Dib, Lina, Daniela Petrelli, and Steve Whittaker. 2010. Sonic souvenirs: Exploring the paradoxes of recorded sound for family remembering. In *Proceedings of the 2010 ACM Conference on Computer-Supported Cooperative Work*, 391–400. New York: ACM Press.

Fickers, Andreas, Susan Aasman, Tim van der Heijden, Tom Slootweg, and Jo Wachelder. 2012. *Changing Platforms of Ritualized Memory Practices: The Cultural Dynamics of Home Movies*. Project funded by the Netherlands. Organization for Scientific Research NWO. http://www .rug.nl/research/arts-in-society/expertisedomeinen/aasmanchangingplatforms?lang=en (accessed November 30, 2014).

Montag, Samuel. 2011. Lawaai. *NRC Weekend*, June 18 and 19, 37.

Oudshoorn, Nelly, and Trevor J. Pinch, eds. 2003. *How Users Matter: The Co-Construction of Users and Technology*. Cambridge, MA: MIT Press.

Silverstone, Roger, and Eric Hirsch, eds. 1992. *Consuming Technologies: Media and Information in Domestic Spaces*. London: Routledge.

Sterne, Jonathan. 2009. On interdisciplinary. http://superbon.net/?p=756 (accessed March 3, 2012).

Sound and Media

The Secret Life of the Motorway, Part II. 2007. BBC Television Documentary, broadcast on August 22. DVD, BBC Archives, Courtesy Andy O'Dwyer.

Tomorrow's World. 1971. BBC Television Program, broadcast on November 26. BBC Archives, Courtesy Andy O'Dwyer.

Woman's Page. 1937. http://www.youtube.com/watch?v=xImnco66C_c (accessed August 25, 2011).

10   Sonic Epistemology

Holger Schulze

At this very moment, you, the reader of this chapter, are enveloped in a whole variety of sonic events and sound waves. Maybe you are sitting at your desk, in a library or on a train, or in your own living room? In any of these cases, you may sense the sounds coming from other human beings around you, other adults or children, even animals; you may hear remote whispers and discussions, movements of bodies and textiles, maybe you hear them using a tablet, a smartphone, or even an old desktop computer. Besides all of that, you are also involuntarily hearing the sounds of the infrastructure around you: the whirring of the air conditioner, the humming of old electric cables or noisy neon lights, the opening and closing of doors, the creaking of wooden floors, of old chairs and tables.

Is there a specific knowledge about sound? How can it be acquired? Could it provide access to genuine forms of knowledge? And last, but certainly not least: what are the limitations of the acquisition of a specific knowledge centered around sound? These questions constitute the foundation of a possible epistemology (Steup 2012) of sound, which I will explore in this chapter.

The Difference between Auditory and Sonic Epistemologies

In the field of sound, it is possible to distinguish between two approaches to epistemologies: (a) *auditory epistemologies* and (b) *sonic epistemologies*. I will focus on the question of sonic epistemologies in this chapter; but to do so it would be helpful to discuss the difference between those two approaches. The approach to (a) *auditory epistemologies* is by now a well-researched and defined field. It encompasses a whole range of publications, studies, and research projects that explore the historical development of specific research approaches, technological innovations, and commodified apparatuses that concentrate on the ear as the listening organ to evaluate certain research findings—from the stethoscope to the headphone, from auditory car analytics to technologies of sonifying scientific data, between *audile techniques* (Sterne 2003) and *sonic skills* (Bijsterveld 2010–2015). A whole range of international researchers, investigative projects, and methodological studies (e.g., Sterne 2003; Bijsterveld 2010–2015; Kursell 2006; Volmar 2012) has explored the methods by which specific bodily trained, technologically refined, and culturally implemented techniques of listening have been developed and implemented into the canon of recognized research practices and apparatuses in Western science culture.

In contrast, any concept of (b) *sonic epistemology* is, frankly speaking, more in its nascent state. Though the more recently proposed shift to a *sonic materialism* (Cox 2011; Cobussen, Meelberg, and Schulze 2013; Schulze 2015) does focus on those specific epistemic qualities of listening and sounding, in scholarly discussions it still seems as if processes of knowledge acquisition not *about* but *by* sound and *by* listening are an invitation to devalue them as being more of a futile or even esoteric illusion. To make this perfectly clear: even today, in the early years of the twenty-first century, conducting historical research on listening as an epistemic form is more widely accepted than actually referring to listening itself as an epistemic form in contemporary cultural research. So, what we observe in research cultures from the distant past is still true today: those repudiations are apparently late symptoms of the fundamental logocentrism in academic research, its original sin (if you enjoy the occasional biblical metaphor). If we try to transcend this form of logocentric angst just for a minute, we might in turn ask ourselves: what could the characteristic research questions be for such a genuinely sonic approach to research? Who would the relevant experts in the field of sound studies be who would work on those approaches? And finally: how could an epistemology actually operate *in the realm* and *by the means* of sound?

The Multiplicity of Sensory Experience

Sonic epistemologies can be found in many sociocultural fields: fields in which practices dominate that have just not (yet) been established as epistemic practices. For the most part, the practices in question lack the quality of reproducibility, and documentable exactitude and distinctiveness, and as such they fall short of the academic standard that is commonly expected from research practices. They are often seen as unintelligible, as subjective, even as esoteric practices which do not justify any further research or theoretical reflection. They may at most be considered a form of skillful craftsmanship, maybe a form of *embodied knowledge* (Gallagher 2005), which could in turn be granted the honor of being a form of *tacit knowledge* (Polanyi 1966). But, in doing so—even in the more symbolic honoring of craftsmanship—the logocentric concept of epistemology still prevails. If sonic epistemologies are to be taken seriously, it is necessary to ascribe to those alternate, thoroughly sonic forms of knowledge the same dignity as forms of knowledge that are more easily transferable to discrete and reproducible, semiotic, and alphanumeric codes that contemporary consumer culture can instrumentalize so perfectly in industrialized research.

If we approach sonic forms of knowledge in this way, coming from a *new materialist* (Howes 2004) perspective, it might be possible to gain specific knowledge and insight into the concrete ways a certain sensory constellation approaches physical and performative reality, and how this creates a thoroughly different but still insightful, sensory representation of the world's physical emanations. This might sound a bit strange, maybe even far-fetched, but only such a broad, sensorily founded definition allows us to speak about the whole of sensory experience in a culturally and historically reflected way: a way that does not—from the viewpoint and sensory dispositives of Western, white, and male-dominated cultures—paternalize the very specific sensory approaches of other cultures, other subcultures, and other individual biographies with their particular sensory profile inscribed and embodied in their flesh. To accept and to acknowledge this rich diversity of sensory dispositives and everyday performativity throughout all history and cultures is just a first, but crucial, step toward also acknowledging the existence of specific sonic forms of knowledge that are simply representative of one particular case, and which are in no way more noble, more subtle, or more lucid than any other sensory constellation (Sterne 2003).

So, in summary, we should at least assume specific, sonically centered forms of knowledge in order to present a specific knowledge that is only or mainly accessible and presentable via the auditory; to enlist a number of distinct, trainable, refinable, and methodologically executable epistemic practices; and to compile a number of epistemic axioms and research interests differing at large from epistemic axioms in well-known logocentric epistemologies, for example, the epistemology of *processing sensory data* or the epistemology of *separated channels of sensory perception* (Serres 1985).

Three Examples of Sonic Epistemologies

To understand the specific characteristics and modes of research, I would like to take a closer look at three relevant and different examples; each of which intends to approach the sonic as an epistemic medium, each of which presents a methodologically trainable practice, and each of which implies at least one epistemic axiom differing from logocentric epistemologies. I am speaking of (a) *human echolocation*, (b) *acoustemology*, and (c) *sonic fiction*.

(a) *Human echolocation*. This research field represents a fairly new focus in the arts and in scholarly research; it is concerned with a perceptual technique that applies virtually all human senses except vision to navigate through an environment and to create representations for a secure orientation in this given environment. In centuries past, this technique was termed *facial sight* or *facial perception* (Supa, Cotzin, and

Dallenbach 1944; Worchel and Dallenbach 1947; Cotzin and Dallenbach 1950; Kellogg 1962), referring somewhat mystifyingly to the perception of the proximity of other people via the facial skin; then the focus lay more on the sounds emanating from the subjects (mostly a specific clicking-practice with the tongue) and reflecting off the target obstacles, sonarlike (Cotzin and Dallenbach 1950).

As a sonic epistemology, human echolocators approach everyday life by means of refined practices of corporeal, auditory perception; these practices can include palatal clicks with the tongue, but also trained kinesthetic senses, passive tactile echolocation via bodily skin perception, especially the discerning of changes in temperature or air flow, ephemeral aural perception, deep listening, and other, more individually developed practices (Kish 1982). So, the epistemic practices are in this case empirical practices exercised by the researchers and their own corporeal-sensory apparatus. As this is often regarded as *only* subjective, it is necessary to step back and to remind ourselves that *every* epistemic practice—in the natural sciences, in the engineering sciences, as well as in social studies and in humanities—relies on the individual intelligence, sensibility, knowledge, inspiration, and perceptual subtlety of a particular researcher and his or her own subjectivity; it does so at least at the crucial point of interpreting, contextualizing, and reasoning (with) a certain corpus of data. The traditional claim of the ubiquitous and ahistoric reproducibility of results regardless of the researcher's subjectivity fails *at least* on this point. Epistemic practices, as in human echolocation, are founded to a large extent on individual and highly subjective perceptual training—and universities as well as research institutions claim to adhere to that in their training, refining, and challenging of precisely that particular ability.

Thus, the stunning and new element in human echolocation as an epistemology is that it leaves logocentric ways of approaching the outside world of everyday life behind to an almost unknown extent, as it concentrates solely on currently barely representable and documentable forms of sensory perception. Its practices assimilate the percepts surrounding the practitioner via the aforementioned multitude of sensory practices. In the strictest sense of the word, we would have to speak of a polysensory perception in contrast to a culturally habitualized sensory restriction that allows us to preferably refer to visual icons and discursive, textual representations as arguments: the dispositives of logocentrism.

In a cultural climate of logocentrism, especially in the realms of commodified research, even only mildly poetic and rich sensory statements can easily provoke angry accusations of being unscientific, off-topic, or just uninformed—even of being insane. For example: "It appears therefore that a major axis of sonic cultural warfare in the twentyfirst [*sic*] century relates to the tension between the subbass materialism of music cultures and holosonic control, suggesting an invisible but escalating micropolitics of frequency that merits more attention and experimentation" (Goodman

2010, 187). In contrast to a rather harsh and antipoetic worldview, the research on human echolocation introduces the entirety of senses as an empirical basis and a sensible approach to the world. Moreover, the practice and theory of human echolocation were hitherto mainly explored by practitioners and researchers who themselves were sight-impaired or blind and developed the sensory apparatus of tactile and corporeal listening (e.g., Daniel Kish, Juan Antonio Martínez, Lawrence Scadden, Ben Underwood, or Tom De Witte). In experimental studies, echolocation trainings, and theoretical explorations of echolocation, the sound studies approach merges with that of disability studies and generates an astoundingly material and corporeal strand of sensory studies. Their research practices are truly polysensory epistemological approaches, and postlogocentric.

(b) *Acoustemology.* In 1996, Steven Feld proposed the following criteria to characterize an acoustemological approach in research: "Acoustemology, acousteme: I am adding to the vocabulary of sensorial-sonic studies to argue the potential of acoustic knowing, of sounding as a condition of and for knowing, of sonic presence and awareness as potent shaping forces in how people make sense of experiences" (Feld 1982, 97). This definition in itself states the fundamental epistemological character of acoustemology quite clearly: by definition, acoustemology refers to all inquiries that approach the outside world via the auditory senses.

Feld's notion of acoustemology refers to an ethnographic field research practice that focuses the attention, the sensibility, and the powers of recollection of the researcher on the auditory senses (Feld 1982). Thus, an acoustemological analysis opens up the social, cultural, and communication practices of a specific culture via the listening researcher and the auditory and sonic practices of the culture in question. An acoustemological analysis is "an exploration of sonic sensibilities, specifically of ways in which sound is central to making sense, to knowing, to experiential truth. This seems particularly relevant to understanding the interplay of sound and felt balance in the sense and sensuality of emplacement, of making place" (Feld 1982, 97). In its practice, it therefore incorporates methodological elements and specific epistemic practices, such as corporeal, deep, or reduced listening, which we also find in human echolocation; and it stretches itself out into a sort of sonic fiction *avant la lettre*, by narrating auditory experiences using *thick description* (Geertz 1973). In this respect, acoustemology is—as the neologism suggests—in and of itself an acoustic approach to epistemology. The question regarding how much of its research relies on an idealized, exoticist idea of an "imagined community" (Erlmann 1998) constitutes the main criticism of this approach—a criticism that might only be denied by research conducted in a less exhilarated and idealizing way. As a consequence, auditory methodologies require the same self-reflective rigidity we demand of other approaches in research: the fact that subjective human audition and interaction is the main tool of research does not mean this tool should be abused in order to superimpose the researcher's projection onto her or his research field. On the contrary, the obvious importance of subjectivity in these research approaches makes researchers more aware of their own interference in epistemic processes than they are in the case of other approaches that tend to present themselves as objective by design.

(c) *Sonic fiction.* The heuristic of sonic fiction as proposed by the DJ and music critic Kodwo Eshun (1998) can be described as a personal, highly suggestive, stylistically innovative, and idiosyncratic narration of an individual auditory experience (Schulze 2013). Eshun relates his presentation of specific auditory experiences to genuine narrations, legends, mystifications, and idiosyncratic associations and imaginations. Whereas acoustemology and human echolocation dive deeply into the sonic cultural practices in everyday life, the sonic fiction approach takes up the semantic aspects of those highly experiential practices. A sonic fiction unfolds the diverse semantic inter-ferences within an auditory experience on the one hand—but it also takes up the pro-vided starting points into the imaginations of this experience as an *affordance* (Gibson 1977) for the elaboration of individual imaginations and fictions. The creation of a (written, visual, programmed, or even musical) sonic fiction is, as such, a way to com-prehend a sonic experience by complementing it with a new genuine artifact and its experience:

So, *More Brilliant Than the Sun* draws more of its purpose from track subtitles than from Techno-Theory, or even science fiction. These conceptechnics are then released from the holding pens of their brackets, to migrate and mutate across the entire communication landscape. Stolen from Sleevenote Manifestos, adapted from label fictions, driven as far and as fast as possible, they mis-shape until they become devices to drill into the new sensory experiences, endoscopes to mag-nify the new mindstates Machine Music is inducing. *More Brilliant Than The Sun*'s achievement, therefore, is to design, manufacture, fabricate, synthesize, cut, paste and edit a so-called artificial discontinuum for the Futurhythmachine. (Eshun 1998, 3)

In the history of musical critique, and in the history of establishing musicology as an academic discipline, such a narrational, even poetic, and highly imaginative and idiosyncratic approach has been repeatedly rejected as being simply the unreflected, excited babbling of laypeople—and as such of no academic interest. Whereas this was true from the perspective of Kantian scientific theory and the intense effort towards a logocentric constitution of research in humanities in the nineteenth century, it has been radically questioned and transformed over the last decades. It has now become quite clear how limited any approach to research would be that denied the relevance of individual listeners' perceptions, imaginations, and articulations. Other, rather deviant approaches and methodologies, including diverging perspectives, tend to be applied and incorporated more and more today into the toolbox of a research culture that is trying hard to leave behind those previously accepted truisms, a step that is long overdue. At this point in academic history, Eshun's approach, derived from the diasporic tradition of Afrofuturism (James 1954; Bernal 1987), becomes an almost irrefutable proposal for transforming and expanding the methodological toolbox of music studies and musicology into sound studies (cf. Goodman 2010; Augoyard and Torgue 2005).

Toward Sensory Epistemologies

The three examined examples demonstrate how sonic epistemologies are currently articulated: first, as newly explored or refined *cultural practices* that prove to be epistemic practices (human echolocation); second, as *comprehensive approaches* that manage to integrate a rich diversity of epistemic practices under one methodological concept (as in acoustemology); and finally, as *forms of representing* the results of epistemic practices that themselves bear strong epistemic potential. In all three examples, the concept of epistemology as it is implemented into our contemporary research cultures is reflected upon, deeply questioned, and remodeled via a thorough consideration of its historicity, culturality, corporeality, and logocentrism. Moreover, these three sonic epistemologies present approaches to us that do indeed propose a *sonocentric* and also a *physiocentric* approach as a kind of counterweight. All in all, they are symptoms of a strong *hypercorporealization* (Schulze 2008) of epistemic practices these days. The research process is rethought and reconceptualized as a corporeal, experiential, situated, and, *horribile dictu*, subjectively guided process. They all more or less leave behind the idea of an abstract and anonymous researcher in the process of researching as being more of an idle illusion of abstraction and negation: negation of the specific research position of researchers' concerning their gender, race, (dis)abilities, age, and economic and political situation. However, while succeeding in promoting this new research field, its main protagonists do also publish scientific articles, and they also augment their arguments with quantified data, statistics, graphs, and algorithmic reasoning—strategies aimed at gaining recognition in the current research environment.

In addition, the experiential aspect of participatory observation is stretched to its farthest limits in all three approaches: the individual auditory or sonic experience of the researcher is the principal means of research and also the main means of presenting the results of this research. Therefore, these new approaches to a sonic or even *sensory epistemology* are joyful manifestations and constitute the exercising of an anthropological potential to transform, to expand, and to develop the abilities and the incorporated sensory practices: this is to be seen as the core experimental method of all three approaches. The concept of the human being in these approaches is therefore not a static and essentialist one, but a concept that allows for an individual and collective plasticity of the human perceptual profile (Wulf 1997, 2009; Kamper and Wulf 1984a, 1984b).

Is it too extreme to imagine a research utopia in the not-too-distant future, in which the results of a project could be presented in the form of a dance, a musical composition, a literary novel, or an architectural concept? Notably *without* any verbal explanation, paratext, or liner notes accompanying it? That would truly be a completely transformed academic culture, in which such a performative demonstration of research findings would be seen not as some strange (at best entertaining and bizarre) exception, as the arts transgressing into the sciences and the humanities—but as a standardized form, which would lead to further projects, further research funding (again: *without* any textual research proposal), and to further research artifacts: *sensory fictions, senseomologies*. I am listening to the sounds of traffic wafting in through the slightly opened window right now, as I sit and finish this chapter. It is late at night, a Sunday evening in December, as I read the final proof.

References

Augoyard, Jean-François, and Henry Torgue, eds. 2005. *Sonic Experience: A Guide to Everyday Sound.* Montreal: McGill-Queens University Press.

Bernal, Martin, ed. 1987. *Black Athena: The Afroasiatic Roots of Classical Civilization.* New Brunswick: Rutgers University Press.

Bijsterveld, Karin. 2010–2015. Sonic skills: Sound and listening in the development of science, technology, and medicine (1920–now). Research project at Maastricht University.

Cobussen, Marcel, Vincent Meelberg, and Holger Schulze. 2013. Towards new sonic epistemologies. Editorial. *Journal of Sonic Studies* 1. http://journal.sonicstudies.org/vol04/nr01/a01.

Cotzin, Milton, and Karl M. Dallenbach. 1950. Facial vision: The role of pitch and loudness in the perception of obstacles by the blind. *American Journal of Psychology* 63 (4): 485–515.

Cox, Christopher. 2011. Beyond representation and signification: Toward a sonic materialism. *Journal of Visual Culture* 10 (2): 145–161.

Erlmann, Veit. 1998. Wie schön ist klein? World Music, Globalisierung und die Ästhetik des Lokalen. *Alternativen: Veröffentlichungen des Instituts für Neue Musik und Musikerziehung* 38:9–23.

Eshun, Kodwo. 1998. *More Brilliant Than The Sun: Adventures in Sonic Fiction.* London: Quartet Books.

Feld, Steven. 1982. *Sound and Sentiment: Birds, Weeping, Poetics, and Song in Kaluli Expression.* Philadelphia, PA: University of Pennsylvania Press.

Gallagher, Shaun. 2005. *How the Body Shapes the Mind.* New York: Oxford University Press.

Geertz, Clifford. 1973. Thick description: Toward an interpretive theory of culture. In *The Interpretation of Cultures: Selected Essays*, 3–33. New York: Basic Books.

Gibson, James Jerome. 1977. The theory of affordances. In *Perceiving, Acting, and Knowing: Toward an Ecological Psychology*, ed. Robert Shaw and John Bransford, 67–82. Hillsdale, NJ: Erlbaum.

Goodman, Steve. 2010. *Sonic Warfare: Sound, Affect, and the Ecology of Fear*. Cambridge, MA: MIT Press.

Großmann, Rolf. 2008. Verschlafener Medienwandel: Das Dispositiv als musikwissenschaftliches Theoriemodell. *Positionen—Beiträge zur Neuen Musik* 74:6–9.

Howes, David. 2004. *The Empire of the Senses*. Oxford: Berg.

James, George G. M. 1954. *Stolen Legacy: The Greeks Were Not the Authors of Greek Philosophy, But the People of North Africa, Commonly Called the Egyptians*. New York: Philosophical Library.

Kamper, Dietmar, and Christoph Wulf, eds. 1984a. *Das Schwinden der Sinne*. Frankfurt am Main: Suhrkamp.

Kamper, Dietmar, and Christoph Wulf, eds. 1984b. *Der andere Körper*. Berlin: Verlag Mensch und Leben.

Kellogg, Winthrop N. 1962. Sonar system of the blind: New research measures their accuracy in detecting the texture, size, and distance of objects "by ear." *Science* 137 (3528): 399–404.

Kish, Daniel. 1982. Evaluation of an echo-mobility training program for young blind people. MA thesis, University of Southern California.

Kursell, Julia. 2006. Epistemologie des Hörens, 1850–2000. Research project, Max-Planck Institut für Wissenschaftsgeschichte Berlin.

Polanyi, Michael. 1966. *The Tacit Dimension*. New York: Doubleday.

Schulze, Holger. 2008. Hypercorporealismus: Eine Wissenschaftsgeschichte des körperlichen Schalls. *PopScriptum* 16 (10). http://www2.hu-berlin.de/fpm/popscrip/themen/pst10/pst10_schulze.htm.

Schulze, Holger. 2013. Adventures in sonic fiction: A heuristic for sound studies. In *Towards New Sonic Epistemologies*, ed. Marcel Cobussen, Vincent Meelberg, and Holger Schulze. Special issue, *Journal of Sonic Studies* 3 (4). http://journal.sonicstudies.org/vol04/nr01/a10.

Schulze, Holger. 2015. Der Klang und die Sinne: Gegenstände und Methoden eines sonischen Materialismus. In *Materialität: Herausforderungen für die Sozial- und Kulturwissenschaften*, ed. Herbert Kalthoff, Torsten Cress, and Tobias Röhl. Munich: Wilhelm Fink.

Serres, Michel. 1985. *Les Cinq Sens: Philosophie des corps mêlés*. Paris: Grasset.

Sterne, Jonathan. 2003. *The Audible Past: Cultural Origins of Sound Reproduction*. Durham, NC: Duke University Press.

Steup, Matthias. 2012. Epistemology. In *The Stanford Encyclopedia of Philosophy* (fall 2012 ed.), ed. Edward N. Zalta. http://plato.stanford.edu/archives/fall2012/entries/epistemology/.

Supa, Michael, Milton Cotzin, and Karl M. Dallenbach. 1944. Facial vision: The perception of obstacles by the blind. *American Journal of Psychology* 57 (2): 133–183.

Volmar, Axel. 2012. Klang als Medium wissenschaftlicher Erkenntnis: Eine Geschichte der auditiven Kultur der Naturwissenschaften seit 1800. PhD diss., Universität Siegen.

Worchel, Philip, and Karl M. Dallenbach. 1947. "Facial vision": Perception of obstacles by the deaf-blind. *American Journal of Psychology* 60 (4): 502–553.

Wulf, Christoph, ed. 1997. *Vom Menschen: Handbuch Historische Anthropologie*. Weinheim: Beltz.

Wulf, Christoph, ed. 2009. *Anthropologie: Geschichte, Kultur, Philosophie*. Cologne: Anaconda.

11   Historicization in Pop Culture: From Noise Reduction to Noise Recording

Bodo Mrozek

In 1964, when the BBC wanted to record the concert of a young band from Liverpool to broadcast it as part of the television series *Juke Box Jury*, it turned into a fiasco. Once the Beatles finally began to play after numerous delays, "due to the crowds of well-wishers and press men on the stage," parts of the female studio audience fell into such a frenzy of ear-piercing screams that the audio technicians could no longer understand one another. Even the band had immense difficulty: in an indignant memo afterward, the BBC production manager complained that "the noise in the theatre was so deafening that The Beatles could not hear their own internal balance, and were therefore singing out of tune" (BBC WAC, T12/644/1). After this and other fundamentally unsuccessful live recordings, the BBC was virtually besieged with letters of protest. For example, one viewer from Hertfordshire County complained that the camera work did not focus on the musicians, instead focusing extensively on the audience, whose screams were "harmful to mind + health" (BBC WAC, N31/5/1). This screaming—which would accompany concerts by the Beatles and other bands from then on—was negotiated under the pathologizing term "Beatlemania," and engaged psychologists, musicologists, and police in a heated debate.

In pop history, the outcome of Beatlemania is well known: increasingly frustrated by their own public, the band gave their last live concert in 1966. In my own work on the emergence of a transnational youth culture in the 1950s and '60s, I analyze the *moral panics*–style debates on the emotional practices of female fans, which refer back to older discourses such as that of hysteria as a typical "female mental illness" or the "Lisztomania" of the nineteenth century (Cohen 1972; Foucault 1972, 351–362; Ehrenreich, Hess, and Jacobs 1997; Noeske 2001). I examined listening protocols from a variety of countries' radio archives, since in the field of what is considered "good form" questions are asked not just about musical dissonance but also about social dissonance. In conjunction with the research imperative presented in this volume, I am mainly interested in the small find that came out of the BBC archive, the complaint filed by an audio engineer. It can be used to exemplarily discuss the relationship between music and noise in pop culture, and to broach fundamental questions about the historicity of sound (cf. Schulze 2007; Morat 2013a; Mrozek 2014a). Using the example of audience noise interference, this essay will demonstrate how variably the borders between music and noise, and/or between the musicians and the audience, have been drawn. Based on that discussion, I will then present an analysis of the connection between sound and historicity using the example of musical performances during the process of German Reunification in 1989, before delineating some further research perspectives in sound studies for historical pop scholarship.

Music, Sound, Noise: Pop History's Shifting Boundaries

Just a few months after the concert fiasco, the BBC audio technicians resorted to trickery in order to be able to obtain recordings in which the music could be heard in spite of the tremendous acoustic noise at the Beatles' concerts. During the filming of a documentary about *Mersey Beat*, they unceremoniously let the "Fab Four" perform in an empty room without an audience so that the music would be audible. The recording went perfectly—but when they listened to the recording, another problem emerged: the concert did not sound real anymore, because the Beatles had never had a concert without the screams, which had become well known through the media. In order to provide the impression of an "authentic" concert in front of an audience, they retroactively mixed a moderately adjusted screaming track into it that they had recorded live at another concert (Frontani 2007, 24). Thus, neither of the sound events was distinguishable as a separate element: not the loud screaming, which was too loud because the music was missing; nor the music alone, because it did not have a live atmosphere. However, the result was so highly artificial and consisted of so many partial elements that it was, in effect, newly created in the end.

This sham, perpetrated by resourceful audio technicians, brought together two things that had always been strictly separated from one another: music and noise—produced by not just the musicians, but the audience as well. In recent years, historical sound studies have fundamentally challenged the separation of these allegedly opposing spheres. While historians used to be interested in music history above all, lately the focus of attention has broadened to sounds and noises of all kinds. Research in music has widened its scope of interest from just the musicians to the audience and thus from the stage to the stands. On the one hand, because it is only in the act of hearing that music is even perceivable—without the sensorial perception of an audience (and the musicians may act as their own audience), it could at most exist as the vibration of air molecules—it cannot exist as sound alone (Sterne 2003, 11; Schulze 2007, 351). And on the other hand, because, as research has shown, the audience's role in music production has become more important than it has ever been: in dialogue with musicians, in the sound design construction of spaces—from the cathedral to the disco—or in the acoustic expressions of the audience itself.

In contrast to music studies, which are interested in aesthetic principles, historiographical sound studies are particular in that they demonstrate a focus on social processes. As a rule, sounds do not interest historians in their form as autonomous works of art, but rather in their complex interaction with society. In the process, they are examining transformation over time: how are sounds socially negotiated? What sounds represent certain groups? Which sounds do people demonstrate for or against (for example, aircraft noise)? And how are they used in demonstrations (for example, caterwauling)? What conceptions of "peace and order" are at the foundation of noise control debates (for example, regarding closing times, restaurants, clubs) and how are those connected to notions of rule and freedom? What are the boundaries of the sayable and the singable in the censuring of concerts or records which infringe upon prevailing norms of decency and morality? What is considered noise, and what is considered art—and at what point in time?

The aforementioned story about the Beatles in the BBC studio is a demonstrative example of how variable the boundaries are between musical sounds on the one hand and noise on the other. For the behavior of the young Beatles' fans in the audience invoked strong opposition in the beginning. Cultural-critical comments in the media melodramatically bemoaned the threat to moral decency (Matheja 2003). Armed with earplugs, special security guards patrolled the aisles in order to stop the fans from getting up from their seats and dancing in the aisles (Kasparek 2000–2001). On the Beatles' tour in the United States, riot police were mobilized against the fans; and when it was reported that some fans had actually lost control of their bladders at the concerts (something that was quite possible given that the guards and police prevented them from going up the aisles to the bathrooms), psychologists, educators, and criminologists speculated on the supposed physically disinhibiting effect of the music, as well as its ability to stimulate sexuality and aggression (Ehrenreich, Hess, and Jacobs 1997). Those kinds of controversies were not new. Even at the time of the introduction of the microphone, there were fears that the electrically amplified voices of crooners such as Frank Sinatra or Pat Boone would expose the teenagers to too much of an "intimate" voice (Frith 2001, 98). In the 1950s, the electrically amplified guitar was thought to stimulate youthful destructiveness, and in the so-called cinema riots, dancing was prevented through the use of physical violence. During Beatlemania, young women now came into focus—shifting it away from the male youths who belonged to subcultures that were stereotyped as aggressive and violent, such as the "teddy boys," "greasers," "beatniks," "blousons noirs," or "nozems." Contrary to their male contemporaries, the women were not stigmatized as dangerous perpetrators of violence, but were instead victimized and said to have succumbed to their own emotions. In both cases, music played an essential role.

While the term *pop* was in the process of development based on a variety of cultural phenomena in the 1960s, not only were new forms of production being established (in the face of culturally conservative opposition in part), but also new forms of sound reception. Pop culture research is as yet divided on this: one side examines the production of sounds under cultural industry conditions (Anand and Peterson 2004), and another places more value on reception (Certeau 1989; Poiger 2000). Often, the two orientations are irreconcilably opposed: the production approach emphasizes the allegedly all-dominant role of an industry in which consumers have little possibility of participation. In contrast, the reception research emphasizes individual will, which is demonstrated by the modes of reception not intended by the producers, such as mix tapes, remixes, or *samizdat*. For it is obvious that the young women did not attend the concert primarily to listen to the Beatles' music, which they proceeded to drown out with their screaming; they went to concerts much more for the purpose of coming together with other young women of their age in order to publicly—and loudly—demonstrate their newly won self-awareness. Seen in that way, publicly living and acting out their feelings together as a bodily practice, or "doing emotions" (Eitler and Scheer 2009), in no way constitutes the moral aberration or even the mental illness that it was depicted as indicating by contemporary media. It was instead the representation of a marginalized group that was publicly distancing itself from the role of future wife and silent sufferer to which it had been assigned.

Detlef Siegfried described the 1960s using a term from the German historian Reinhart Koselleck, "*Sattelzeit*," a term denoting a particular transition period in which important shifts in values and politics occurred, even before the free speech movement in the United States and the German and French student protests in the late '60s (Siegfried 2008, 10). Using the example of the Beatles in the BBC studio, this kind of shift becomes audible: at the beginning of the decade, that wave of shrill excitement was supposed to be filtered out using sensitive recording techniques, or by drowning them out with more effective amplifiers—but over the course of just a few months, those screams had become an accompanying noise to the music, so that the BBC had to mix them in for the recording to attain authenticity. That original background noise, which had so upset viewers that they sent in "listeners' responses" to protest against the acoustic affront, had now become part of the concert experience, thereby losing its character of noise.

This example points to a fundamental shift occurring in the twentieth century, which constituted a reversal of earlier processes (at least in subdomains). In the eighteenth and nineteenth centuries, the concert audience turned increasingly silent (as classical studies already demonstrated in the 1990s), and music halls transformed into quasi sacred spaces, in which no noise was allowed to disturb the performance, and no sneezing or heckling to adulterate the work of sound art (Gay 1996, 1–36; Johnson 1995; Weber 1997). As a part of that process, music transformed from a background medium into a medium of attention. This shift, which still affects the classical concert at the beginning of the twenty-first century, had admittedly not included the field of popular music. At folk music festivals, fairs, in ballrooms, and the *caves du jazz*, movement and conversation have always been permitted. At jazz concerts, calling out and applauding during the performance was such an integral part of the show that the musicians would explicitly encourage their audience to do so; and they would then incorporate those verbal and nonverbal expressions into the set. This also became the case at rock 'n' roll concerts in large arenas, the Alan Freed Show for instance, or the first international tours of musicians such as Louis Armstrong or Bill Haley, where frenetic applause that degenerated into riots was a common occurrence in some parts. While the rock music of the 1950s was very generation and class specific, the *Mersey Beat* opened up a cross-section of youths across all levels of income and education. From the very beginning, the Beatles were presented as a band that was also palatable to the upper classes, demonstrated by their act of giving a concert for members of the British Royal Family (Thomson and Gutman 2004, 53–56). For that reason, pop music increasingly infiltrated into the established halls as well, such as Carnegie Hall in New York and the Royal Albert Hall in London or the Budokan in Tokyo. In those concert houses, the modes of reception had been established since the nineteenth century: remain calmly seated and listen silently. The clash with the new reception practices of teenagers led to conflicts with the concert house managers, security guards, and repeatedly with the police as well. After a massive riot following the breakup of a Rolling Stones concert in an open air theater in Berlin that was made for a seated audience and not sufficiently secured, the German police conducted an internal analysis of Beat concerts in order to understand the puzzling behavior of the fans—they even ended up educating their officers in jazz music (Mrozek 2014b). As conflicts between the police and street musicians in Munich led to days-long riots, the first position for a police psychologist was created, which also led to the education of the police in music psychology (Sturm 2006). After some brawling at an open air concert in France, similar debates were held (Tamagne 2014). Most of them ended with the insight that the audience's modes of reception for pop concerts required different spaces and different methods beyond those of police force.

Media Historicization of Sound Events: The Live Album

The audio technicians' recording of audience noise in the early Beatles' recording marked a tonal shift: the importance of the concert for pop music increases if the excitement of the masses becomes acoustically audible. In the end, that created a medium all of its own. In the first half of the century, there were already some gramophone records that included not only sounds of applause, but also call-outs from the audience and vendors, whistles, and even "salacious verbal exchanges between the sexes" were recorded (Morat 2013b, 114). However, it was only with the live album that the juxtaposition of music and noise, musicians and audience, was provided with a suitable medium. Live albums make it possible to hear the acoustic expressions of the audience alongside the musical performance. That is what emphasized the event-character of the concert: based on its special mood, the so-called live atmosphere, the recording distinguished itself from the studio recordings, in that they were deemed too perfect or even aseptic in comparison. Noises, glass-clinking, conversational buzz, foot shuffling, clapping, and heckling were not noise interference in this case, but rather an integral component of the sound experience.

With studio technology and audio technicians' growing influence on the recording process, music increasingly lost its singular event quality, its production had trans-formed from the simple documentation of an event to a time-transcending process (Schmidt-Horning 2013, 6). Above all, owing to audio tapes' multiple tracks, it was also possible and common to record musicians' joint concerts—musicians who had never actually played at the same time and place—and to achieve artificial results through mixing and editing techniques that would never have been possible to reproduce on the stage under live conditions, just as the Beatles were forced to discover in the mid-'60s (ibid., 171–207). The live album also constituted a backlash to the trend toward the industrialization of pop music production. By reproducing one individual stage event, it once again approached the origins of music recording in an almost conservative way. The audience noises—recorded by microphones designed just for that purpose and carefully modulated—additionally served to authenticate the character of singular sound events, for example a particular stage or a specific historical situation. So-called bootlegs (often illegally made recordings that would be secretly exchanged and disseminated) acquired particular importance to collectors.

Frequently, live albums historicize performances, transforming them into events that situate musicians within certain contexts and can lend social or political mean-ing to a recording—as in the case of the country singer Johnny Cash's concert for the inmates at Folsom Prison (Cash 1968). Some concerts are also attributed with histori-cal importance as the culmination point of an entire historical process. For instance, in 1969, Woodstock was stylized as a transnationally significant musical high point of a political protest movement with cross-generational shaping power. This concert also received its protest character based on a particular articulation of the audience, which, in response to Vietnam veteran Joseph "Country Joe" McDonald's call-out, used a *four-letter-word* as part of a comment on the war. And thus, a concert, which many did not experience as particularly meaningful while sludging around in the mud and rain, ended up constituting a decidedly political articulation of political-historical importance (Various Artists 1970).

More and more, live albums were planned in advance and staged as spectacles: in 1973, the Elvis comeback, *Aloha from Hawaii*, was the first global satellite transmis-sion. It was watched by a billion people and documented in the form of a film as well as an album (Parks and McCartney 2007, 254). In the end, that exchange with the audience—which was so essential to the live album—had generated its own music genre, so-called arena rock. This genre attempts to structurally integrate the massive contribution from the audience into the music from the outset, for instance with call-and-response (originally a Gospel music practice), in which the a capella singing of individual passages will stimulate thousands of voices, or where the rhythm anticipates in advance the participation of as many people as possible clapping or stamping along with the music. In that way, a mass aesthetic of the spectacle emerges that oscillates between the egalitarian democratization of the music and a totalitarian aesthetic of overpowerment.

The Sonification of History: Sonorous Cesuras

A further field of research for the sound history of popular sound worlds is not only the historicization of sound events on well-known recordings (for instance)—but also the reversed process: the sonification of historical events through sonic productions. To the extent to which it became increasingly self-evident to listen to those massive concerts as historic culmination points of particular eras, generations, or processes, the need to performatively and musically shape political events also grew—if they appeared historically important based on contemporaneous expectations. One instance of this can be exemplified by the musical forms at the fall of the Berlin Wall in November 1989, and I would like to discuss two of those events here.

One day after the fall of the Berlin Wall, German politicians gathered on Kennedy Platz, the place where the American president had expressed his now famous statement "Ich bin ein Berliner" twenty-six years earlier. The German Chancellor at the time, Helmut Kohl, had underestimated both the Germans' patriotism as well as their singing power, when he spontaneously (or deliberately) called out for them to intone the national anthem together—something that was still taboo for a lot of leftists owing to concerns about German nationalism: the concert went awry similarly to the Beatles' recording discussed above. However, it did not founder beneath shrieks of excitement, quite the contrary: it was met with boos and a chorus of disapproving whistles. Two days later, on November 12, the broadcast service Radio Free Berlin organized a rock concert in the western part of the city with musicians from both parts of Germany, among them individuals who had been affected by the bans on leaving East Germany (or entering it, as the case may be), such as the West German rock singer Udo Lindenberg, who wrote songs critical of East Germany; or the punk singer Nina Hagen, who had left East Germany in protest in the 1970s. 50,000 audience members from both parts of Germany attended the eleven-hour-long concert. The moderator's announcements of the further easing of border restrictions—which had only been open for a few days and was still the official state border inside of Germany—repeatedly provoked cheers from the audience. The concert took place in West Berlin in an arena called the Deutschlandhalle, but citizens from both of the German states were called on to put their radios out onto their window sills so that the concert could be broadcast throughout the streets of East and West Berlin—and thousands of people did just that. So, although the East and the West were still politically separated, a collaborative sound space was created in this way, a sound space that spread out from the performance venue over the divided city and out beyond into the territories of both states. And even though it was highly symbolic, this sound act went beyond the purely representational: a concrete space of experience was created, one that was not to be constricted readily ever again—through the reclosing of the border, for example, which was certainly still in the realm of possibility during the turbulent events of those days. In this respect, the sound act contributed to shaping the departure of a political process that was contingent upon this historical point in time: as far as the auditory perspective is concerned, the members of the population of two German states created one *soundscape* through the act of collective listening, applauding, and singing along. And from a political perspective, they also created an *imagined community*, as defined by Benedict Anderson (Sterne 2013).

In the process, they were able to refer to the differing (but nevertheless still collective) world of experience of an internationalized pop culture that the division of states had not been able to effectively prevent. From a pop historical perspective therefore, the concert demonstrates a further historical development: the wide establishment of the electrified, Anglo-American rock and/or pop music (on pop history, see Geisthövel and Mrozek 2014). Forty years after Beatlemania, which had also been controversial in Germany, pop music was so well established that it could function as an appropriate expression of a political process of transformation of historical scope. Perhaps, based on its mass appeal, it was better able to do justice to the democratic character of the event than a string concert shaped by the musical tastes of an educated middle-class elite (although this demand was not completely answered, as a classical concert was simultaneously broadcast on television from the Berliner Philharmonie). One irony of (sound) history lies in the fact that the only live record of the event that exists is a vinyl single released as satire by the leftist newspaper *die tageszeitung*, which consists of the chorus of whistles let out by Reunification skeptics and Kohl critics at Kennedy Platz. Thus the Konzert für Berlin, planned as a huge historical event of epic dimension, was supposed to be released on record but never was, because the participants were not able to agree on the rights.

The different case studies discussed in this essay—from the BBC recording to the live album to the Berlin concert—demonstrate how sonic events can be interpreted as historical events in which the boundaries between melody and noise are newly adjusted, groups are constituted as communities, and public life is newly balanced and partitioned. Both the historicization of musical events and the sonic background to political change are of interest to sound studies and historical research alike. Both inquire into questions of how to deal with historical turning points, and explore audio signatures pertaining to particular eras. Those could concern everyday shifts in the construction of gender, generation, class, or emotions—as became clear in the example of Beatlemania—but can also aid in the understanding of great historical events such as the fall of the Berlin Wall. In between both examples lies a gradual process in which the audience noisily creates a place for itself within musical performances and recordings, in which it establishes new forms of sounds, and in which it has, at least partially, reversed the historical process of earlier eras described as the silencing of the audience.

Translated by Jessica Ring

References

Bibliography

Anonymous. 1963a. Handwritten letter, December 10; BBC Written Archives, Reading (BBC WAC), N31/5/1.

Anonymous. 1963b. Louder Beatles. *The Record Retailer and Music Industry News,* November 28, 29.

Anand, Narasimhan, and Richard A. Peterson. 2004. The production of culture perspective. *Annual Review of Sociology* 30:311–334.

Certeau, Michel de. 1989. *Kunst des Handelns*. Berlin: Merve.

Cohen, Stanley. 1972. *Folk Devils and Moral Panics: The Creation of the Mods and Rockers*. London: MacGibbon & Kee.

Colehan, Barney. 1963. Report from "The Beatles" Recordings December 7; BBC Written Archives, Reading (BBC WAC), T12/644/1.

Ehrenreich, Barbara, Elisabeth Hess, and Gloria Jacobs. 1997. Beatlemania: A sexually defiant consumer subculture? In *The Subculture Reader*, ed. Ken Gelder and Sarah Thornton, 523–536. London: Routledge.

Eitler, Pascal, and Monique Scheer. 2009. Emotionengeschichte als Körpergeschichte: Eine heuristische Perspektive auf religiöse Konventionen im 19. und 20. Jahrhundert. *Geschichte und Gesellschaft (Vandenhoeck & Ruprecht)* 35:282–313.

Foucault, Michel. 1972. *Histoire de la folie à l'âge classique*. Paris: Gallimard.

Frith, Simon. 2001. Pop music. In *The Cambridge Companion to Pop and Rock*, ed. Simon Frith, Bill Straw, and John Street, 93–108. Cambridge: Cambridge University Press.

Frontani, Michael R. 2007. *The Beatles: Image and the Media*. Jackson, MS: University Press of Mississippi.

Gay, Peter. 1996. *The Naked Heart*, vol. 4 of *The Bourgeois Experience: Victoria to Freud*. New York: W. W. Norton.

Geisthövel, Alexa, and Bodo Mrozek eds. 2014. Einleitung. In *Popgeschichte*, vol. 1: *Methoden und Konzepte*, 7–31. Bielefeld: transcript.

Johnson, James. 1995. *Listening in Paris: A Cultural History*. Berkeley: University of California Press.

Kasparek, Jonathan. 2000–2001. A day in the life: The Beatles descend on Milwaukee. *Wisconsin Magazine of History* 84:14–23.

Matheja, Bernd. 2003. *"Internationale Pilzvergiftung": Die Beatles im Spiegel der deutschen Presse 1963–1967*. Hambergen: Bear Family.

Morat, Daniel. 2013a. Der Klang der Zeitgeschichte: Eine Einleitung. *Zeithistorische Forschungen/ Studies in Contemporary History* 8 (2): 172–177.

Morat, Daniel. 2013b. Sport und Vergnügungskultur: Der Sportpalastwalzer (Wiener Praterleben). In *Sound des Jahrhunderts: Geräusche, Töne, Stimmen 1889 bis heute*, ed. Gerhard Paul and Ralph Schock, 112–115. Bonn: Bundeszentrale für Politische Bildung.

Mrozek, Bodo. 2014a. Écouter l'histoire de la musique: Les disques microsillons comme sources historiques de l'ère du vinyle. *Le Temps des Médias: Revue d'histoire* 22:92–106.

Mrozek, Bodo. 2014b. Die Spur der Steine: 1965 drohte en Beat-Verbot im Westen. http:// pophistory.hypotheses.org/1498 (accessed May 26, 2015).

Noeske, Nina. 2001. Die Geburt der Virtuosität aus dem Geiste der Hysterie? Zur "Lisztomanie" als weibliches Phänomen. *Die Tonkunst* 5–4 (10): 495–505.

Parks, Lisa, and Melissa McCartney. 2007. Elvis goes global: Aloha! Elvis live via satellite and music/tourism/television. In *Medium Cool: Music Videos from Soundies to Cellphones*, ed. Roger Beebe, and Jason Middleton, 252–268. Durham, NC: Duke University Press.

Poiger, Uta. 2000. *Jazz, Rock, and Rebels: Cold War Politics and American Culture in a Divided Germany*. Berkeley: University of California Press.

Schmidt-Horning, Susan. 2013. *Chasing Sound: Technology, Culture, and the Art of Studio Recording from Edison to the LP*. Baltimore, MD: The Johns Hopkins University Press.

Schulze, Holger. 2007. Wissensformen des Klangs. Zum Erfahrungswissen in einer historischen Anthropologie des Klangs. *MusikTheorie. Zeitschrift für Musikwissenschaft* 22:349–356.

Siegfried, Detlef. (2006) 2008. *Time Is on My Side: Konsum und Politik in der westdeutschen Jugendkultur der 60er Jahre*, 2nd ed. Göttingen: Wallstein.

Sterne, Jonathan. 2003. *The Audible Past: Cultural Origins of Sound Reproduction*. Durham, NC: Duke University Press.

Sterne, Jonathan. 2013. Soundscape, landscape, escape. In *Soundscapes of the Urban Past: Staged Sounds as Mediated Cultural Heritage*, ed. Karin Bijsterveld, 181–194. Bielefeld: transcript.

Sturm, Michael. 2006. "Wildgewordene Obrigkeit"? Die Rolle der Münchner Polizei während der "Schwabinger Krawalle." In *"Schwabinger Krawalle": Protest, Polizei und Öffentlichkeit zu Beginn der 60er Jahre*, ed. Gerhard Fürmetz, 59–105. Essen: Klartext.

Tamagne, Florence. 2014. La Nuit de La Nation: Jugendkultur, Rock' n' Roll und *moral panics* im Frankreich der sechziger Jahre. In: *Popgeschichte*, vol. 2: *Zeithistorische Fallstudien 1958–1988*, ed. Bodo Mrozek, Alexa Geisthövel, and Jürgen Danyel, 41–62. Bielefeld: transcript.

Thomson, Elizabeth, and David Gutman, eds. 2004. *The Lennon Companion*. Cambridge: Da Capo Press.

Weber, William. 1997. Did people listen in the 18th century? *Early Music* 25:678–691.

Sound and Media

Cash, Johnny. 1968. *At Folsom Prison*. LP, Columbia Records, CS 9639.

Presley, Elvis. 1973. *Aloha from Hawaii via Satellite*. LP, RCA Victor, 6089.

Various Artists. 1970. *Woodstock: Music from the Original Soundtrack and More*. LP, Cotillion, SD 3-500.

12 Sound and Media Studies: Archiving and the Construction of Sonic Heritage

Carolyn Birdsall

In an article written over a decade ago, Germany's central storage facility Barbara-Stollen was criticized for its limited definition of national historical and cultural heritage: "The current commission maintains an archival concept that is almost entirely oriented towards the State, as critics have complained. This means that anything the State has not produced and preserved itself is not included in the [Barbara-Stollen] facility" (Marek 2003). This underground UNESCO-protected site was established outside of Freiburg in the 1950s, several years after the official formation of the West German Federal Republic, in the context of escalating Cold War tensions and fears of nuclear attack. Kept secret until 1990, today there are around 30 million meters of microfilm duplicates (including copies of East German archival records) selected for what has been described as the "Super Archive of German History and Culture."[1]

Film and sound recordings are absent from the national heritage selection at the Barbara-Stollen.[2] However, in the case of sound recordings, we can note similar past efforts to not only archive and preserve recorded sound, but also to protect it from potential wartime destruction. The national broadcasting authority, Reichs-Rundfunk-Gesellschaft (RRG), started making recordings as early as the late 1920s, with Berlin station director Hans Flesch initiating the first radio sound archive in 1930.[3] Contemporary commentators praised the establishment of radio production archives, with a distinct emphasis on their potential development as national sound archives dedicated to Germany's shared sonic heritage (see Tasiemka 1930, 4; Vertun 1932, 1055–1056).

In the context of National Socialism, party officials lamented the perceived selectiveness of the Weimar-era archive, while archivists boasted about their own achievements in the preservation of recordings of historical significance after 1933 (Brauchitsch 1934, 294–298; Valentyn 1941, 234–239). In 1942, however, the escalation of aerial attacks on Berlin led to the decision to bring recordings to German-controlled Graz, Wrocław, and Prague for safekeeping, thereby separating historically valuable recordings from those used for everyday program production. In the 1950s, archivists continued to appraise broadcast recordings of politicians, artists, and public events as historically significant, and a selection of these recordings were, in turn, held separately in a Frankfurt bank vault in case of fire.[4]

My motivation for drawing attention to sound preservation and the institutional establishment of archives is to emphasize the historical formation of heritage discourses and archival practices which have influenced the types of broadcast programming—such as those, for instance, from the National Socialist era—that are available in the historical sound archive today. In previous research, I have joined in sound studies' critique of textual and visual archives, and insisted on a critical reinterpretation of available sources in order to approach past soundscapes and histories of listening (Birdsall 2012, 13, 26–28; Smith 2004). Here, I will offer reflections on the social negotiations involved in the construction of heritage, with particular attention to the preservation and restricted distribution of Nazi-era radio and film.[5]

This chapter will begin by acknowledging existing work and theoretical concepts for the understanding of the archiving of sound, before focusing on case studies of how sound circulates through institutional archives and other less formal processes of reuse and remediation. With these perspectives in mind, I will revisit my own analysis of recorded and reused archival sound (in the form of special wartime announcements on German radio), before reviewing where and how archival sound from the National Socialist era is made available in the present day. In doing so, I will consider how a renewed examination of the archive may be productive in order to reveal the institutional and material formation of sound archives from the 1930s onward. At the same time, a more nuanced understanding of the social construction of sonic heritage may help us to reassess the relationship between sound history and contemporary developments connected to sound-based heritage.

Archiving Sound: Remediation and Recycling

As noted above, the question of cultural heritage in Germany is a complex one, particularly given the histories of dictatorship, persecution, and war during the twentieth century. The normative bent of much heritage discourse would suggest—in the case of Germany—a preference for preserving music, literature, and other (high) art forms. Against such tendencies, recent scholarship has elaborated on "conflict heritage" as an important field of inquiry, particularly as related to commemoration practices in the built environment (see Forbes et al. 2009; Bassanelli and Postglione 2013). The history of German heritage preservation and archival practices has also been examined, with a view to the exclusive definitions of national identity, history, and memory during National Socialism (Koshar 2000; Kretzschmar and Eckert 2007).

More generally, reflections on heritage, preservation, and presentation have been extensive in numerous fields of cultural studies research, including in particular sub-fields such as music and media studies.[6] The literature on the (digital) preservation and curation of moving images is ever growing, with a related concern about how various sociopolitical, industry, and technological shifts have shaped media archival practices (Amad 2010; Frick 2011; Fossati 2009; for television, see Dwyer 2008; Noordegraaf 2010). One blind spot in this scholarship is that "audiovisual heritage" is usually used as shorthand for discussions of image preservation, thereby neglecting sound archival materials in the scope of these discussions (Müske 2010; Campanini 2014).

Sound studies itself has been adept at examining the popular circulation of mediated sound, and has elaborated on the relationship between sound, technology, and memory (Bijsterveld and van Dijck 2009). In sound studies, and more noticeably in radio history, however, there has not been an integrated or sustained account of the institutional and noninstitutional processes of organizing, preserving, and presenting sound.[7] The field, on the whole, has also been slower to develop a critical response to recent developments in digital heritage, and as a result, it has also failed to fully investigate both the formal and informal processes of sound archival practice, both past and present (two early exceptions include Evens 2005; Sterne 2009). There are two concepts—namely remediation and recycling—with which sound studies could productively engage in order to better attend to the ways that sound is reframed and circulated through media culture.

The first term, *remediation*, was introduced by media scholars David Bolter and Richard Grusin, in their book, *Remediation: Understanding New Media* (2000). This concept was introduced as a means to grasp the processes by which the contents of one medium are re-presented in another, or how new media borrow from or imitate the appearances of existing media, such as print, radio, film, or television.[8] Each individual medium and its form of remediation depends on a network of technological, social, and economic factors. In other words, this process is not always evenly distributed across mediums, as Bolter and Grusin point out, since online media "refashion the newer perceptual media of radio, television, and telephone more aggressively than they refashion print" (ibid., 200). While their account investigates a number of hybrids produced in new media, such as computer games, digital photography, television, and the Internet, the analyses contain little elaboration on the remediation of sound, whether in other sound-based media or in audiovisual media.

The second concept of recycling has been used mainly in relation to compilation and found footage films, particularly for television history documentaries that reuse archival film materials. Film scholar Frances Guerin, for example, has studied the ways in which amateur film and photography from the National Socialist era have been

Figure 12.1
Berliner Rundfunk Sound Archive, Berlin 1946 (German Federal Archive picture 183-Z1105-501/
Schwahn/CC-BY-SA, Wikimedia Commons).

recycled in contemporary print media, exhibitions, film, television, and online. Guerin
draws attention to the archival function of these recyclings, namely that of preserving
them for researchers today, but also the potential for "transformations in perception,
in use, and in meaning of the images across time" (Guerin 2012, xviii). Guerin observes
how television documentaries often depend on voice-over commentary and musical
soundtracks that seek to contain recycled National Socialist images in the narrative,

but also acknowledges a range of experimental films that open up these same images to questions regarding interpretation, memory, and the archive. As such, Guerin's approach involves the analysis of film and photographic material in both their archival and recycled forms, while asking how these iterations facilitate particular ways of remembering and forgetting the past (see also Rigney and Erll 2012). These two terms provide conceptual tools for thinking about the (digital) media that frame and remediate archival sound material, as well as the cultural narratives and media conventions that have developed in the presentation of archival material, as, for instance, from the National Socialist era.

Radio Archives and Sound History

In my own research, I discovered that one of the most remembered radio programs from World War II was the special announcements (*Sondermeldungen*) (Birdsall 2009). These announcements are quite difficult to research: given their status as interruptions to the usual broadcast schedule, they did not appear in program magazines, nor did such publications discuss this program format (for more detailed discussion of the *Sondermeldung* genre, see Birdsall 2012, 113–119, 161–162). What is striking, however, is just how many recordings of these special announcements have been preserved, with almost 150 listed in the database of the Deutsches Rundfunkarchiv (DRA) in Frankfurt.

In terms of their media composition, these special announcements contained a mixture of live and prerecorded elements. The prerecordings consisted of musical intro and brass fanfare sequences, the two national anthems, and war campaign songs. Both the sound effects and short musical sequences demonstrate a strong similarity to the prerecorded materials held in broadcast production archives, as suggested by official archival catalogs for the years up to 1939.[9] *Sondermeldungen* not only circulated in public and private life through radio loudspeakers, but the campaign songs attained the status of catchy hit songs (*Schlager*), featuring in gramophone recordings, feature films, and newsreels, particularly in the first three years of the war.

What makes the sound archive interesting here is that radio production archives played a crucial role in facilitating the reusage of short segments from sound effects libraries (*Geräuscharchive*) as well as the remediation and circulation of recorded music through radio. The difficulty in studying this production process is in part due to the National Socialist investment in the "liveness" of *Sondermeldung* segments (Favre 2007). Official discourse stressed their immediacy (*Unmittelbarkeit*) and the notion that the whole nation was listening attentively and simultaneously. Propaganda Minister Joseph Goebbels was said to be furious when the magazine *Die Woche* published a picture of the sound effects record used for fanfare sounds in *Sondermeldungen*; such evidence of prerecorded material was understood as detracting from an understanding of these broadcasts as live, spontaneous, and unrepeatable (Boelcke 1966, 383). From the perspective of transparent remediation, *Sondermeldung* programs (and campaign theme songs) were presented and commented upon in fictional and nonfictional film genres, and even featured in television program production throughout 1941.[10]

While there may have been ambivalence about the extent to which listeners should be aware of prerecorded sound, in trade magazines and other publications, we can observe how National Socialist–era archivists and radio officials discussed the importance of recording technology for program production, and for the development of national sound archival collections (Valentyn 1941, 234–39). It remains difficult to ascertain where and when particular recordings were placed in storage, yet it would seem that the sizable presence of *Sondermeldungen* in the archive today is due to their perceived historical value during National Socialism and afterward. While many recordings were lost or destroyed after 1945, it can be estimated that a significant portion of these were selected for safe-keeping in occupied cities, or were among those confiscated by various Allied forces in 1945.[11] Either way, during the postwar period, *Sondermeldungen* did not appear to the same extent that other iconic sound clips (such as air-raid sirens) or radio programs (for example, *Wunschkonzert fur die Wehrmacht*) did, even though these segments appear to be among the most remembered programs from this period, connected as they were to, first, the celebration of initial German military successes and, later, the growing sense of the war being lost in the East and on other fronts.[12]

This case offers a reminder of the role played by institutional archiving in the organization, remediation, and recycling of recorded sound in National Socialist radio. The decision to record such segments for the historical archive also reflects a choice concerning their perceived historical and political significance, and potential for future reuse. It is crucial for a critical sound studies project to be able to identify National Socialist era practices of sound archiving, as well as the choices made regarding their selection, storage, and preservation in subsequent decades. In what follows, I will reflect on some different ways in which archival sound from the National Socialist era is now available to both researchers and the general public, including the alternative sound archives of this period made available on digital platforms.

Archives of Sound

The "archives" of sound from the National Socialist period comprise not only recorded sound, but also the remediation and visualization of sound in other media, including photography, film, and print media (Schmidt 2014). In this vein, I have examined a compilation film that seeks to uncover the audio and audio-visual archive of National Socialism in the present. *Hitler's Hit Parade* (2003) is striking as a documentary film, since it reveals the heterogeneous nature of archival materials, in contrast to the

canonical selections of images that are usually recycled in contemporary media representations of National Socialism. Its usage of twenty-five hit songs from the National
Socialist era, each played in full, seems like a "Best of" album of sentimental and catchy
tunes from this period. The film combines hit songs from the sound archive with the
(audio)visual archive of this period, ranging from feature and advertising films to amateur home movies and images of persecution. In doing so, the film reopens debates
about National Socialist cultural production and the status of the image as "evidence"
of this period. At the same time, the hit song soundtrack raises questions about the
ongoing circulation of hit songs from the 1930s and 1940s in commercially sold CD
compilations, as supposedly nonpolitical artifacts from that era.[13] The circulation of
archived sound in the present requires some investigation into how and where these
materials are presented, and the interfaces—or "scripted spaces"—that structure the
remediation of such materials (Chamberlain 2011).

In terms of conventional audio publishing, I have not yet ascertained a date for
the earliest commercial publications of radio programming.[14] While Germany and
many other European nations have restrictions on the commercial publication of
National Socialist era sound and film materials classified as propaganda, we can
observe the commercial distribution in places such as Luxembourg, Liechtenstein,
and the United States, which have not banned the trafficking of Nazi memorabilia.
One example of illegal distribution is Documentary Series Est., which distributed
vinyl copies of National Socialist radio recordings through Liechtenstein in the
1970s. Despite their appearance as documentary releases, it seems that this series of
approximately 100 LPs was produced by German Nazi sympathizers whose operation was eventually shut down by government authorities (see Anonymous 1976;
how such a large number of radio recordings came into the possession of this organization remains unclear). While this distribution was eventually prevented, library
catalogs suggest that these recordings are widely available in public libraries around
the world.[15]

As for more official publications, in 1995, the German DRA initiated a CD
series *Stimmen des 20. Jahrhunderts* (Voices of the Twentieth Century) in cooperation with the Deutsches Historisches Museum, Berlin. Starting with a compilation
of radio broadcasts on the German capitulation in 1945, this series has since put
out thirty-five releases, most of which deal with political speeches, live events, and
other cultural productions.[16] One of the most interesting releases is *Das Verbrechen
hinter den Worten* (The Crimes Behind the Words), which seeks to challenge the omissions of the National Socialist archive, and to find statements that serve as evidence
of the Holocaust and others forms of persecution. Each of the releases in the CD
series includes an online listening sample, and the DRA also frames archival sound
through its online exhibition pages, such as the "Die Olympischen Spiele 1936 im

NS-Rundfunk." Such pages include contextual framing and visual documentation, with digitally remediated audio delivered through an MP3 player embedded on the site. Other official archives, for example the British Library, also list copies of National Socialist era recordings in their online database. All of these sound recordings are available online, and are suggestive of dual holdings shared with the DRA; some of these materials also potentially extend to programs recorded by UK propaganda monitors during World War II.[17]

The unofficial circulation of National Socialist era audio materials is vast and is in part connected to a broader memorabilia culture and collectors' market, whose online presence in Europe has been a point of controversy in recent years.[18] In this process, even collection specialists at large institutions like the Deutsches Historisches Museum have been called to reassert their faith in the pedagogical benefits of including original artifacts in exhibition displays (Steffes 2014). On download forums and other platforms for sharing digital content, there is noticeable popularity for original sound recordings, particularly of military marches and wartime songs, along with National Socialist propaganda films and newsreels. Aside from unofficial channels, there are also open access archival platforms like the nonprofit Internet Archive (www.archive.org). On the site's Community Audio section, we can find numerous recordings of National Socialist radio, primarily of speeches by National Socialist party leaders Adolf Hitler and Joseph Goebbels. While it is possible to see who provided these recordings, very little contextual or technical information is given. For example, one collection is titled "Wartime Radio (1932–1946)," and was put together in 2011 using donations from "Ad Bongers-Scheltema Verlag" and "Clauds Historical Record and Recording Collection." Although some technical information is supplied about recording formats, it remains very difficult to track down the collectors or find out more about the provenance of these collections.

More generally, one of the other key means of accessing archival material is through their reusage in radio, television, and film. In the case of radio, much of these efforts have been linked to the perceived pedagogical use of the voices of Hitler and Goebbels for the understanding of German history. Indeed, during the 1950s, Hamburg-based NWDR (later NDR) and other stations were preoccupied with securing copies of recordings, as part of an effort to close the gaps in the archived sound history for "The Weimar period up to 1945."[19] These materials, as I pointed out earlier, should be examined critically as to their structural absences and omissions. It should therefore be remembered, as wartime propaganda analyst Ernst Gombrich pointed out, that crucial information about Jews and Eastern European slave labor remained all but absent from wartime German radio.[20]

The reuse of historical clips in West German radio appears to have been common practice, as suggested by correspondence in the mid-1950s when requests for historical spoken word clips tripled between 1955 and 1956 (Anonymous 1955–56, 3). During a

1958 radio feature, marking "60 Years of Sound Documents," a Frankfurt reporter takes a tour of the national sound archive. The archive is presented as providing the listener with access to "the voices of the past" and performing "contemporary sonic history" through sound recordings. In the program, there is a pedagogical introduction to German history from 1918 to 1945, with the inclusion of Reichstag parliamentary recordings, communist and Nazi party anthems, and speeches by Adolf Hitler and Joseph Goebbels. In this reuse of historical recordings, radio's limited definition of the public sphere is confirmed through the particular selections of sound recordings, which are mainly restricted to the voices of male political leaders and well-known events (Lacey 1996). Moreover, it allows former National Socialist radio functionaries like archivist Hans-Joachim Weinbrenner to ignore their own role in earlier propaganda production. Instead, Weinbrenner's statements are consistent with the selective remembering of the 1950s, and he interprets these historical recordings according to the values of the West German democratic state (see *60 Jahre Tondokumente*, 1958; Moeller 2001). Within the general trend of remixing found footage and other forms of reuse from the 1960s to the present day, the ongoing presence and appropriation of National Socialism in contemporary digital culture is also worth noting (Tofts and McCrea 2009; Rosenfeld 2015).

In summary, there is a range of ways in which archival sound from the National Socialist era has been preserved (in various media and formats), archived (in institutional and noninstitutional frameworks), and made accessible (through various forms of publishing and access). The remediations and recycling of such material have been emphasized as important operations in the ongoing negotiation and production of German sonic heritage. As this chapter has shown, media scholars have called for a more nuanced understanding of "archival footage" as not only restricted to "physical materials stored in archives controlled by state or other institutions, collections officially sanctioned as authoritative repositories of audiovisual evidence about the past. However, this definition is problematic in that it simply refers to a location in which certain documents, whose contours are determined by variously informed acts of inclusion and exclusion, are stored" (Baron 2014, 7). Such observations recognize that an archival document is constituted not only by virtue of its professional appraisal or storage in a particular location, but also through its circulation, reuse, and the experiences of media audiences. Through examining conditions of archiving, preservation, presentation, and circulation of archival sound, sound studies research will be better situated to offer critical insights into the constitution of sound-based heritage, both past and present.

To conclude, then, it is evident from the previous discussion that a critical perspective on archiving and the social production of sonic heritage is crucial for the understanding of sound in media culture. Using conceptual tools from media studies and developing comparative approaches to media archiving may offer a means with which researchers can grasp the specific processes by which mediated sound is made available for the purposes of analysis today. This awareness does not require sound scholars to abandon their critical assessments of the limits to visual and textual archives, nor to fetishize the sound archive; rather it requires us to further inquire into the ways in which sound materials are preserved and presented in the present, as well as the historical formation of sounds from the past. An awareness of such processes—in combination with critical perspectives drawn from archive and media theory—is essential in order to reach a more nuanced understanding of the constitution of mediated sound.

Notes

1. See the official brochure for the facility, published on the website of Germany's Federal Office of Civil Protection and Disaster Assistance (Bundesamt für Bevölkerungsschutz und Katastrophenhilfe, BBK): http://www.bbk.bund.de/SharedDocs/Downloads/BBK/DE/Publikationen/ Broschueren_Flyer/Flyer_Barbarastollen.pdf?__blob=publicationFile (accessed June 2, 2014).

2. This omission may be in part due to the existence of other organizations involved in audiovisual preservation, the most prominent being the German Film Institute (Frankfurt), Deutsche Kinemathek (Berlin), German Music Archive (Leipzig), and the public broadcasting funded Deutsches Rundfunkarchiv (Frankfurt and Potsdam).

3. Of course, sound archiving began earlier, following the invention of the phonograph in the 1870s, although primarily in the context of academic research. Here, however, I mainly discuss archives that were perceived in shared national terms.

4. In 1950, the *Lautarchiv des Deutschen Rundfunks* (now *Deutsches Rundfunkarchiv*), was announced as the central historical archive for all stations of the West German ARD broadcast organization (*Arbeitsgemeinschaft der öffentlichen-rechtlichen Rundfunkanstalten der Bundesrepublik Deutschland*). For the use of bank vault storage, see "60 Jahre Tondokumente: Besuch im Lautarchiv des Deutschen Rundfunks," Hessischer Rundfunk, November 2, 1958 (Reporter: Käthe Beckmann; interviews with Kurt Magnus, Hans Weber, Hans-Joachim Weinbrenner). Sound recording held at Deutsches Rundfunkarchiv, Frankfurt.

5. For discussions of the dangerous nature of certain National Socialist films, and postwar restrictions on their distribution and exhibition, see Leiser 1974 and Kelson 1996.

6. In popular music studies, for instance, recent research has shown increasing interest in heritage discourses and institutions such as museums, music tourism, fan practices, and online archiving initiatives. See, e.g., Bennett 2009, 474–489; Brandellero et al. 2013.

7. In radio history, none of the major national history volumes has devoted significant attention to archival practice in broadcasting. Two recent exceptions that discuss the archiving of radio include Dolan 2003 and Jensen 2012.

8. They concede that this is a longer-term process, citing examples of how television has remediated film since the early 1950s. However, Bolter and Grusin do not, perhaps, sufficiently acknowledge the differences between the television culture of the United States and other global contexts.

9. For the first catalog of the Weimar period, see *Schallaufnahmen der deutschen Rundfunkgesellschaften in den Jahren 1929/1931* (Berlin: Hesse, 1931). For the two subsequent catalogs, see *Schallaufnahmen der Reichs-Rundfunk GmbH von Ende 1929 bis Anfang 1936* (Berlin: Reichs-Rundfunk-GmbH, 1936); *Schallaufnahmen der Reichs-Rundfunk GmbH von Ende 1936 bis Anfang 1939* (Berlin: Schallarchiv der Reichs-Rundfunkgesellschaft, 1939).

10. See, e.g., Miltner 1940; Anonymous 1941a; Anonymous 1941b.

11. For German-language radio recordings held in Russia, see the DRA publication, Tischler 1997.

12. For a rare example of a compilation that included *Sondermeldungen*, see *Aus Dem Führerhauptquartier Teil 1 + 2*, vinyl LP, catalog no. DS 353, Documentary Series Est. Schaan (Liechtenstein), year unknown. I discuss this LP series in more detail in the following section.

13. An illustrative example is the CD release *Heimat Deine Sterne*, which includes digitally remastered film hit songs from the period 1930–1942, thus spanning both Weimar and National Socialist cultural production.

14. Against the background of limited explicitly political content on radio (prior to 1933), the National Socialist party did produce gramophone releases, such as a 1932 speech by Adolf Hitler as part of election campaigning. *Völkischer Beobachter* (July 15, 1932), cited in Paul 1992, 198.

15. For a full listing of the series and library holdings, see the discography available at Discogs (http://www.discogs.com) or the bibliographic listings offered by WorldCat (http://www.worldcat.org).

16. Titles directly relevant to the National Socialist era include historical overviews like "Frauenstimmen, 1908–1997," "Hymnen der Deutschen," "Prosit Neujahr!," "Hundert deutsche Jahre 1900–2000," and "Von Pan-Europa zur Europäischen Union." Titles that are more specific to National Socialism include "Weimar—Das Scheitern einer Demokratie," "1933—Der Weg in die Katastrophe," "Die XI. Olympischen Sommerspiele" and "Der Nürnberger Prozess" (the full list can be accessed http://www.dra.de/publikationen/cds/).

17. More generally, the British Library Sound Archive has had global scope and ambition since its inception as the British Institute of Recorded Sound in 1955. See Saul 1956, 173.

18. A French court ruled in 2000, for instance, that Yahoo! was responsible for the accessibility of Nazi memorabilia to the French public, in violation of the French criminal code. More recently, the availability of Hitler's *Mein Kampf* online and preparations for a critical edition of this book have been the subject of further debate. See, e.g., Brändlin 2014; Range 2014.

19. German original: "Zeitgeschichte von der Weimarer Zeit bis 1945." See the annual report for the DRA-forerunner in Frankfurt: "Jahresbericht des Lautarchivs des deutschen Rundfunks 1955/56," 4, NWDR/NDR Sammlung, file 1381, held at the Hamburg Staatsarchiv. In 1955, the BBC allowed NWDR archivist Dietrich Lotichius to spend a week in London in order to register and copy a selection of approximately 4,500 disks with Reichs-Rundfunk-Gesellschaft broadcast content. In earlier archivist correspondence, an estimate was given that 2,000–3,000 disks held in London needed to be copied to avoid imminent deterioration. See internal correspondence from March 1955, NWDR/NDR Sammlung, file 1381, held at the Hamburg Staatsarchiv.

20. Gombrich, an Austrian art historian with a Jewish family background, worked for the BBC Listening Post between 1939 and 1945, monitoring German-language broadcasting. As he recalled, "the concentration camps were not mentioned on the Home Service except very late in the war when they were described in a talk as places of harsh corrective labor, presumably both to intimidate and to reassure the home population" (Gombrich 1970, 23).

References

Bibliography

Anonymous. 1941a. Die neue Wochenschau: Kampf und Sieg von Finnland bis zum Schwarzen Meer. *Film-Kurier*, July 23, 1.

Anonymous. 1941b. Deutsches Rundfunk Schrifttum: Bibliographische Beilage. *Rundfunkarchiv* 14 (September): 9.

Anonymous. 1955–56. Jahresbericht des Lautarchivs des deutschen Rundfunks 1955/56 [Annual report for the DRA-forerunner in Frankfurt]. NWDR/NDR Sammlung, file 1381, Hamburg Staatsarchiv.

Anonymous. 1976. Liechtenstein: Eine Treuhand wäscht die andere. *Spiegel* 34, August 16. http://www.spiegel.de/spiegel/print/d-41147305.html.

Amad, Paula. 2010. *Counter-Archive: Film, the Everyday, and Albert Kahn's Archives De La Planète*. New York: Columbia University Press.

Baron, Jaimie. 2014. *The Archive Effect: Found Footage and the Audiovisual Experience of History*. New York: Routledge.

Bassanelli, Michela, and Gennaro Postglione, eds. 2013. *Re-enacting the Past: Museography for Conflict Archaeology*. Siracusa: LetteraVentidue.

Bennett, Andy. 2009. Heritage rock: Rock music, representation, and heritage discourse. *Poetics* 37 (5–6): 474–489.

Bijsterveld, Karin, and José van Dijck, eds. 2009. *Sound Souvenirs: Audio Technologies, Memory, and Cultural Practices*. Amsterdam: Amsterdam University Press.

Birdsall, Carolyn. 2009. Earwitnessing: Sound memories of the Nazi period. In *Sound Souvenirs: Audio Technologies, Memory, and Cultural Practices*, ed. Karin Bijsterveld and José van Dijck, 169–181. Amsterdam: Amsterdam University Press.

Birdsall, Carolyn. 2012. *Nazi Soundscapes: Sound, Technology, and Urban Space in Germany, 1933–1945*. Amsterdam: Amsterdam University Press.

Boelcke, Willi A. 1966. *Kriegspropaganda, 1939–1941: Geheime Ministerkonferenzen im Reichspropagandaministerium*. Stuttgart: Deutsche Verlags-Anstalt.

Bolter, Jay David, and Richard Grusin. 2000. *Remediation: Understanding New Media*. Cambridge, MA: MIT Press.

Brandellero, Amanda, Susanne Janssen, Sara Cohen, and Les Roberts, eds. 2013. *Popular Music as Cultural Heritage*. Special issue, *International Journal of Heritage Studies* 20 (3).

Brändlin, Anne-Sophie. 2014. Hitler's "Mein Kampf" becomes an online bestseller. *Deutsche Welle*, January 11. http://www.dw.de/hitlers-mein-kampf-becomes-online-bestseller/a-17355880.

Brauchitsch, Konrad von. 1934. Schallaufnahme und Schallarchiv der Reichs-Rundfunk-Gesellschaft. *Rufer und Hörer* 6–7:294–298.

Campanini, Sonia. 2014. Film sound in preservation and presentation. PhD diss., University of Amsterdam.

Chamberlain, Daniel. 2011. Scripted spaces: Television interfaces and the non-places of asynchronous entertainment. In *Television as Digital Media*, ed. James Bennett and Niki Strange, 230–254. Durham, NC: Duke University Press.

Dolan, Josephine. 2003. The voice that cannot be heard. *Radio Journal: International Studies in Broadcast and Audio Media* 1 (1): 63–72.

Dwyer, Andy. 2008. European television archives and the search for audiovisual sources. In *A European Television History*, ed. Jonathan Bignell and Andreas Fickers, 257–262. Malden, MA: Wiley-Blackwell.

Evens, Aden. 2005. *Sound Ideas: Music, Machines, and Experience*. Minneapolis: University of Minnesota Press.

Favre, Muriel. 2007. Goebbels "phantastische Vorstellung": Sinn und Zweck des O-Tons im Nationalsozialismus. In *Original/Ton: Zur Medien-Geschichte des O-Tons*, ed. Harun Maye, Cornelius Reiber, and Nikolaus Wegmann, 91–100. Konstanz: UVK.

Forbes, Neil, Robin Page, and Guillermo Perez, eds. 2009. *Europe's Deadly Century: Perspectives on 20th-Century Conflict Heritage*. Swindon: English Heritage.

Fossati, Giovanna. 2009. *From Grain to Pixel: The Archival Life of Film in Transition*. Amsterdam: Amsterdam University Press.

Frick, Caroline. 2011. *Saving Cinema: The Politics of Preservation*. New York: Oxford University Press.

Germany's Federal Office of Civil Protection and Disaster Assistance. (Bundesamt für Bevölkerungsschutz und Katastrophenhilfe, BBK): http://www.bbk.bund.de/SharedDocs/Downloads/BBK/DE/Publikationen/Broschueren_Flyer/Flyer_Barbarastollen.pdf?__blob=publicationFile (accessed June 2, 2014).

Gombrich, Ernst H. 1970. *Myth and Reality in German War-Time Broadcasts*. London: Athlone Press.

Guerin, Frances. 2012. *Through Amateur Eyes: Film and Photography in Nazi Germany*. Minneapolis: University of Minnesota Press.

Jensen, Eric Granly. 2012. Access and history: The digitisation of the Danish Broadcasting Archives and its cultural heritage. *International Journal of Media and Cultural Politics* 8 (2–3): 305–316.

Kelson, John. 1996. *Catalogue of Forbidden German Feature and Short Film Productions*. [Zonal Film Archives, Hamburg, Germany.] Ed. K. R. M. Short. Westport, CA: Greenwood Press.

Koshar, Rudy. 2000. *Germany's Transient Pasts: Preservation and National Memory in the Twentieth Century*. Chapel Hill: University of North Carolina Press.

Kretzschmar, Robert, and Astrid Eckert, eds. 2007. *Das deutsche Archivwesen und der Nationalsozialismus: 75 Jahre Deutsche Archivtag 2005 in Stuttgart*. Essen: Klartext.

Lacey, Kate. 1996. *Feminine Frequencies: Gender, German Radio and the Public Sphere, 1923–1945*. Ann Arbor: University of Michigan Press.

Leiser, Erwin. 1974. *Nazi Cinema*. London: Secker & Warburg.

Marek, Michael. 2003. Schatzkammer der Nation. *Deutsche Welle*, May 19. http://www.dw.de/schatzkammer-der-nation/a-864249.

Miltner, Heinrich. 1940. Filmaufnahme mit—*Sondermeldung*. *Film-Kurier*, July 9, 3.

Moeller, Robert G. 2001. *War Stories: The Search for a Usable Past in the Federal Republic of Germany*. Berkeley: University of California Press.

Müske, Johannes. 2010. Constructing sonic heritage: The accumulation of knowledge in the context of sound archives. *Journal of Ethnology and Folkloristics* 4 (1): 37–47.

Noordegraaf, Julia. 2010. Who knows television? Online access and the gatekeepers of knowledge. *Critical Studies in Television: Scholarly Studies in Small Screen Fictions* 5 (2): 1–19.

Paul, Gerhard. 1992. *Aufstand der Bilder: Die NS-Propaganda vor 1933*. Bonn: Dietz.

Range, Peter Ross. 2014. Should Germans read "Mein Kampf"? *New York Times*, July 7. http://www.nytimes.com/2014/07/08/opinion/should-germans-read-mein-kampf.html?_r=0.

Rigney, Anne, and Astrid Erll, eds. 2012. *Mediation, Remediation, and the Dynamics of Cultural Memory*. Berlin: de Gruyter.

Rosenfeld, Gavriel D. 2015. *Hi Hitler! How the Nazi Past Is Being Normalized in Contemporary Culture*. Cambridge: Cambridge University Press.

Saul, Patrick. 1956. The British Institute of Recorded Sound. *Fortes Artis Musicae* 3 (2): 170–173.

Schmidt, Michael J. 2014. Visual music: Jazz, synaesthesia, and the history of the senses in the Weimar Republic. *German History* 32 (2): 201–223.

Smith, Mark M., ed. 2004. *Hearing History: A Reader*. Athens, GA: University of Georgia Press.

Steffes, Annabelle. 2014. The right approach to Nazi memorabilia. *Deutsche Welle*, April 16. http://www.dw.de/the-right-approach-to-nazi-memorabilia/a-17573542.

Sterne, Jonathan. 2009. The preservation paradox in digital audio. In *Sound Souvenirs: Audio Technologies, Memory, and Cultural Practices*, ed. Karin Bijsterveld and José van Dijck, 55–65. Amsterdam: Amsterdam University Press.

Tasiemka, Hans. 1930. Ein Funkarchiv für die Ewigkeit. *Der Deutscher Rundfunk* 8 (30): 4.

Tischler, Carola. 1997. *Inventar der Quellen zum deutschsprachigen Rundfunk in der Sowjetunion (1929–1945): Bestände in deutschen und ausländischen Archiven und Bibliotheken*. Potsdam: Verlag für Berlin-Brandenburg.

Tofts, Darren, and Christian McCrea. 2009. Editorial: Remix. *Fibreculture* 15:n.p. http://fifteen.fibreculturejournal.org/.

Valentyn, Eduard van den. 1941. Die Entwicklung der Schallaufzeichnung im Grossdeutschen Rundfunk. *Reichsrundfunk*, August 31, 234–239.

Vertun, Hans. 1932. Konservierte Geschichte. *Die Sendung* 49 (9): 1055–1056.

Sound and Media

60 Jahre Tondokumente: Besuch im Lautarchiv des Deutschen Rundfunks. 1958. Hessischer Rundfunk, November 2. Reporter: Käthe Beckmann; Interviews with Kurt Magnus, Hans Weber, Hans-Joachim Weinbrenner). Sound recording held at Deutsches Rundfunkarchiv, Frankfurt.

Aus Dem Führerhauptquartier Teil 1 + 2. N.d. LP, Documentary Series Est. Schaan, Liechtenstein, DS 353.

Das Verbrechen hinter den Worten: Tondokumente zum nationalsozialistischen Völkermord. 2001. CD, Deutsches Rundfunkarchiv in cooperation with the Deutsches Historisches Museum, Berlin.

Die Olympischen Spiele 1936 *im NS-Rundfunk*. Web portal, Deutsches Historisches Museum Berlin. http://1936.dra.de/ (accessed June 2, 2014).

Heimat Deine Sterne. 2003. CD Series: Deutsche Tonfilmschlager, The International Music Company AG, Hamburg.

Hitler's Hit Parade. 2003. Directors: Oliver Axer and Susanne Benze. Documentary, C. Cay Wesnigk Filmproduktion, Berlin.

Schallaufnahmen der deutschen Rundfunkgesellschaften in den Jahren 1929/1931. Berlin: Hesse, 1931.

Schallaufnahmen der Reichs-Rundfunk GmbH von Ende 1929 bis Anfang 1936. 1936. Berlin: Reichs-Rundfunk-GmbH.

Schallaufnahmen der Reichs-Rundfunk GmbH von Ende 1936 bis Anfang 1939. 1939. Berlin: Schallarchiv der Reichs-Rundfunkgesellschaft.

Stimmen des 20. Jahrhunderts. 1995. CD Series, Deutsches Rundfunkarchiv in cooperation with the Deutsches Historisches Museum, Berlin. http://www.dra.de/publikationen/cds/ (accessed June 2, 2014).

13   Soccer Stadium as Soundscape: Sound and Subjectivity

Jochen Bonz

In a study about the songs sung by soccer fans at stadium games, musicologists Reinhard Kopiez and Guido Brink contrasted the creativity expressed by those songs with the mere yelling that takes place at the events (Kopiez and Brink 1998, 7). However, it is precisely that yelling, the insults and the cries of desperation and jubilation which articulate—through the medium of noise—what Norbert Elias and Eric Dunning described in 1969 as the cultural function inherent to soccer: in a culture based on "excitement control" ([1969] 1986, 63), football and other leisure activities constitute "an enclave for the socially approved arousal of moderate excitement behaviour in public" (ibid., 65). For Elias and Dunning, soccer enthusiasm is synonymous with the desire and search for intensive and extraordinary emotional experiences.

The concept that modern soccer stadiums are architecturally structured to increase the intensity of emotions is consistent with Elias and Dunning's discussions. For that purpose, the inner stadium is shut off from the outside world to focus attendants on the inside space and increase noise: "The sound of the goal cheer does not just disappear into nothingness; it in fact echoes back through the inside space with even more force" (Alkemeyer 2008, 92; see also Prosser 2002; Schäfer and Roose 2010).

Garry Robson (2004) provides an alternative explanation for the noisiness of soccer fans in a study on the fan culture of the southeast London team, Millwall FC. Robson discusses the "Millwall Roar,"[1] a collective roar often lasting several minutes in which the syllable "Miiiiill" is drawn out almost interminably, thus making it possible to experience it as a material sound event.[2] According to Robson, this is how the fans secure their collective community, and how they provide their fan culture's proletariat values with a presence in today's world. In this way, the community of fans and their culture come alive and are able to be experienced emotionally on an individual level: "The roar brings the collective and its world alive, and can overwhelm both participants and observers" (Robson 2004, 183).

Robson's analysis confirms the widely held view that soccer fans are essentially engaged in identity development. At the same time, however, he expands on the traditional understanding of identity construction. Today, it is mainly understood through the mechanism of "Othering" as a mode of identification in which the self is conceived of as a member of a group and the identity of the group is asserted through an idealization directed inward, and the devaluation and invalidation of an Other located on the outside. In the construction of Otherness, this Other is experienced simultaneously as a threat and as an affirmation of the self through the Other's supposed inferiority (see Fabian 1983; Said [1978] 2003).

Robson demonstrates that there is a different mode of identity construction in the realization of identifications that are inherent to the subject. In Robson's analysis, he determines that the Millwall FC fans reference British working class values. For the participants, those values become perceptible as a part of their own self in the emotionally laden atmosphere of the soccer game: they become deeply moved. An already existing subject of Millwall fandom is, in a way, performatively awoken and can be experienced for the person as self.[3]

His interpretation of the Millwall Roar as constituting an expansion of presence from the already available identifications enables Robson to formulate a research

Figure 13.1
Ultra fans of FC Wacker Innsbruck at Hanappi stadium, Vienna (photo courtesy of Faninitiative Innsbruck, Verein zur Förderung der Fankultur).

approach that makes it possible to focus on identifications, and the objects at which they are directed, in the context of soccer enthusiasm. These are identifications around objects of analysis that are difficult to nail down for research purposes—these identification objects or forms of identification are not just located within the subject's field of perception. In fact, it is only through an identification that this kind of field of perception, this desire—in other words, a subject—is generated in the first place. As Robson's understanding of the Millwall Roar demonstrates, however, objects of identification can emerge within cultural phenomena and be made accessible to analysis. But in that bellowed "Miiiiiil!!," the identification not only becomes present for the fans themselves—the objects of identification also gain presence for fan culture research, and thus an existence in the here and now within the dimension of noise.

Identifications in the Soundscape of the Stadium

Robson's study does no more than establish a starting point for a sound-oriented approach to the research of soccer enthusiasm as an event of identification. It is undercomplicated in that Robson implicitly defines a specific situation: how the events that occur in the stadium on the day of the game are perceived from the perspective of a very particular form of fandom. As a series of recent studies on manifestations of soccer enthusiasm have shown, one must assume greater situational variety (see Giulianotti 2002; Porat 2010; Schmidt-Lauber [2003] 2008). Therefore, I suggest looking at soccer in general as a spectator sport and at the stadium situation in particular as metasituations that encompass diverse situations in which the fans' objects of fandom acquire presence.

Robson's proposed analysis is confirmed by research into the subculture of the so-called ultras in my ethnographic analysis of manifestations of soccer enthusiasm concerning the German soccer club SV Werder Bremen: the fans' behavior references a subject position with which they identify. I will not go into further details here, but will instead concentrate on other sound events and the forms of identification linked to them.

The large majority of sound phenomena at Bremen's Weser Stadium articulate an object of identification that is both unambiguous and completely vague: I am speaking of the home team fans' relationship with their team and/or club and everything that they represent. In Bremen—where, because of SV Werder Bremen's great success, soccer plays so large a role that the club is a central topic of conversation in local politics as well as in everyday conversations between individuals—SV Werder Bremen represents the city of Bremen itself. A community is articulated in many ways through the dimension of sound: in the singing of the club's songs at the beginning of the game, in the yelling out of the home-team players' last names in answer to the stadium announcer's calling out of the first names for the lineup, in the collective goal celebration. Without a doubt, it has the characteristics of an "imagined community" (Anderson [1983]

2006), which can be experienced as a "performative community" (Alkemeyer 2008, 88) by the subject in the stadium situation.

Imagined, performatively experienceable communities also consist of smaller, divided sections within the stadium. Collective call-outs, which Mery Kytö has described as "call and response" (2011, 127), go back and forth between the different bleacher sections across the stadium. In Bremen's Weser Stadium, for example, thousands of people occupy each side of the stadium, the eastern and the western stands: in the middle of the game, and generally during a more boring phase, a call echoes out over the playing field: "Heeel-lo eastern stands!" To which the fans in the eastern stands answer: "Heeel-lo western stands!" Whereas in this case the back and forth is limited to a one-time exchange that may perhaps be repeated at a later point, another call-and-response game usually repeats itself several times: from somewhere in the stadium, a resounding call of "Werrrr-derrrrrr!" sounds out, followed by an immediate response from another area: "Breeeeee-mennnnn!" As a rule, this call-out occurs when the game is at a particularly dramatic highpoint; it constitutes a cheer of encouragement. In contrast to the eastern-western stands call-out, the addressees are less clearly defined here and the bellowed-out names echo less distinctly from one side of the stadium to the other. That leads to a roar, the wildness of which signalizes support of the home team players, and enables the roarers to feel that they are a part of their team's game. Accordingly, Bremen fans are greeted as the "twelfth player" during the game's opening program. An imagined, performative, and fleeting experience of the self as part of a community thus occurs in this case as well.

The Humming of the Peer Groups

Imagined collectives are articulated and generated through the sound events discussed above. Beyond that, the soundscape of the stadium is also determined through sound phenomena whose initiators and addressees are more concrete: individual call-outs; groaning; whooping; cheering; clapping; chanting by a few fans that surge up and then ebb back down; dull rumblings, such as would be the result of kicking metal barriers or seats; and last but not least, conversations or conversation sounds and fragments of conversations. Together, those noises create a humming that can also be understood as the articulation of the specific situation of soccer enthusiasm, which Jan Jirat (2007) identified as "peer groups" in his ethnography of the Swiss soccer club FC Schaffhausen's fan stands. Jirat defines them in the following way: "[Peer groups are] associations of people … who have approximately the same position, status and age in relation to one another" and "who have been a part of the [fan stands] by the dozens, mostly for many years already. These peer groups differ starkly in their age structure as well as in their size, they range from small groups (3 to 4 people) to groups of over 20 people" (Jirat 2007, 111). Alongside having a general "interest in the game and in the fate of the association, the social factor plays a large role" for these peer groups (ibid.). Jirat elaborates on this point:

The stadium is not only both the site of the game and the platform for the acting out of fandom—it is primarily a meeting place that provides the possibility of making and nurturing social contacts within a precisely determined cycle. For many peer group members, important reasons for going to a game are: collective discussion, commenting on the plays, living out collectively felt emotions and experiencing the game with a familiar group in which one can move around completely naturally and even be emotional on occasion. (ibid.)

In Frank Müller's (2010) autoethnographical reflections about a group of friends in the standing room only section of Bremen Weser Stadium, he clarifies the meaning of "commenting on the plays" and "living out collectively felt emotions." At the time of the study, the group had existed for over fifteen years and had met at every home game. The core of the group was made up of four people, including one person who vocally expressed himself: "He enjoys screaming out his commentary about the game and often provokes the other spectators who find his allegations to be either funny, wrong, or disruptive" (93). For the other group members, that opens up other possibilities for action that would contribute to the humming. Müller writes about his own behavior: "It is fun for me to relativize Bernd's often polemical contributions, to reinforce them or just to entertain people with some corny joke. I often defend him, and, interestingly, his complaints about the Bremen team's game are seldom wrong" (ibid.).

The peer group situation comprises essential references to the events of the game, to individual players of both one's own team and the opposing team, as well as to other fans, individuals, or groups. For that reason, on the one hand, the group of friends can be an identification object for the peer group situation. On the other hand, a plurality of identification objects are obviously available here, namely objects that are contained in all of the references and experienceable for this situation's subject.

An example of this is the referencing of individual players or parts of the team and their actions. That clearly involves a desire to have an influence on the game, to play with them, to be a part of shaping the events of the game. This identification is demonstrated by the stadium's soundscape, in call-outs such as those cited by Müller: "C'mon, run, run, run, faster, don't make it so complicated, pass it … pass it over now, man, no shoot—oh man, he's really kicking it in" (101).

As a phonetic utterance, these kinds of statements contribute to the humming of the situation. The humming, however, reveals what the situation's function is for the participants. In their encyclopedia of sound phenomena, Jean-François Augoyard and Henry Torgue (2009) describe the envelopment sound effect as a sound situation in which the subject is embedded in sounds. Their origin and semantic content is irrelevant—what arises instead is a "feeling of being surrounded by a body of sound that has the capacity to create an autonomous whole, that predominates over other circumstantial features of the moment" (47). Augoyard and Torgue describe the accompanying experience as very pleasant for the subject: "The accomplishment of this effect is marked by enjoyment, with no need to question the origin of the sound" (ibid., 22). In my opinion, if the sound situation is understood in that way, it can be transferred to the peer group situation: their humming swirls around the subjects in the same way that it encompasses the subject in the form of a situation. It provides support, is considered pleasant, gives security, and enables and encourages the subject to realize their desire. Les Back correspondingly characterizes the soundscape of the stadium as an "atmosphere of sociability rather than communication" (2003, 320).

The Affective Impact of the Stadium Situation and the Fans' Identity Work

Today, soccer enthusiasm is plausibly explained by the aesthetic qualities that are inherent to the game. Thomas Alkemeyer (2008) thus emphasizes how the stadium situation's high degree of affectivity results from the improbability of successfully coordinating the ball, the feet, and the players. Hans Ulrich Gumbrecht (2012) understands successful plays as "forms," and the emergence of such a form is experienced as an epiphany by the spectators. In addition, my sound ethnographic interpretation of the soundscape of the stadium points to how subjective perception and desire are dependent upon the identifications that constitute perception and desire in the first place. Beauty and a high degree of affectivity do not exist per se. It is the subject who is enthusiastic about soccer—the subject who has an identification with soccer—who finds it. As demonstrated by the approach to the soundscape of the soccer stadium undertaken here, there is a diversity of available identifications in the stadium alongside the diversity of situations. A great variety of identification objects appear to be in existence here—and they articulate themselves through the medium of sound and are thus made accessible to interpretation. As Hermann Bausinger, one of the fathers of German-language European ethnology, puts it: "Soccer is [therefore] one of the few topics and areas which adapt to the pluralization of lifeworlds (there are many different ways of experiencing soccer, and many different orientations), but which, on the other hand, bridge that plurality, scaling back the fragmentation of society a bit" (2000, 56).

The diversity of identification objects and their perceivability in the metasituation of the stadium engender a similarly diverse desire, which agglomerates within the metasituation and results in an atmosphere of affective force. In the more recent past, the sound of the stadium atmosphere's emotionality has been deployed in pop music in a variety of ways as an aesthetic medium—for instance, by Panda Bear in his piece "Benfica" (*Tomboy*, 2011), the Chromatics in "The Page" (*Kill for Love*, 2012), and Frank Ocean in "Pink Matter" (*Channel Orange*, 2012).

The affective force of the stadium atmosphere is not static, but dynamic. It also results from the state of play: the suffering when the opposing team scores; the relief at a successful defense; the joy of a winning goal; the beauty of a successful play, etc. The soundscape also provides important access to the exploration of this dynamic, as Jirat (2007) impressively illustrated by means of a documentation of songs, call-outs, and mood trends during a soccer game: from aggression to despondency to protest to disillusion, various emotional atmospheres are articulated within the soundscape and therefore made tangible for fandom subjects and analyzable for research purposes.

Just as in other manifestations of fandom, soccer enthusiasm also serves as identity work, the function of which is the generation of a subject. On the one hand, this occurs in situations of soccer enthusiasm in which preexisting and enduring identifications come to life or in which fleeting identifications are newly created. On the other hand, it also occurs when the subject itself becomes completely present in the intensity of the affective force—a force to which the identification processes and the physical copresence of the other fans and of the material sound contribute.

Translated by Jessica Ring

Notes

1. I am not the first to use the Robson study, and in particular the Millwall Roar, as a main reference point for further considerations; see also Les Back (2003) for his arguments based on this phenomenon.

2. See, e.g., the Millwall Roar at Ipswich April 21, 2012: http://www.youtube.com/watch?v=46L1xiMksUk (accessed May 9, 2013).

3. Robson's logic of the presence expansion of an already existing subject identification comes close to canonical concepts such as Althusser's ([1969] 1973) concept of interpellation and Butler's ([1990] 2007) understanding of performance. In their case, however, the actualization of identifications reference a hegemonic symbolic order, whereas Robson's reidentification concerns a subcultural community.

References

Bibliography

Alkemeyer, Thomas. 2008. Fußball als Figurationsgeschehen: Über performative Gemeinschaften in modernen Gesellschaften. In *Ernste Spiele: Zur politischen Soziologie des Fußballs*, ed. Gabriele Klein and Michael Meuser, 87–111. Bielefeld: transcript.

Althusser, Louis. (1969) 1973. Ideologie und ideologische Staatsapparate. In *Ideologie und ideologische Staatsapparate: Aufsätze zur marxistischen Theorie*, 108–153. Berlin: VSA.

Anderson, Bendict. (1983) 2006. *Imagined Communities: Reflections on the Origin and Spread of Nationalism*. London: Verso.

Augoyard, Jean-François, and Henry Torgue. 2009. *Sonic Experience: A Guide to Everyday Sounds*. Montreal: McGill-Queen's University Press.

Back, Les. 2003. Sounds in the crowd. In *The Auditory Culture Reader*, ed. Michael Bull and Les Back, 311–327. Oxford: Berg.

Bausinger, Hermann. 2000. Kleine Feste im Alltag: Zur Bedeutung des Fußballs. In *Über Fußball: Ein Lesebuch zur wichtigsten Nebensache der Welt*, ed. Wolfgang Schlicht, and Werner Lang, 42–58. Schorndorf: Hofmann.

Butler, Judith. (1990) 2007. *Gender Trouble: Feminism and the Subversion of Identity*. London, New York: Routledge.

Elias, Norbert, and Eric Dunning. (1969) 1986. The quest for excitement. In *Quest For Excitement: Sport and Leisure in the Civilizing Process*, 63–90. Oxford: Blackwell.

Fabian, Johannes. 1983. *Time and the Other: How Anthropology Makes Its Object*. New York: Columbia University Press.

Giulianotti, Richard. 2002. Supporters, followers, fans, and flaneurs. *Journal of Sport and Social Issues* 26 (1): 25–46.

Gumbrecht, Hans Ulrich. 2012. *Präsenz*. Berlin: Suhrkamp.

Jirat, Jan. 2007. Der zwölfte Mann—die Schaffhauser Bierkurve: Ethnografie einer Fussball-Fankurve. *Schweizerisches Archiv fur Volkskunde* 103:105–131.

Kopiez, Reinhard, and Guido Brink. 1998. *Fussball-Fangesänge: Eine FANomenologie*. Würzburg: Königshausen & Neumann.

Kytö, Meri. 2011. "We are the rebellious voice of the terraces, we are Carsi": Constructing a football supporter group through sound. *Soccer and Society* 12 (1): 77–93.

Müller, Frank. 2010. Lebenslang grün-weiß. In *Fans und Fans: Fußball-Fankultur in Bremen*, ed. Jochen Bonz , 90–105. Bremen: Edition Temmen.

Porat, Amir Ben. 2010. Football fandom: A bounded identification. *Soccer and Society* 11 (3): 277–290.

Prosser, Michael. 2002. "Fußballverzückung" beim Stadionbesuch: Zum rituell—festiven Charakter von Fußballveranstaltungen in Deutschland. In *Fußball als Kulturphänomen: Kunst, Kultur, Kommerz*, ed. Markwart Herzog, 260–292. Stuttgart: Kohlhammer.

Robson, Garry. 2004. *"No One Likes Us, We Don't Care": The Myth and Reality of Millwall Fandom*. Oxford: Berg.

Said, Edward. (1978) 2003. *Orientalism*. London: Penguin.

Schäfer, Mike S., and Jochen Roose. 2010. Emotions in sports stadia. In *Stadium Worlds: Football, Space and the Built Environment*, ed. Sybille Frank and Silke Steets, 229–244. London: Routledge.

Schmidt-Lauber, Brigitta, ed. (2003) 2008. *FC St. Pauli: Zur Ethnographie eines Vereins*. Hamburg u. Münster: Lit.

Sound and Media

Chromatics, the. 2012. "The Page." *Kill for Love*. CD, Italians Do It Better, B00FY3US6O.

Ocean, Frank. 2012. "Pink Matter." *Channel Orange*. CD, Def Jam, B00FZ0W5OY.

Panda Bear. 2011. "Benfica." *Tomboy*. CD, Paw Tracks, B004MGMJ3E.

II Formations of Listening: Popular Culture by Ear

Making History by Ear

14   The Invention of the Listener: An(other) History

Veit Erlmann

Let us begin with a painting. Or, rather, a description of a painting:

In 1847, the American artist William Sidney Mount solidified his reputation as a genre painter with a canvas he called *The Power of Music*. It depicts a shirt-sleeved fiddler playing in a barn to an attentive audience of two men. But outside, hidden from the improvised concert by the half-closed barn door, he has a third listener, a black man, hat in hand, rapt. His axe and jug beside him, he has paused to drink in the sounds reaching him at his humble vantage point. (Gay 1996, 11)

This passage opens the first chapter on "The Art of Listening" in Peter Gay's *The Naked Heart*, the fourth in a series of five volumes entitled *The Bourgeois Experience: Victoria to Freud* (1996). There are several ways to "read" *The Power of Music*, Gay argues. On the surface, the painting may be seen as a "gently amusing set piece, inviting a benevolent smile"—a reading that echoes Mount's opinion that the inclusion of black people added to the "humor of the scene" (Gay 1996, 11). On quite a different level, however, the work might "serve as a commentary on a society half slave and half free: though quite unaware of it, the amateur violinist is performing for a segregated audience" (ibid.). Ultimately, though, Gay concludes his interpretation of *The Power of Music* by asserting that the work yields a deeper meaning: the silence of the white listeners and the "beatific smile" on the black man's face attest to the "ascent of inwardness in the Victorian age" (ibid.).

Gay may well be the first scholar to put forward the audacious claim of an affinity between Victorian "inwardness" and race. But he is hardly alone in arguing for the recognition of an intimate relationship between listening, subject formation, and race. Nor is he the only scholar who simultaneously evokes racial difference as a rationale for the invention of modern listening while silencing the discourses and sonic worlds of racialized listening subjects by subsuming them under the hegemonic auditory regime of an era, however that is defined. Thus, it did not occur to Gay that the black eavesdropper's "rapt" silence was the posture that corresponded to the slaveholders' understanding of social order: "slaves at rest were, ideally, quiet slaves, calm slaves,

Figure 14.1
William Sidney Mount (American, 1807–1868), *The Power of Music*, 1847. Oil on canvas, 43.4 × 53.5 cm. The Cleveland Museum of Art, Leonard C. Hanna, Jr. Fund 1991.110, © The Cleveland Museum of Art.

composed and obedient" (Smith 2007, 51). (Just as singing slaves in the cotton fields fed the planters' conceit about slavery as the most efficient way to organize labor.) In fact, the black listener is made to relapse into silence all over again: Gay does not take up the issue of race and slavery in the remainder of *Naked Heart*. Instead he turns to the "ascent of inwardness," recounting with great flourish the story of how E. T. A. Hoffmann, Beethoven, and a host of other "romantic" figures "invented" the listener. Listening to music, they asserted, was more of a mental act requiring attentive silence than a sensuous distraction.

The argument put forward in this essay is the following: The emergence of the (modern) listener is one of the key topics of sound studies. Yet while this figure has been and continues to be defined in predominantly "neutral," nongendered, and non-racial terms, the Western aural self is deeply caught up with the history and politics of difference. Race in particular has been one of the most recurrent and at the same time parenthetical themes in the story of modern aurality.

This essay consists of three sections. After a brief review of the musicological "invention" of the modern listener, the first section suggests that the construction of this listener rests on a hidden foundation, one in which the figure of the modern listener is crafted from a narrative of difference and Otherness. The second section is about the reverse process, namely the construction of auditory alterity in opposition to hegemonic colonial and postcolonial orders in the author's home disciplines of anthropology and ethnomusicology. The final section provides a cursory review of a variety of alternative attempts to conceptualize forms of aurality in which non-Western listeners are neither marginalized as racialized Others nor folded back into homogeneous cultural formations based on fixed racial, gendered, or social identities.

Music's "Outer Beauties": "Savages" and the Origin of Attention

By now, the story of how the silent, inward-turned "listener" was invented has become a familiar one, having been told numerous times both before and after the publication of Gay's book. Attention and thus, implicitly, an auditory orientation toward music's supposedly metaphysical rather than material core looms large in these accounts. Thus, as early as 1882 German musicologist Hugo Riemann argued for a listening stance whose roots resided in the "nature of the perceiving mind" and could therefore only be understood in psychological rather than physiological terms. Following in Riemann's footsteps, Heinrich Besseler pioneered an approach in the mid-1920s in which certain periods of musical history (along with their key genres: Renaissance motet, Baroque oratorio, etc.) are said to correspond to specific auditory practices (Besseler 1926). Thus, the fifteenth-century motet required a merely passive "*Vernehmen*," an absorbing of music in its objectlike given status, whereas during the Baroque era listening advanced to an intellectual activity that comprehended, in an act of cognitive mastery, music's structural totality. More famously (albeit as alarmist in tone as "traditional" musicological approaches), in 1938, Theodor W. Adorno diagnosed a "regression of listening" resulting from the "fetishization" of music in the capitalist entertainment industry (Adorno 2002a). Corresponding to such a regression is a typology of listeners that ranges from the infantile jazz fan to the radio ham to the sexually inhibited loner. Eventually, by 1995, and concurrent with the publication of Gay's book, in *Listening in Paris* (1995) historian James H. Johnson wonders why audiences in post-Revolutionary France increasingly fell silent, instead of engaging in the raucous, noisy behavior common during the *ancien regime*. He argues that this shift toward a deeper and more focused engagement on the part of the bourgeois listener was in part a consequence of the new musical styles of the period—styles that demanded more attention to "indescribable feelings and urges" than did the baroque styles with their depictions of storms, birds, and battles. Other factors, such as new architectural settings for performance, also facilitated this shift (Johnson 1995, 3).

While in those earlier texts the "listener" is figured as a subject whose auditory stance is determined by broader historical shifts in social identities and musical aesthetics, current constructions of the modern listener complicate this narrative by exploring the interplay of attentive listening and audio technology or by disrupting conventional historical periodizations. Thus, in *The Audible Past* (2003), Jonathan Sterne sees the sweeping transformations of sound and hearing that occurred during the nineteenth and twentieth centuries as being intertwined with the rise of modern sound reproduction. As such, the notion that technologies such as the stethoscope, the phonograph, and the headset used to harness, modify, and shape the auditory perception of individuals "in the service of rationality" differs markedly from earlier ontologies of listening that pit "spherical hearing" against "directional" or affect-driven hearing against intellectual vision or from typologies such as the one advanced by Besseler or Adorno. For Sterne, the modern "listener" is neither the product of technological progress, nor is he or she alienated or divorced from a "natural" faculty. Music historian Andrew Dell'Antonio, for his part, upsets conventional chronology by tracing the origins of "attentive" listening to a period well before the turn of the nineteenth century (Dell'Antonio 2011). Moreover, contrary to standard musicological accounts, it was not the secular, bourgeois practices of eighteenth-century concert-goers that contributed to the rise of taste as the primary basis for a new form of attentive listening, but the spiritual practices of the aristocratic elite in seventeenth-century Rome (ibid.). Finally, in *Reason and Resonance: A History of Modern Aurality* (Erlmann 2010), the present writer examines a wide range of scientific, literary, and philosophical texts suggesting that, starting in the early modern period, auditory perception rivaled vision as a major site of the discursive construction of modern subjectivity.

In many of those writings, however, race maintains an eerie presence, always lurking below the discursive surface. But how did race become the shadow of the discourse about the emergence of the modern listener? David Howes has shown that the roots of the racialization of the senses lie in nineteenth and early twentieth-century anthropologists, missionaries, and colonial administrators' fascination with the physical and sensory properties of the colonial Other (Howes 2008, 4–5). Ironically, however, the same discipline that sought to justify the West's claims to superiority—and thus explicitly or implicitly the colonial venture on the whole—in the association of the colonizers with rationality and the colonized with sensuality also produced a wealth of sensuous ethnographic detail, providing "a tantalizing indication of the vitality and sophistication of sensory symbolism across cultures" (ibid., 6).

And not only across cultures—within the colonizing culture as well. As Constance Classen reminds us, the inverse relation between sense and intellect was held to be true not only of the colonized but also of Europe's own internal Others: the marginalized, lower class, or mentally impaired members of society were said to be beholden to "primitive" senses such as olfaction (Classen 1998, 118–121).

Clearly, the specter of the Other is the ground upon which the Western normative musical experience (and its study) has been erected (Radano and Bohlman 2001, 1–2). For instance, as early as the late eighteenth century, Johann Nikolaus Forkel devoted considerable attention to the music of "savages" in his massive *Allgemeine Geschichte der Musik*—one of modern musicology's foundational texts. Yet much like other Enlightenment and post-Enlightenment thinkers—and in contrast to Jean-Jacques Rousseau who dreamed of a return to nature and for whom the history of music (and especially French music) was one of decline—Forkel saw the music of the ancient Egyptians, Greeks, Romans, the "American savage," and other "raw, uncultivated nations"—although no doubt beautiful to their own members—utterly lacking in "order and beauty" (Forkel 1788, vol. 1, xiv). The propensity for music's "outer beauties"—that is, timbre and rhythm—was merely a primitive, albeit inevitable, phase in the evolution of music toward ever greater perfection, beyond mere sensory gratification. Proper listening, accordingly, was teleological listening, moving up from raw sensuousness to higher levels of transcendence.

A little over a century later, at the apogee of European imperialist expansion, the focus shifted from Forkel's dismissal of "savage" auditory practices to a heightened sense of anxiety about a different albeit equally "strange" set of practices at the heart of contemporary industrial society. Besseler, for instance, decried the "intrusion" of the "new rhythms and sounds of the nigger jazz band" that required a "vitally rhythmic" collective merging with the music, instead of the individual listener placing herself or himself vis à vis the heard music (Besseler 1926, 38). Besseler's concept of the jazz fan locked into an undifferentiated, primordial collectivity of listeners is echoed in Adorno's almost contemporaneous critique of jazz. Yet in true dialectic fashion, Adorno cautions that the listener who is longing for the expression of the authentic and archaic in jazz recovers only a "regression through suppression; there is nothing archaic in jazz but that which is engendered out of modernity through the mechanism of suppression" (Adorno 2002b, 478).

As these examples illustrate, the invention of a modern, attentive, "inward" listening as the culmination of a long history of cultural "achievement" (Gay's "ascent") is deeply interwoven with the Othering of the senses within the larger project of Western colonial domination. What is at issue in this process is less the difference between attentive and inattentive forms of listening than the fact that the sensing body becomes the passive ground on which (racial) difference is inscribed.

Postcolonial Continuities

The inscription of race—or the sonification of race, as one might call textual inscription's sonic twin brother—on the body of the Other shapes musical scholarship to this day. Strikingly, this colonial legacy endures even in the face of the growing acceptance of postcolonial critiques that query conventional understandings of cultural identity. And it survives in ethnomusicology and anthropology's (and in part cultural history and popular music scholars') sustained investment in constructions of the "sensuous" listening practices and aesthetic choices of Africans, Asians, or African Americans as fundamentally "different" from those of the racially unmarked "listener." Although such constructions do not usually involve any racial (let alone racist) categories as such, they do reflect certain institutional legacies and imbalances. Thus, the emergence of area studies in the United States (and its subsequent proliferation to other parts of the world) has led to the formation of distinct "cultures" and their isolation as discrete objects of study, which has in turn hindered the development of critical tools to deal with the complexities of difference and sameness across racial and ethnic divides. At the same time, the invention of discrete "areas" has marginalized the scholarly study of the discursive construction of race even further, leaving it to those performers and listeners (and a small group of scholars) from marginalized communities who value their musical traditions to define and defend their preferences by essentializing them in racial terms or by opting for what Ronald Radano calls an "assertive, affirmative scholastics" (2003, 32).

A prominent example of the racialization of difference are the almost ritual evocations of rhythm (along with cognate phenomena such as "drumming" and "repetition") as quintessential markers distinguishing the sensibilities of Western listeners from those allegedly prevalent in Africa and among African Americans. Thus, in "non-literate," "non-Western" societies communication is said to be "overwhelmingly sonic," involving whistles, horns, and above all "drum languages" to convey information across large distances (Smith 2007, 46–47). Needless to say, such "historical" communicative technologies are not only held to be "complicated and subtle," they often serve important purposes such as defining political and social territory or, as among the New World's enslaved populations, resisting bondage and coordinating revolts.

This sort of narrative (here offered by a historian of the senses) is not uncommon, even in musical scholarship. And as such, not surprisingly, it became the target of incisive criticism. For instance, in a discussion of the Harlem Renaissance, Houston Baker argued that African American modernism, unlike the modernism of the likes of James Joyce and T. S. Eliot, which drew its main impetus from the destruction of traditional form, rests on a subtle interplay between deformation and a desire to master form. But this dialectic of formal mastery and deformation is embedded in a complex process of filtering African American traditions rooted in slavery through what Baker (1987) calls the minstrel mask. To advance their own critical modernism—or as Baker puts it, "crafting a voice out of tight places" (ibid., 33)—African American modernists had to convert the nonsense of racist minstrelsy into "good sense, or, sense intended for a common black good," that is, a black nation (ibid., 32). Sound and differently tuned ears are key for this mastery of form through the minstrel mask, whether the template for such cultural work is provided by the gibberish of a "Negro" dialect, the sounding horns of maroons (runaway slaves of the Caribbean), or, more ominous, "deep-rooted African sound" (ibid., 49).

Investing music and musical listening with some inherent power to transcend the very social divisions they are said to be based on is of course highly redolent of nineteenth-century romantic aesthetics—and thus deeply problematic. Yet Baker is not the only black intellectual struggling to distance himself from reductionist, Afrocentrist claims to authenticity as a way of forging a countermodernity that is mediated by distinctive creative practices rooted in black experiences such as sound and music. Paul Gilroy's highly acclaimed book *The Black Atlantic* (1993) bears powerful testimony to this strand of deconstructing racial determinations of listening while at the same time affirming them. The power of African-American music, Gilroy argues in a poststructuralist, anti-essentialist vein, does not reside in some racial core such as blues poet Amiri Baraka's "changing same," but in the terrifying experience of slavery. Born from and bearing the scars of oppression, African-American music betrays an almost modernist, Enlightenment desire for transformation. But it does so by partially transcending modernity, "constructing both an imaginary anti-modern past and a postmodern yet-to-come"; and thus it is that, in a lower frequency undetectable to the white overseers, the "formation of a community of needs and solidarity … is magically made audible in the music itself" (Gilroy 1993, 37). One can only presume that there is also a corresponding "magically" listening subject to the absolute object thus made "magically" audible.

A completely different kind of critique has been offered by Kofi V. Agawu. The Ghanaian-born music theorist and ethnomusicologist berates Western (and African) scholars' fixation on rhythmic complexity as a hallmark of African music and thus, implicitly, of Africanness (Agawu 2003). To counter this fetishization and Othering of Africans by the ear, as it were, Agawu argues for a radical sameness. For instance, by applying a Schenkerian perspective to a highlife tune by Ghanaian veteran composer E. T. Mensah, Agawu explores the possibility of engaging with the musics of the African continent from multiple vantage points, which often include an unabashedly aggressive emphasis on things such as formal analysis, "structural listening," or context-free aesthetics—all of which are said to be preoccupations (or defects) of an older brand of musicology and as such cut against the grain of much postcolonial theory.

Listening Awry, Listening through Writing—Alternative Concepts of the Listener

As we have seen, postcolonial discourses have been unable (or unwilling) to completely disengage themselves from concepts of irreducibly racial or ethnic difference and in so doing frequently invoke sonic and auditory practices to support their claims. Alternatives to this sort of dogmatism are emerging, however. They are especially noticeable in those projects focusing on forms of listening that openly resist or playfully undermine the Othering of normative (Western or non-Western) listening protocols by foregrounding or taking pleasure in indeterminate, in-between, and hybrid listening experiences. As we shall see, such projects are increasingly being instigated from some unexpected quarters.

Musicology and music theory are two disciplines in which alternative histories of listening are beginning to be written as histories of listeners as Others. In *The Singing of the New World* (2007), music historian Gary Tomlinson, for instance, seeks to reconstruct the sensory experience of the indigenous populations of Mexico and Latin America in the early phases of Spanish colonization from a range of written and visual evidence, such as the well-known Aztec *cantares*. Much like in Tomlinson's earlier work on Renaissance music, this project of listening to long-faded soundscapes requires a certain dissociation of the false sense of familiarity centuries of colonial and postcolonial mastery of the Other have instilled in us. However, Tomlinson points out, this attempt to "hear" these songs does not only mean rescuing them "from the logocentric prison-house guarded over by speech" (Tomlinson 2007, 11). Making the supposedly familiar strange, ironically, also entails the "nudging of singing into alliance with writing, over against speech and its venerable privilege" (ibid.). In short, it means constantly disrupting the tendency to use listening as the site of pure identity-making, through what French deconstructionist Jacques Derrida calls *différance*.

Another approach is inspired by theories of creolization. For instance, in her book *Sonic Spaces of the Karoo* (2011), Marie Jorritsma examines the sacred music of a "colored" (mixed race) community in a small South African town marked by centuries of oppression and violence but also "cultural encounters" across racial and ethnic divides. Jorritsma argues that a listening that is guided by notions of hybridity, creolization, or syncretism and uncovers hidden layers of African influence on four-square musical structures typical of mainstream European Christian missions might be well intentioned in seeking to undermine apartheid's infatuation with racially "pure" communities. Ultimately, however, such an auditory stance would reinforce stereotypes of "colored" identity as ambiguously situated in a no-man's land, at the "frontier" between white and black, colonizer and colonized. More useful might thus be the metaphor of the "seam," or better still, of a "listening at the seam"; in a space of fusion and healing for sure, but also in a space that retains "the marks of unsuccessful attempts to flatten out difference" (Jorritsma 2011, 28).

To conclude this section, I should mention an interesting body of work that probes forms of listening that, though not primarily concerned with race, "other"—that is, reify—music and its listeners by locking them into fixed relationships between text and context. In *Listening Awry* (2006), music theorist David Schwarz, for example, proposes a different approach toward listening inspired by Lacanian psychoanalysis—a "listening awry." Instead of focusing the listener's attention on the objective, internal properties of a work in a "straight," unidirectional sense, listening awry is a listening that bypasses the work's cohesion, as it were. Listening awry means being attuned to the social spaces that make it possible for such principles of cohesion to come into being by marginalizing other practices on the basis of their supposed lack of cohesion—such as the music and sounds of "the Jew"—or by just generally suppressing the uncanny layers making up the acoustic unconscious (Schwarz 2006, xii).

Conclusion

In closing, let us return to Mount's *The Power of Music* and allow the work to raise a different set of questions about modern listening. Instead of asking what sense the black listener might have made of the sounds he was eavesdropping on, and what feelings the music—merry music, presumably—might have evoked in a man likely to have experienced mostly pain and sorrow, might it be more productive to attend to the entanglement of racialized (white and black) subjects in power relationships mediated by sound and hearing? How, for instance, can hegemonic music-making such as the planters' fiddling make listeners "unaware" of power imbalances?

References

Adorno, Theodor W. 2002a. On the fetish-character in music and the regression of listening. In *Essays on Music*, ed. Richard Leppert, trans. Susan H. Gillespie et al., 288–317. Berkeley: University of California Press.

Adorno, Theodor W. 2002b. On jazz. In *Essays on Music*, ed. Richard Leppert, trans. Susan H. Gillespie et al., 470–495. Berkeley: University of California Press.

Agawu, Kofi V. 2003. *Representing African Music: Postcolonial Notes, Queries, Positions.* New York: Routledge.

Baker, Houston A. 1987. *Modernism and the Harlem Renaissance.* Chicago: University of Chicago Press.

Besseler, Heinrich. 1926. Grundfragen des musikalischen Hörens. *Jahrbuch der Musikbibliothek Peters* 32:35–52.

Classen, Constance. 1998. *The Color of Angels: Cosmology, Gender, and the Aesthetic Imagination*. London: Routledge.

Dell'Antonio, Andrew. 2011. *Listening as Spiritual Practice in Early Modern Italy*. Berkeley: University of California Press.

Erlmann, Veit. 2010. *Reason and Resonance: A History of Modern Aurality*. New York: Zone Books.

Forkel, Johann Nikolaus. 1788. *Allgemeine Geschichte der Musik*. Leipzig: Schwickert.

Gay, Peter. 1996. *The Naked Heart*. Vol. 4 of *The Bourgeois Experience: Victoria to Freud*. New York: W. W. Norton.

Gilroy, Paul. 1993. *The Black Atlantic: Modernity and Double Consciousness*. Cambridge, MA: Harvard University Press.

Howes, David. 2008. *Sensual Relations: Engaging the Senses in Culture and Social Theory*. Ann Arbor: University of Michigan Press.

Johnson, James H. 1995. *Listening in Paris: A Cultural History*. Berkeley: University of California Press.

Jorritsma, Marie. 2011. *Sonic Spaces of the Karoo: The Sacred Music of a South African Coloured Community*. Philadelphia: Temple University Press.

Radano, Ronald M. 2003. *Lying Up a Nation: Race and Black Music*. Chicago: University of Chicago Press.

Radano, Ronald M., and Philip V. Bohlman, eds. 2001. *Music and the Racial Imagination*. Chicago: University of Chicago Press.

Riley, Matthew. 2004. *Musical Listening in the German Enlightenment: Attention, Wonder, and Astonishment*. Aldershot: Ashgate.

Schwarz, David. 2006. *Listening Awry: Music and Alterity in German Culture*. Minneapolis: University of Minnesota Press.

Smith, Mark M. 2007. *Sensing the Past: Seeing: Hearing, Smelling, Tasting, and Touching in History*. Berkeley: University of California Press.

Sterne, Jonathan. 2003. *The Audible Past: Cultural Origins of Sound Reproduction*. Durham, NC: Duke University Press.

Tomlinson, Gary. 2007. *The Singing of the New World: Indigenous Voice in the Era of European Contact*. Cambridge: Cambridge University Press.

15   Sonic Modernities: Listening to Diasporic Urban Music

Carla J. Maier

This chapter deals with the ways in which certain forms of music listening can constitute a critique of conceptions of culture and modernity that are based on binary oppositions of traditional and modern, East and West, urban and tribal. These binarisms are problematic because they are based on essentialist representations of (cultural) identity and difference (Hall 1996) and tend to construct a hegemonic relationship between Western cultures (as allegedly stable and homogeneous entities) and "other" cultures (as marginalized and exoticized ethnic cultures). In contrast, this article will emphasize that constructions of modernity have also been shaped by processes of globalization and migration. It therefore investigates how the thriving musical cultures that have emerged from the South Asian diaspora in the UK continuously transgress the boundaries of Eurocentric musical discourses and national culture(s).

Based on an analysis of the track "A History of Now" (2011) by Asian Dub Foundation (ADF), I will elaborate on the new modes of listening that have been (and could be) developed—modes that both incorporate and require new ways of thinking about concepts of culture and sonic modernities.

I chose ADF's title track "A History of Now" to anchor my analysis of diasporic music in the actual material of a music track and to assist in a reflection on conceptualizations of sonic modernities.[1] Investigating new modes of listening in relation to this particular example is important for a sonic reconsideration of culture and modernity, since ADF has always explored musical as well as political ideas in order to develop new ways of thinking about essentialist or racist notions in regard to South Asians in the UK. This article will therefore investigate how particular modes of listening to the band's sampling and remixing sound practices—and their usage of Indian as well as Western instrumentation and technology—challenge representations of what supposedly "sounds Indian." This close listening aims to work toward demystifying Asianness as the "exotic other" (see Sharma et al. 1996; Hutnyk 2000) of Western modernity.

The problem with the exotic tag often being attached to all things Asian is that it promotes a mode of listening that ignores the music's diversity and reinscribes notions of a traditional culture that is somehow distinct from modern culture. From British Asian musicians of the late 1980s such as the band Alaap (one of the first British Asian bands to gain popular attention in the UK for their mixing of South Asian sounds with British pop music instrumentation), to British Sri Lankan musician and producer Maya Arulpragasam, a.k.a. M.I.A., who became popular in the second half of the 2000s with her politicized mix of rap, Indian, and Brazilian samples, and London club sounds—these musics have persistently been perceived as primarily representing some sort of "authentic" Indianness. Although Alaap was primarily promoted and marketed as a world music act, M.I.A. was often reduced to her Sri Lankan heritage and her upbringing as the daughter of a member of the Liberation Tigers of Tamil Eelam, a militant organization fighting for an independent state for the Tamil people. These journalistic approaches tend to deal with these artists primarily in terms of their ethnic origin, and, consequently, M.I.A.'s music has been filed in record stores' "world music" section—a categorization that ignores the fact that this diasporic music is firmly based in the UK's electronic dance music practices. Therefore, a mode of listening that focuses on the Asian markers prevails here, thereby favoring a listening for ethnic otherness. Although Alaap's music was influenced by various musical styles and traditions and incorporated Bhangra music from the Punjab, British pop music, as well as Spanish and Middle Eastern styles, the Spanish and Middle Eastern sounds in particular were silenced in journalists' commentaries because they could not be authenticated as Asian. As Nabeel Zuberi aptly points out with regard to recent diasporic urban musicians in the UK, such as Dizzee Rascal, Sway, and M.I.A., as well as dubstep producers:

While it has been highly mobile, much of the commentary on this music in journalism and websites centers on "placing it," situating it, marking its boundaries, and making critical claims about its authenticity, success or failure based on its belonging to a place or places. Popular music discourse, inside and outside academia, manifests a territorializing imperative and desire for enclosure that is integral to the boundary-making of music cultures. (Zuberi 2010, 179)

There is a clear connection between the territorialization of diasporic music and the marking of it as premodern in terms of its cultural heritage that reestablishes and reinforces binary oppositions between the modern and the traditional, or the West and the East. In his book *tracks 'n' treks* (2011), Johannes Ismaiel-Wendt elaborates an analytical framework, which he terms TRX studies, that synchronizes a subjective mode of close listening, a reflection of postcolonial theories of spatial deterritorialization, and notions of transcultural transformation. With what he calls "postcolonial ear training," Ismaiel-Wendt challenges the established causal relations between sound (*Lautbild*) and idea (*Vorstellungsinhalt*) in order to open up new perspectives on diasporic music as "performative spaces" (ibid., 25), in which sampling, layering, instrumentation, and production techniques charge our perception with imagined geographies that are saturated with folkloristic associations, or train our ears to discern and critique exoticizing sonic strategies to arrive at a mode of listening that acknowledges the transatlantic and transcultural links between these musics (ibid., 53).

To elaborate on the new modes of listening that create alternative concepts of modernity, which are incorporated into and demanded (in terms of analytical practice) by the music of bands such as ADF, M.I.A., and Dizzee Rascal, the analysis of ADF's track "A History of Now" will serve as an example. In the interest of a concise analysis, the main focus will be on the Bhangra sample, which structurally frames the track and also permeates its sonic fabric.

Three different modes of listening that aim to concretize sonic particularities of diasporic urban music and refine the notion of listening as an analytical practice are proposed in the following section: listening to musical *diversity*, listening to the sound's *materiality*, and listening to the music's *multimodality*.

Acknowledging Musical Diversity

As announced on their official website, ADF defines its musical style as a mixture of rapcore, dub, dancehall, and ragga, but this list can easily be extended to include punk rock and Bhangra. The track "A History of Now" is based on a sped-up Bhangra loop featuring a dhol drum and a tumbi. The dhol drum is a double-barreled drum played with two sticks—one for the bass and one for a higher note. A rolling and uplifting rhythm is played with this drum that is distinctive to Punjabi folk dance songs. The accompanying tumbi is a wooden, plucked instrument that provides a high-frequency, repetitive melody.[2] Looping and speeding up this Bhangra sample thus results in a repetitive and highly danceable rhythmic pattern. After a number of repeats, the bass guitar kicks in, introducing the full force of ADF's instrumental fabric of distorted guitars, a three-chord punk rock riff, a thumping bass line, and a straightforward beat on the drum set. The tumbi melody keeps running through the track, providing a repetitive hookline in the instrumental parts. The band's distinct sound is complete only when the singer comes in, articulating his lyrics in a mixture of rap and ragga style that progresses like a pulsating sonic-verbal flow. This short description of the track clearly reveals that it cannot be easily categorized as "traditionally Indian" or captured under the generalizing umbrella term *world music*.[3] Throughout the band's history, ADF has recombined and refabricated sounds in order to explore different culturally charged forms of music—and in doing so, the band has constantly reshaped what actually constitutes popular music in the UK.

Listening to ADF's music in this way challenges an exoticizing mode of listening. Instead, the mode of listening proposed here acknowledges musical diversity, thus overcoming essentializing categorizations.

Listening to the Sound's Materiality

Listening to the sound's materiality means paying attention to the sound practices of looping and layering that characterize ADF's "A History of Now" and how those sound practices incorporate new musical ideas. Significantly, the Bhangra elements, such as the dhol drum beats and the tumbi melody, are used in this track as sonic raw material that is geared toward further manipulation and recombination.

As mentioned in the previous section, the high-pitched and amplified tumbi melody is recurringly featured in the track as a persistent hookline. However, listening to this sound as an authentic expression of Indian culture simply misses the point. In the context of this track, the tumbi sound directly corresponds to the electric guitar's melody lines in the parts of the track that sound, first and foremost, like straightforward punk rock. Later, the tumbi melody is cut up into smaller sections to form part of an accelerating and repetitive segment in which the phrases "you can't download the sun" and "you can't download the sea" are sung alternately until the whole structure detonates into the final call "you'll never download me." At this point, the full force of electric guitars and tumbi melody unite together with the other instruments to create the collective noise of the track's final segment.

Listening to how the Bhangra sounds are used as sonic raw material enables alternative versions of this musical style to emerge that are perceived in their materiality—contrasting the bass, emphasizing the electric guitar, fragmenting, dissolving in and merging with the track's multitude of sounds. The sound's materiality therefore incorporates new sonic and musical ideas that can no longer be captured in clear-cut musical styles or rigidly defined cultural identities.

Opening Up a Multimodal Listening Perspective

Listening to "A History of Now," the perception of the Bhangra sample is twisted once more when played against the ragga lyrics of ADF's track. Although Bhangra music is commonly associated with Hindi lyrics about love, dancing, and marriage, the lyrics of "A History of Now" that come in after the Bhangra sequence deal with a dystopian vision of information overload and the danger of losing oneself in the sped-up rhythms of globalized media culture. The song is also a fervent expression of finding a way to cope with the experience of mental overload caused by the buzzing input of digitalized information. The vibrant effect created by the sound tapestry of rapped words and impulsive beats is full of determination and urgency.

The Bhangra sample, which comes in again as a bridge, further decontextualizes its traditional connotations. In the final section, the instrumental and vocal parts are mixed and merged to generate an accelerating and repetitive, sonic and rhythmic figure that makes the track resound as a form of collectivized noise. This multimodal perspective fosters an agitated mode of listening in which the Bhangra sample counteracts and amplifies the meaning of the lyrics as they become an urgent and collective invitation, or even demand, to "live in a history of now" (see ADF's lyrics to "A History of Now").

Theorizing Sonic Modernities

Against the backdrop of this analysis, it becomes paramount to ask how established concepts of modernity need to be reconsidered in the light of diasporic urban music.

The "territorializing imperative" with regard to South Asian music in the UK to which Zuberi refers (Zuberi 2010, 179) is inextricably linked to a critique of the idea of Western modernity as opposed to an allegedly premodern global South. In the last couple of decades, a number of scholarly works have challenged the idea of a specifically Western modernity and have started to rethink modernity as a concept under the new conditions of the globalizing tendencies of contemporary culture. Terms such as "reflexive modernization" (Beck et al. 1994), "liquid modernity" (Eisenstadt 2000), "alternative modernity" (Gaonkar 2001), and "multiple modernities" (Baumann 2000; Welz 2004) have started to circulate in the realms of sociology, cultural anthropology, and literary and cultural studies.

What we can learn, therefore, from the diversity, materiality, and multimodality of diasporic music is that musical ideas are not fixed to one place, nation, or community but are instead in a "flow," and that cultural formations are fundamentally influenced by these transcultural dynamics. Dilip Parameshwar Gaonkar emphasizes the decentering tendency of modernity without fixity:

Modernity today is global and multiple and no longer has a governing centre or master narrative to accompany it. ... Modernity has travelled from the West to the rest of the world not only in terms of cultural forms, social practices, and in situational arrangements, but also as a form of discourse that interrogates the present. (Gaonkar 2001)

The complexity of the relationship of the global flows of cultural production and the local practices of (re)appropriation indicate that cultural space is not only a geographically locatable space, but rather one that consists of multiple, context-related connections and transcultural practices (Hannerz 1996).

Taking into account concepts such as multiple modernities, diasporic urban music not only responds to the emerging global condition of modernity, but also plays an important role in constituting new modernities. As Weheliye notes: "Sound … holds out more flexible and future-directed provenances of black subjects' relation to and participation in the creation of western modernity" (2005, 11). Therefore, what has been conceived of as Western modernity is not thinkable without black culture—including literature, art, theater, and, of course, music—a fact that causes Weheliye to propose an alternative concept of modernity that puts sound technology, black cultural production, and modernity into the same realm: "Overall, it is sound, especially in its ties to modern technologies, that allows these diverse laborers in the realm of culture to 'mess' with the strict cadence of western modernity in order to present us with a disjointed, singular, and 'mixed-up' modernity: 'sonic Afro-modernity'" (ibid., 105). Applying Weheliye's proposed "grooves of sonic Afro-modernity" (ibid., 16) to the analysis of "A History of Now," ADF has consistently used music technology to demystify South Asian sounds—and ADF's music's politicized resonance has been transferred to diverse contexts, thus creating a form of collective noise across cultural, social, and political boundaries.

Listening to the *recontextualization* of their music can be described as the fourth mode of listening. How ADF's sound as a form of collective noise is reappropriated in different contexts becomes manifest in the different versions of the "A History of Now" music video. The track's initial video addresses the information overload of digital communication technologies and depicts a speed-ridden journey through a city from the perspective of a boy who is following a race through virtual reality. However, there are subsequent versions of this music video that have been recontextualized using the political protests on Tahrir Square in Cairo and in Tunisia in the context of the so-called Arab spring of 2011, events which coincided with the release of ADF's album. Jimmy Cauty (founding member of KLF and The Orb) produced a series of visual interpretations of ADF's track in which "A History of Now" becomes the background for the struggles of the people on the streets to achieve the resignations of Mubarak and Ben Ali, respectively. The experience of this form of recontextualization of ADF's distinctive sound has the potential to generate transnational trajectories for the struggle against hegemonic power structures in contemporary Egypt and Tunisia. In this process, ADF's sound becomes part of a sonic modernity that incorporates ideas of righteousness and people power across and beyond the boundaries of race and ethnicity.

To sum up, the notion of listening as an analytical practice developed in this chapter requires at least four different modes of listening: listening to *diversity* involves an openness toward a multiplicity of musical styles; listening to the sound's *materiality* entails paying attention to the multilayered sonic fabric of a musical track, which affects its interpretation and meaning; listening to a track's *multimodality* implies a consideration of the mutual impact of sound and lyrics (or other forms of textual, sonic, or visual representation, for that matter); and, finally, listening

Figure 15.1
Video stills of Jimmy Cauty's version of ADF's "A History of Now (Chapter 1)" music video (courtesy of Jimmy Cauty).

to examples of the *recontextualization* of the music's performance (e.g., in a music video, a live gig, or played on the car radio) acknowledges the complex material, as well as the cultural and critical capacity and power, of diasporic urban music. Applying these modes of listening to ADF's music demonstrates how their sound practices incorporate and demand modes of listening that involve agency—leading to the highlighting of transcultural dynamics and the emergence of multiple sonic modernities.

Notes

1. In my PhD thesis (Müller-Schulzke 2012), I investigate sound practices such as sampling, remixing, and sonic manipulation with regard to some earlier ADF tracks, as well as the music of other British Asian artists such as Apache Indian, M.I.A., Nathan "Flutebox" Lee, and British producers Dusk + Blackdown.

2. A sound that became popular with Panjabi MC's hit "Mundian To Bach Ke."

3. Since the 1980s, world music has largely been communicated as part of a marketing strategy to sell "non-Western" music to a white mainstream audience (see Frith 2000, 305). This concept was criticized by a number of scholars, such as Ashwani Sharma, who stated that "in World Music marketing practices, a pervasive strategy has been to promote specific artists as representations of authentic ethnic musical cultures" (Sharma 1996, 23).

References

Bibliography

Baumann, Zygmund. 2000. *Liquid Modernity*. Cambridge: Polity.

Beck, Ulrich, Anthony Giddens, and Scott Lash, eds. 1994. *Reflexive Modernization: Politics, Tradition, and Aesthetic in the Modern Social Order*. Cambridge: Polity Press.

Eisenstadt, Shmuel N. 2000. Multiple modernities. *Daedalus* 129:1–29.

Frith, Simon. 2000. The discourse of world music. In *Western Music and Its Others: Difference, Representation, and Appropriation in Music*, ed. Georgina Born and David Hesmondalgh, 305–322. Berkeley: University of California Press.

Gaonkar, Dilip Paramshwar, ed. 2001. *Alternative Modernities*. Durham, NC: Duke University Press.

Hall, Stuart. 1996. Who needs identity? In *Questions of Cultural Identity*, ed. Stuart Hall and Paul du Gay, 1–17. London: Sage.

Hannerz, Ulf. 1996. *Transnational Connections: Culture, People, Places*. London: Routledge.

Hutnyk, John. 2000. *Critique of Exotica: Music, Politics, and the Culture Industry*. London: Pluto.

Ismaiel-Wendt, Johannes. 2011. *tracks 'n' treks: Populäre Musik und Postkoloniale Analyse*. Münster: Unrast.

Lipsitz, George. 1994. *Dangerous Crossroads: Popular Music, Postmodernism, and the Poetics of Place*. London: Verso.

Müller-Schulzke, Carla J. 2012. Transcultural sound practices: South Asian sounds and urban dance music in the UK. PhD diss., Goethe-Universität Frankfurt.

Sharma, Ashwani. 1996. Sounds Oriental: The (im)possibility of theorizing Asian musical cultures. In *Dis-Orienting Rhythms: The Politics of the New Asian Dance Music*, ed. Sanjay Sharma, 15–31. London: Zed Books.

Sharma, Sanjay, John Hutnyk, and Ashwani Sharma, eds. 1996. *Dis-Orienting Rhythms: The Politics of the New Asian Dance Music*. London: Zed Books.

Sharma, Sanjay. 2004. The sounds of alterity. In *The Auditory Culture Reader*, ed. Sanjay Sharma, 409–418. Oxford: Berg.

Weheliye, Alexander G. 2005. *Phonographies: Grooves in Sonic Afro-Modernity*. Durham, NC: Duke University Press.

Welz, Gisela. 2009. Multiple modernities: The transnationalization of cultures. In *Transcultural English Studies: Theories, Fictions, Realities*, ed. Frank Schulze-Engler and Sissy Helff, 37–57. Amsterdam: Rodopi.

Zuberi, Nabeel. 2010. Worries in the dance: Post-millennial grooves and sub-bass culture. In *Britpop and the English Music Tradition*, ed. Andy Bennett and Jon Stratton, 179–192. Farnham: Ashgate.

Sound and Media

Asian Dub Foundation. 2011. "A History of Now." *A History of Now*. CD, Cooking Vinyl, COOKCD532.

Asian Dub Foundation. 2011. *A History of Now (Chapter 1). Artistic version of Asian Dub Foundation's official music video of "A History of Now."* Music video (dir. Jimmy Cauty), https://www.youtube.com/watch?v=fqrmFEKWBtQ (accessed January 19, 2015).

16   On the Modern Listener

Marta García Quiñones

Modernity and Aurality: Some Preliminary Remarks

As is well known, the name of the protagonist of this chapter—the "modern listener"—has often been considered to be almost an oxymoron. At least until the beginning of this century, the identification of modernity with visuality—what Martin Jay (1992) has called "Cartesian perspectivalism," attested by the abundance of metaphors of light in philosophical and scientific language and by the central role of the gaze in different social practices—seemed to condemn the ear to a process of revaluation as the "other" sense that too often took on suspiciously lyrical tones or simply fell into contradiction. As Jonathan Sterne observed in the introduction to *The Audible Past* (2003), the result of this dualistic approach was an "audiovisual litany" that could not provide any solid foundation for sound studies. Sterne's research into the conditions of possibility of sound reproduction underlined instead the importance of certain ways of listening—specific cases of what he called "audile technique"—that developed around older sound technologies such as the stethoscope, the telegraph, or the telephone. His work on the archeology of audio technologies and their social contexts, together with the work of other scholars who have focused on a variety of subjects related to the history of sound, such as the acoustics of spaces (Thompson 2004) or the regulation of noise (Bijsterveld 2008), have raised attention to aurality as a fundamental aspect of the modern project.

Yet, although the vindication of the centrality of sound in modernity is quite a recent phenomenon, the identification of a certain *musical* listener (or rather, a certain ideal of listening) with some of the principles of modernity—particularly if we consider modernity in its broadest sense as almost a synonym for the Western way of thinking and acting in the world—has been broadly accepted and actively promoted worldwide until not long ago. I am referring here to the ideal listener of the classical music tradition and to the "apparatus" (in the Foucauldian sense; Foucault 1980) formed by the notion and practices of listening (silent listening, immobile attention) adopted since the mid-eighteenth century in European concert halls, the disposition of those spaces, the musical works of the classical music tradition—more precisely, since circa 1800, a canonical selection of them (Weber 2008)—and the critical idiom and analytical procedures that developed accordingly as tools for musical comprehension (Campos and Donin 2005). Within this framework, the claim that music should be heard "for itself" went hand in hand with the institution of certain material conditions for listening, as well as with the introduction of new listening attitudes that were conceived at the same time as an almost natural response to the excellence of musical "masterworks." Within this framework, music developed its purported autonomy (Dalhaus 1989), whereas the specific modes of music and music culture acquisition, which were linked to social conditions at least as much as to individual aesthetic choices, became obscured (Bourdieu 1984; Botstein 1992; Gramit 2002; Love 2004; Leppert 2004).

At the beginning of the twentieth century, (classical) music had established itself as a closed code that demanded attention and competence on the part of cultivated listeners, to whom it promised in exchange not only a refined form of aesthetic pleasure, but also intellectual (Bonds 2006) and even moral rewards. At that point in Western history, and well into the 1980s, the political task of forming good citizens—that is, self-conscious individuals in control of their own bodies and emotions—seemed to coincide with the pedagogical project of training attentive listeners to music that was considered "of the highest kind." Therefore, the ideal of attentive listening born in the concert hall was incorporated both into the national school curricula (in many countries of Europe and America, but also in other regions worldwide, often at the expense of local music traditions) and to programs of popular education devised by the new mass media.

Auditory Styles in Music History: The Emergence of the Question of Listening

However, parallel to those efforts to broaden the appreciation of music, it also became evident that there had been different listening styles related to different music styles or genres in the history of Western music, as well as to the social contexts in which those styles or genres had developed. As it is generally acknowledged (Wegman 1998), this notion was introduced by German musicologist Heinrich Besseler in his 1925 essay "Grundfragen des musikalischen Hörens" (1978; "Fundamental Issues of Musical Listening," 2011) and later in "Das musikalische Hören der Neuzeit" (Musical Listening in the Modern Era, 1959). In these works, Besseler expanded the notion of listening beyond the limits of musical classicism and the ideology of aesthetic autonomy—an expansion that was not only a consequence of the boost that historical and comparative musicology had received during the previous decades, but that was also linked to the contemporaneous debate about the lost community-building power of concert music (Pritchard 2011). These questions are particularly present in Besseler's 1925 essay, which deals with the question of the "possibilities of access" to different types of music, and with the "social formations" derived from those possibilities (Besseler 2011, 50). In contrast, Besseler's 1959 essay stresses the link between listening attitudes and musical material, as does Adorno's later typology of listeners presented in the first pages of his *Einleitung in die Musiksoziologie* (1973; *Introduction to the Sociology of Music*, 1976): though both authors focus on the social role of music, albeit in different ways, the music itself seems to be the most determinative factor for the way in which it is listened to.

While Adorno was a pioneer in discussing how broadcasting could affect the experience of listening to a symphony (Adorno 1941, 1984, 2002) and in critically examining popular initiatives to bring classical music to mass audiences (Adorno 1994), he condemned jazz and popular music for the kind of listening they supposedly encouraged (Adorno 1990, 2002). Contrarily, his notion of "structural listening" may be interpreted as a form of expertise, a professional practice akin to the ones classified by Jonathan Sterne under the concept of "audile technique": like those professional listeners trained to create meaning out of subtle acoustic differences (as in stethoscopy) or to interpret acoustic signs as quickly as possible (as in telegraphy), isolated from their surroundings and from their own bodies, structural listeners are defined as much by their auditory and interpretive skills as by their limitations. As Rose Rosengard Subotnik (1996) has remarked (for further developments see Dell'Antonio 2004), attention to musical structure commonly entails inattention to sound and style—more specifically, to qualities such as timbre or the spatiality of sound—whereas the ability to follow thematic developments is not always compatible with the kind of passionate involvement that so many music lovers long for.

In the last decades, we have gained a better understanding of how models of music listening evolved throughout history—that is, not only how the institution of the concert began and was consolidated (Weber 2004, 2008), but also how other important musical institutions and genres for which performance and the body are key—notably, opera (Johnson 1995) and vocal music in general, but also the art of instrumental virtuosi, especially during Romanticism (Kramer 2002; Gooley 2004)—developed in dialogue with their audiences. Thus, music historians have described and discussed other notions of listening, such as those that preceded the classical era, which some have anachronistically characterized as distracted or disengaged (Weber 1997), or the literary elaborations of music listening that flourished in Romanticism. More recently, Veit Erlmann's research on the centrality of aurality to the modern project, *Reason and Resonance* (2010), has recovered and examined a wealth of historical discourses on sound and music related to what he defines as "the collapse of the boundary between perceiver and perceived" (10). In doing so, he has unveiled a more complex picture of the modern listener: one who is not so self-assured and who is less confident about her (more often, his) autonomous character than the bourgeois concertgoer. Though the prospects of delineating a history of the various auditory "epistemes" (Foucault 1970) or "regimes of listening" (Szendy 2008) are probably dim, so far it seems safe to affirm that the ideal of listening disseminated by the classical music tradition—the modern listener—was neither as well defined nor as hegemonic as it presented itself to be.

Listening to Music, Today

Yet the question of music listening can by no means be considered as exclusive to the classical music tradition, as it has also been analyzed in relation to the disciplines of ethnomusicology and popular music studies. Actually, it is an issue of interest to the entire field of music studies and to sound studies more generally. In particular, in the last decades, popular music scholars have explored the experience of listening to popular music (Negus 1996; Frith 1998; Moore 2001; Gracyk 2007), to specific popular music genres (Fabbri 1982, 2012; Berger 1999; Henriques 2011; to name just a few), to the various generations of recording and playback technologies (Sterne 2003; Katz 2004; Milner 2009; Bull 2000, 2007), mass media (Goodwin 1992; Douglas 1999; Baade 2012; Lacey 2013), and sound formats (Sterne 2012; Papenburg 2013). While popular music is often associated with forms of physical engagement (typically dance) that go beyond the notion of listening (Wicke 2001), and in spite of the aforementioned attempts to portray popular music listening as distracted or disengaged, situations of extremely focused listening—most commonly to recordings (see, for instance, Novak 2008)—are far from rare among popular music audiences. Indeed, many of them are as keen to listen to records or attend live performances as to discuss those experiences, and their discussions may also deal with the more adequate modes of listening to certain musics: advice on the best disposition of hi-fi systems for listening to stereophonic recordings (see Fabbri, "Concepts of Fidelity," this vol.), thoughts on the convenience of consuming specific substances (alcohol, drugs) to enhance the experience of listening to specific records or artists, or considerations about which songs are more apt to intensify or counterbalance certain moods while listening in motion (Bull 2000, 2007). Yet material aspects, such as the importance of rhythm in many popular music genres, the complex sound textures created in the recording studio, the audiovisual imaginary that has become such a central part of the popular music experience, as well as the variety of contexts where listening takes place nowadays, prompt us to think of music listening in new ways.

The expansion of auditory media—which began in the mid-twentieth century but which has become particularly intense since the 1990s, when a new digital compression format, the MP3, was adopted as the standard and exploited commercially—has resulted in the ubiquitous presence in public and private spaces of all types of sounds and music. This has created a range of relatively new listening situations, to which Anahid Kassabian (2013) has referred as "ubiquitous listening." This term includes, among others: the commercial use of background music to create emotional states in potential consumers (North and Hargreaves 2009) or to keep some of them away from certain spaces (Sterne 2013), the presence of music to accompany specific activities such as exercising (DeNora 2000; Facci 2013) or driving (Bull 2004; Fabbri 2013), the whole gamut of circumstances in which music can be listened to through portable players (Bull 2000, 2007; García Quiñones 2007, 2013), or the everyday interaction with audiovisual materials. In most of these contexts, the implicit listener (if she could even be described as a "listener") can hardly be identified with any of the avatars of the modern listener, as she is sometimes a captive listener but is often an inattentive or intermittent one (Stockfelt 1997; DeNora 2000; Kassabian 2002, 2013; García Quiñones, Kassabian, and Boschi 2013). As Jonathan Sterne (2012) has argued, conditions of low-fi or intermittent listening seem to be inscribed in the very procedures of perceptual coding on which digital compression formats are based.

One aspect that I have studied (García Quiñones 2007, 2013) is the experience of listening to portable digital players, which, in spite of being commonly represented as an intensively private, even isolating practice, poses many questions about the relationship of the (often moving) body to the music and the surrounding context. Categories such as "mood," so frequently mentioned by users, or "affect" may be helpful in discussing that zone of awareness that is not yet consciousness, as well as those complex overlappings of feelings, emotions, and thoughts that music can trigger even when we are not deeply engaged in it or when we are just accidentally exposed to it. Not surprisingly, an increasing interest in the power of music to affect us and, more generally, to interact with our minds and bodies at different levels—conscious and unconscious, emotional and rational, but also purely sensual and physical (Austern 2002; Kennaway 2010, 2012)—is noticeable among popular music and sound studies scholars today. This is attested to by the exploration of notions such as vibration (Goodman 2010; Trower 2012) and, generally, by the need to push the study of music beyond the threshold of its aesthetic status, covering also its applications to health and well-being (Gouk 2000; Horden 2000; DeNora 2013) and its ability to create unrest, trigger violence, or even be used as a weapon (Johnson and Cloonan 2008; Goodman 2010; Volcler 2013).

Although these explorations may be considered a novelty, there is actually a link between them and the traditional reflections on the "effects of music" that can at least be dated back to Plato's *Republic*, in which music (actually, the mix of poetry, music, and dance that Ancient Greeks called *mousiké*) was described as a force that should be kept under control by banning particular modes. Concerns about what music can do to us have in fact appeared periodically throughout Western history, associated, for instance—as Gary Tomlinson (1993) has showed with reference to the Renaissance conception of music—with natural magic, but also with contemporary technologies

such as the gramophone, which was used at the beginning of the twentieth century
to test the physiological and emotional effects of music (Schoen 1927; Selfridge-Field
1997; Hui 2012; Grajeda 2013). Indeed, the two notions of music defined here, namely
the notion of music as something to be listened to actively as it evolves over time and

Figure 16.1
Dancer at the Boiler Club, Barcelona, 2004 (photo courtesy of José Carlos Soto López).

the idea that music may have some strong effects on people even if they do not pay attention to it, do not seem to correspond to specific moments in history, but rather to contrasting approaches that have been present in discourses on music at different times, with different intensities.

Though introducing those two principles cannot act as a substitute for the necessity of studying each historical period and every new (and old) listening situation, I think that the historical dynamics of the notions of music and listening may be articulated around them, at least provisionally. In my opinion, any attempt to build an aesthetics of popular music (or rather, a popular aesthetics of music) should aim at reconciling those two principles. For that purpose, future research must not only try and address the variety of musical situations that we may encounter in the everyday; it must also be open to considering ways of creating meaning that do not necessarily crystallize into discourse, since undercurrents of affect and bodily sensations always travel with the flow of musical meaning. At the end of the day, what we need to understand is the continuity between the "modern listener" and the "modern hearer"—and also, perhaps more crucially, between those and the "modern dancer."

References

Adorno, Theodor W. 1941. The Radio Symphony. In *Radio Research 1941*, ed. Paul F. Lazarsfeld, and Frank N. Stanton, 110–139. New York: Duell, Sloan & Pearce. Repr. in Theodor W. Adorno, *Essays on Music*, ed. Richard Leppert, trans. Susan H. Gillespie et al., 251–270 (Berkeley: University of California Press, 2002).

Adorno, Theodor W. 1973. Einleitung in der Musiksoziologie. In *Gesammelte Schriften*, vol. 14: *Dissonanzen: Einleitung in der Musiksoziologie*, ed. Rolf Tiedemann, 169–420. Frankfurt: Suhrkamp.

Adorno, Theodor W. 1976. *Introduction to the Sociology of Music*. Trans. E. B. Ashton. New York: Seabury Press.

Adorno, Theodor W. 1984. Der getreue Korrepetitor: Lehrschriften zur musikalischen Praxis. In *Gesammelte Schriften*, vol. 15: *Komposition für den Film: Der getreue Korrepetitor*, ed. Rolf Tiedemann, 157–368. Frankfurt: Suhrkamp.

Adorno, Theodor W. 1990. On popular music. In *On Record: Rock, Pop, and the Written Word*, ed. Simon Frith, and Andrew Goodwin, 301–314. London: Routledge. Repr. in Theodor W. Adorno, *Essays on Music*, ed. Richard Leppert, trans. Susan H. Gillespie et al., 437–469 (Berkeley: University of California Press, 2002).

Adorno, Theodor W. 1994. Analytical study of the NBC "Music Appreciation Hour." *Musical Quarterly* 78 (2): 325–377.

Austern, Linda Phyllis, ed. 2002. *Music, Sensation, and Sensuality*. London: Routledge.

Baade, Christina. 2012. *Victory through Harmony: The BBC and Popular Music in World War II*. New York: Oxford University Press.

Berger, Harris M. 1999. *Metal, Rock, and Jazz: Perception and the Phenomenology of Musical Experience*. Hanover, NH: Wesleyan University Press.

Besseler, Heinrich. (1925) 1978. Grundfragen des musikalischen Hörens. In *Aufsätze zur Musikäesthetik und Musikgeschichte*, ed. Peter Gülke, 29–53. Leipzig: Reclam.

Besseler, Heinrich. 1959. *Das musikalische Hören der Neuzeit*. Berlin: Akademie-Verlag.

Besseler, Heinrich. 2011. Fundamental issues of musical listening. Trans. Matthew Pritchard, with Irene Auerbach. *Twentieth-Century Music* 8 (1): 49–70.

Bijsterveld, Karin. 2008. *Mechanical Sound: Technology, Culture, and Public Problems of Noise in the Twentieth Century*. Cambridge, MA: MIT Press.

Bonds, Mark Evan. 2006. *Music as Thought: Listening to the Symphony in the Age of Beethoven*. Princeton, NJ: Princeton University Press.

Botstein, Leon. 1992. Listening through reading: Musical literacy and the concert audience. *19th-Century Music* 16 (2): 129–145.

Bourdieu, Pierre. 1984. *Distinction: A Social Critique of the Judgement of Taste*. Trans. R. Nice. Cambridge, MA: Harvard University Press.

Bull, Michael. 2000. *Sounding Out the City: Personal Stereos and the Management of Everyday Life*. Oxford: Berg.

Bull, Michael. 2004. Automobility and the power of sound. *Theory, Culture, and Society* 21 (4–5): 243–260.

Bull, Michael. 2007. *Sound Moves: iPod Culture and Urban Experience*. London: Routledge.

Campos, René and Nicolas Donin. 2005. La musicographie à l'oeuvre: écriture du guide d'écoute et autorité de l'analyste à la fin du XIXe siècle. *Acta musicologica* 77 (2): 151–204.

Dahlhaus, Carl. 1989. *The Idea of Absolute Music*. Trans. Roger Lustig. Chicago: University of Chicago Press.

Dell'Antonio, Andrew, ed. 2004. *Beyond Structural Listening: Postmodern Modes of Hearing*. Berkeley: University of California Press.

DeNora, Tia. 2000. *Music in Everyday Life*. Cambridge: Cambridge University Press.

DeNora, Tia. 2013. *Music Asylums: Wellbeing through Music in Everyday Life*. Farnham: Ashgate.

Douglas, Susan J. 1999. Listening. In *Radio and the American Imagination*. New York: Times Books.

Erlmann, Veit. 2010. *Reason and Resonance: A History of Modern Aurality*. New York: Zone Books.

Fabbri, Franco. 1982. What kind of music? *Popular Music* 2:131–144.

Fabbri, Franco. 2012. How genres are born, change, die: Conventions, communities, and dia-chronic processes. In *Critical Musicological Reflections*, ed. Stan Hawkins, 179–191. Aldershot: Ashgate.

Fabbri, Franco. 2013. Taboo listening (or, What kind of attention?). In *Ubiquitous Musics: The Everyday Sounds That We Don't Always Notice*, ed. Marta García Quiñones, Anahid Kassabian, and Elena Boschi, 161–173. Aldershot: Ashgate.

Facci, Serena. 2013. An anthropology of soundtracks in gym centers. In *Ubiquitous Musics: The Everyday Sounds That We Don't Always Notice*, ed. Marta García Quiñones, Anahid Kassabian, and Elena Boschi, 139–160. Aldershot: Ashgate.

Foucault, Michel. 1970. *The Order of Things: An Archaeology of the Human Sciences*. London: Tavistock.

Foucault, Michel. 1980. The confession of the flesh. In *Power/Knowledge: Selected Interviews and Other Writings, 1972–1977*, ed. and trans. Colin Gordon et al., 194–228. Brighton: Harvester Press.

Frith, Simon. 1998. *Performing Rites: Evaluating Popular Music*. Oxford: Oxford University Press.

García Quiñones, Marta. 2007. "Listening in shuffle mode": Lied und populäre Kultur/Song and popular culture. *Jahrbuch des Deutschen Volksliedarchivs* 52:11–22.

García Quiñones, Marta. 2013. Body and context in mobile listening to digital players. In *Ubiqui-tous Musics: The Everyday Sounds That We Don't Always Notice*, ed. Marta García Quiñones, Anahid Kassabian, and Elena Boschi, 107–118. Aldershot: Ashgate.

García Quiñones, Marta, Anahid Kassabian, and Elena Boschi, eds. 2013. *Ubiquitous Musics: The Everyday Sounds That We Don't Always Notice*. Aldershot: Ashgate.

Goodman, Steve. 2010. *Sonic Warfare: Sound, Affect, and the Ecology of Fear*. Cambridge, MA: MIT Press.

Goodwin, Andrew. 1992. *Dancing in the Distraction Factory: Music Television and Popular Culture*. Minneapolis: University of Minnesota Press.

Gooley, Dana. 2004. *The Virtuoso Liszt*. Cambridge: Cambridge University Press.

Gouk, Penelope, ed. 2000. *Musical Healing in Cultural Contexts*. Aldershot: Ashgate.

Gracyk, Theodore. 2007. *Listening to Popular Music, or How I Learned to Stop Worrying and Love Led Zeppelin*. Ann Arbor: University of Michigan.

Grajeda, Tony. 2013. Early mood music: Edison's phonography, American modernity and the instrumentalization of listening. In *Ubiquitous Musics: The Everyday Sounds That We Don't Always Notice*, ed. Marta García Quiñones, Anahid Kassabian, and Elena Boschi, 31–47. Aldershot: Ashgate.

Gramit, David. 2002. *Cultivating Music: The Aspirations, Interests, and Limits of German Musical Cul-ture, 1770–1848*. Berkeley: University of California Press.

Henriques, Julian. 2011. *Sonic Bodies: Reggae Sound Systems, Performance Techniques, and Ways of Knowing*. New York: Continuum.

Horden, Peregrine, ed. 2000. *Music as Medicine: The History of Music Therapy since Antiquity*. Aldershot: Ashgate.

Hui, Alexandra. 2012. Sound objects and sound products: Standardizing a new culture of listening in the first half of the twentieth century. *Culture Unbound* 4:599–616.

Jay, Martin. 1992. Scopic regimes of modernity. In *Modernity and Identity*, ed. Scott Lash and Jonathan Friedman, 178–195. Oxford: Blackwell.

Johnson, Bruce, and Martin Cloonan. 2008. *The Dark Side of the Tune: Popular Music and Violence*. Aldershot: Ashgate.

Johnson, James H. 1995. *Listening in Paris: A Cultural History*. Berkeley: University of California Press.

Kassabian, Anahid. 2002. Ubiquitous listening. In *Popular Music Studies*, ed. David Hesmondhalgh and Keith Negus, 131–142. London: Arnold.

Kassabian, Anahid. 2013. *Ubiquitous Listening: Affect, Attention, and Distributed Subjectivity*. Berkeley: University of California Press.

Katz, Mark. 2004. *Capturing Sound: How Technology Has Changed Music*. Berkeley: University of California Press.

Kennaway, James. 2010. From sensibility to pathology: The origins of the idea of nervous music around 1800. *Journal of the History of Medicine and Allied Sciences* 65 (3): 396–426.

Kennaway, James. 2012. Musical hypnosis: Sound and selfhood from mesmerism to brainwashing. *Social History of Medicine* 25 (2): 271–289.

Kramer, Lawrence. 2002. *Musical Meaning: Toward a Critical History*. Berkeley: University of California Press.

Lacey, Kate. 2013. *Listening Publics: The Politics and Experience of Listening in the Media Age*. Cambridge: Polity Press.

Leppert, Richard. 2004. The social discipline of listening. In *Aural Cultures*, ed. Jim Drobnick, 19–35. Toronto: YYZ Books.

Love, Harold. 2004. How music created a public. *Criticism* 46 (2): 257–271.

Milner, Greg. 2009. *Perfecting Sound Forever: An Aural History of Recorded Music*. London: Faber & Faber.

Moore, Allan F. 2001. *Rock: The Primary Text: Developing a Musicology of Rock*. Aldershot: Ashgate.

Negus, Keith. 1996. *Popular Music in Theory: An Introduction*. Cambridge: Polity Press.

North, Adrian C., and David J. Hargreaves. 2009. Music and consumer behaviour. In *The Oxford Handbook of Music Psychology*, ed. Susan Hallam, Ian Cross, and Michael Thaut, 481–490. Oxford: Oxford University Press.

Novak, David. 2008. 2.5 × 6 metres of space: Japanese music coffeehouses and experimental practices of listening. *Popular Music* 27 (1): 15–34.

Papenburg, Jens Gerrit. 2013. Soundfile: Kultur und Ästhetik einer Hörtechnologie. *Pop Kultur und Kritik* 2:140–154.

Pritchard, Matthew. 2011. Who killed the concert? Heinrich Besseler and the inter-war politics of gebrauchsmusik. *Twentieth-Century Music* 8 (1): 29–48.

Schoen, Max, ed. 1927. *The Effects of Music*. London: Kegan Paul, Trench, & Trubner.

Selfridge-Field, Eleanor. 1997. Experiments with melody and meter, or the effects of music: The Edison-Bingham music research. *Musical Quarterly* 81 (2): 291–310.

Sterne, Jonathan. 2003. *The Audible Past: Cultural Origins of Sound Reproduction*. Durham, NC: Duke University Press.

Sterne, Jonathan. 2012. *MP3: The Meaning of a Format*. Durham, NC: Duke University Press.

Sterne, Jonathan. 2013. The nonaggressive music deterrent. In *Ubiquitous Musics: The Everyday Sounds That We Don't Always Notice*, ed. Marta García Quiñones, Anahid Kassabian, and Elena Boschi, 121–137. Aldershot: Ashgate.

Stockfelt, Ola. 1997. Adequate modes of listening. In *Keeping Score: Music, Disciplinarity, Culture*, ed. David Schwarz, Anahid Kassabian, and Lawrence Siegel, 129–146. Charlottesville: University Press of Virginia.

Subotnik, Rose Rosengard. 1996. Toward a deconstruction of structural listening: A critique of Schoenberg, Adorno, and Stravinsky. In *Deconstructive Variations: Music and Reason in Western Society*, 148–176. Minneapolis: University of Minnesota Press.

Szendy, Peter. 2008. *Listen: A History of Our Ears*. Trans. Charlotte Mandell. New York: Fordham University Press.

Thompson, Emily. 2004. *The Soundscape of Modernity: Architectural Acoustics and the Culture of Listening in America, 1900–1933*. Cambridge, MA: MIT Press.

Tomlinson, Gary. 1993. *Music in Renaissance Magic: Toward a Historiography of Others*. Chicago: University of Chicago Press.

Trower, Shelley. 2012. *Senses of Vibration: A History of the Pleasure and Pain of Sound*. New York: Continuum.

Volcler, Juliette. 2013. *Extremely Loud: Sound as a Weapon*. New York: New Press.

Weber, William. 1997. Did people listen in the 18th century? *Early Music* 25 (4): 678–691.

Weber, William. 2004. *Music and the Middle Class: The Social Structure of Concert Life in London, Paris, and Vienna between 1830 and 1848*, 2nd ed. Aldershot: Ashgate.

Weber, William. 2008. *The Great Transformation of Musical Taste: Concert Programming from Haydn to Brahms*. New York: Cambridge University Press.

Wegman, Rob C. 1998. "Das musikalische Hören" in the Middle Ages and Renaissance: Perspectives from pre-war Germany. *Musical Quarterly* 82 (3–4): 434–454.

Wicke, Peter. 2001. Sound-Technologien und Körper-Metamorphosen: Das Populäre in der Musik des 20. Jahrhunderts. In *Handbuch der Musik im 20. Jahrhundert*, vol. 8: *Rock-und Popmusik*, ed. Peter Wicke, 11–60. Laaber: Laaber-Verlag.

Listening Materialities and Techniques

17 Listening and Digital Technologies

Anahid Kassabian

It is rare that scholars stop to think about how digital technologies, as a group, have interacted with listening practices. While there is outstanding scholarship on particular technologies (e.g., Sterne 2012) or particular practices (e.g., Katz 2012), and there is also a body of writing on analog vs. digital in music and elsewhere (e.g., Massumi 2002, under "On the Superiority of the Analog," chapter 5), the relationship between listening and digital technologies generally is not an area that has had a great deal of scholarly attention. It would be impossible to redress that absence in this brief essay, but what follows will sketch at least the major areas one would need to account for in a larger work on this topic.

Impossible Starting Points: Technology or Practice

First, as is always the case in studying mediation, it is necessary to note the impossibility of establishing a starting point. All listening technologies come from listening practices, and listening practices are developed in relation to technologies. Every beginning of a story is thus a false one, and yet necessary, since one cannot begin with the birth of the human (which itself has an unclear starting point). It is crucial to remember, always, that practices and technologies are inseparably intertwined.

Given the recognition that any starting point is in some sense arbitrary and in some sense results from the choice to ignore what comes before it, the MP3 would certainly be one plausible starting point in this discussion.[1] The compression algorithm has caused a great deal of hand-wringing; audio specialists, hobbyists, musicologists, and professional musicians worry about the loss in the MP3's particular "lossy compression," concerned that listeners will lose the ability (that it is presumed they once had) to hear the full range of audio information available in live music, taken as the "correct" "origin" of all music. So one important touchstone in the discussion of listening and digital technologies will always be compression formats and algorithms.

But MP3s are compressed because of the possibilities that compression offers to listeners, which is most easily conceived as portability. The compression allows listeners to store many more files in the same amount of space, which is, not coincidentally, also an amount of information, or digital "real estate," that can be transported easily as email attachments, downloads, and so forth. This portability has drastically changed listening practices, making playlists for specific activities, for example, possible to carry around in one's pocket. The portability of MP3s has created a whole new set of uses for music, or perhaps conceptions of music, which have had consequences for CD packaging, Internet radio marketing, and so on. (These issues will be discussed further below.)

Digital technologies have also made possible a host of editing and recording technologies that are themselves the results of imaginative listening processes and that have had dramatic consequences for listening. These programs, from Cubase to the Reactable mobile app, allow you to input information for when and how to make a particular sound and then layer the sounds together to make a song. But even for listeners who have never made sounds before in their lives, the understanding that that is how the song was made leads to profound changes in their listening. Additionally, some of them will become producers themselves, on any range of levels— from making birthday songs for family members, to actually working with musicians to make more complex tracks, to working hard enough and long enough to manage to sell something. In this way, listening through digital technologies has become an impetus for new kinds of creative practices, which, in turn, also underpin changes in listening practices.

At the very least, listeners are now well trained to think of songs, or tracks, as being made up of bits that can be rearranged to make other songs. This kind of "training" began at least with hip hop; listeners learned to listen for samples from other tracks by other artists, not always even recent ones.[2] In the race to produce tracks and albums, producers work with samples from all across music's past, though some will limit their "hunting and gathering" to examples that address the same kinds of issues they want to raise. Whatever the case, both audiences and producers listen to music quite differently now, making all sorts of new forms of creating and listening possible.

For just one example, let us turn to mashups. By listening to albums and tracks as a series of separate units, music producers can then use them to create something entirely new. There are many, many mashup artists; one of the most famous is Danger Mouse, who released *The Grey Album* in 2004. By matching up an a cappella version of Jay-Z's *The Black Album* with the instrumentals from the Beatles' *White Album*, Danger Mouse created a smash hit. All music-making requires a keen ear and good musicians, but this kind of practice certainly requires a very close, attentive, and creative listening.

In return, mashups invite an attentive, creative listening on the part of their listeners. By being created from other songs, they invite the listener to identify all the fragments of songs from which they are composed. In this sense, mashups are just another step down the road first created by the use of samples in hip hop, which has taught its listeners a wide range of African-American music history, but also a fair amount of other music histories as well.

Digital technologies, as it is already possible to see, have had many and profound consequences for, or perhaps relationships with, listening. For example, a flash mob—a group that comes together via messages spread around among people through various social media and e-communication tools—appears to be a dance practice. And largely it is. But it also has a range of consequences for kinds of listening: a flash mob that comes together to dance can also come together to make music. So, for example, one response to the raising of student fees from £3,000 to £9,000 per year in the UK was a national, orchestral flash mob called the Guerilla Orchestra that played the Mission Impossible theme around the country, mainly in train stations, on one particular night in 2009 at a specific time. Reports were that many people were somewhat stunned by the event and had a range of reactions from bemused to confused. However, once the notion of a flash mob was created—dependent as it was on digital information technologies and social media platforms—then other social groupings were free to take on the form and pretend to have a flash mob. There are too many examples of this to recount, including corporate advertising, but what might be of interest are examples like the Opera Company of Philadelphia taking popular favorites such as Verdi's "Libiamo ne' lieti calici" (more commonly known as "Brindisi") from *La Traviata* and Handel's "Hallelujah Chorus" from *The Messiah* to an everyday shopping situation in an enclosed public market in Philadelphia, with the company's singers acting as ordinary customers, or the Gothenburg Symphony Orchestra launching four music chairs (in which one can sit and see and hear past GSO performances) in Landvetter Airport with a pop-up orchestra playing *Also sprach Zarathustra* when people sat in the new chairs. In other words, digital technologies enable new forms of listening outside of specific engagements with the technologies themselves.

Listening as Creating and Managing Music

But if these are examples of events that were made possible by our engagements with digital technologies without relying on such technologies (or at least not obviously and directly—people probably used email and mobile phones to arrange the events), they are also examples of how listening to and creating music are coming ever closer together. Another example of this phenomenon is one of Improv Everywhere's events (or, in truth, many of them). "The Cell Phone Symphony" took place in 2006 in the Strand Bookstore in New York City. Half of the participants were instructed to leave their mobile phones in their bags and check them at the front desk of the shop. Then they gave their numbers to one of the people who had not deposited their bags, and they were sorted by phone manufacturer. At particular cues, all of the Samsung, or

Motorola, or Nokia phones were called, so that the people in the store, and especially the staff at the front desk, were baffled by these coordinated events in which ten bags rang all at once and played the same ringtone.

Another way that the consuming and producing of music[3] cross over is in small music-making apps. There are thousands of these, but one interesting example is MadPad, made by Smule, who also created the Ocarina app and a handful of others. MadPad is a grid of twelve boxes that can either record or play whatever you want. You can then create a mix of the sounds you assigned to the boxes and directly upload that to YouTube (or your photos folder). To make MadPad work, you have to be listening to the world around you in order to choose the sounds you want, then record them, then imagine how to make the sounds work together. At that point, you can either mix them live or record them. Apps like this one also make listening and creating cross over each other—one listens to the world, imagines what one might record, records it, mixes it, and then either listens as it is being mixed or afterward to a recording, which may or may not be saved for future listening.

If recording made it possible to listen differently and thus to develop compositional practices like *musique concrète* and soundscape composition, then apps like MadPad are massifying what once were niche approaches to music-making. Digital technologies have also made it easier to codify music along different parameters. Whereas once one may have thought of music as belonging to an artist, a genre, and a period, now it is possible to add all sorts of other tags to tracks and organize one's musical world in new ways.

One of the most common is according to mood. Although there have been collections of music by mood since more or less the beginning of recording (collections for filmmakers, anthologies of music for romance, etc.), there has been an explosion of music for mood management over the past ten or so years. There are endless CD collections organized for specific weather (*… for a Rainy Day*), for relaxation (*Keep Calm and Relax*), and for working out (*Fitness Beats [The Running Mix 2014]*). There are apps to sort your music collection by a range of moods, so that you can get a customized playlist from your own collection (e.g., the Moodagent app), or you can go to Internet radio provider Songza, whose tagline is "Listen to Music Curated by Experts" (see Jarman 2013; Meier 2011; Baade 2014). Songza provides playlists for your every need, dividing and subdividing like a well-fed bacterium into all sorts of categories; today's playlists, for example, include a "Funky Dance Party for Prince's Birthday," "Inmate Playlists" in honor of the second season of *Orange Is the New Black*, and "Songs to Take a Dump To." (And I did not even skip any or take them out of order. Today is June 11, 2014.)

But these are only a very few of the possibilities of mood music. *…for a Rainy Day* is just one in a long list of "Music for …" albums, and *Keep Calm and Relax* is one in another long list of "Keep Calm and …" albums. There are hosts of mood music apps, such as Stereomood, Relax Music, and Mood of the Day, and there are similar hordes of sites, such as Musicovery, AUPEO!, and thesixtyone, all of which will help one find music to suit a mood.

The mood music industry and its affective marketing and branding include a range of activities—from scented candles and brand scents to color schemes to the repackaging of recorded music into mood collections, often with multiple CDs in a set—that is allowing contemporary capitalism to explore new approaches and especially new sensory regimes (see, for some examples, Thrift 2004).

Control Society, Ubiquitous Listening, Distributed Subjectivity

The very wide (and very incomplete) range of changes that have occurred with the development of digital technologies for listening that I have discussed here leads to the ubiquity of listening and to distributed subjectivity, as I have argued elsewhere (Kassabian 2013). Listening is no longer necessarily an event; it still can be, but it is more and more difficult to make it so (see, for example, the frequent use of visual materials in traditional concert settings). Because listening is everywhere and everywhen, it feels "natural" to do other things while we are engaging in listening; this, I believe, is in direct relation to developments in digital technologies, even though the phenomenon begins with radio and Muzak in the early twentieth century. Moreover, listening has become a constitutive feature of what I have called distributed subjectivity: a form of subjectivity that does not engage individuals as individuals, but rather both human and nonhuman, part objects and part subjects, in fields of activity that create ebbs and flows of subjectivities without discrete individuals as specific subjects. As we move more and more toward "control society," where subjects are no longer engaged in a Gramscian "struggle for consent" or in Althusserian "hails" and "interpellation," or even labor power as we once understood it, this form of subjectivity becomes more and more important to recognize and theorize. Although it makes agency more difficult, if not impossible, to posit as a political response, that path has in any case already been blocked by the elements of control society. Distributed subjectivity offers a way of thinking through how sound and music and listening can become a response to those new—and very grim—political developments. All that is needed now are ways to imagine and develop those responses.

Notes

1. Along with the MP3, other potential starting points are music's digitization, the popularization of the CD, the invention of the browser, and the establishment of file sharing systems and

sites. These would also connect—as does the MP3, for that matter—to the radical shifts the music industries have undergone in the past decade or two. To begin with any of these would tell a slightly different but very similar story.

2. Oddly enough, serialist composition may have the closest listening practices to hip hop, insofar as one is required to hear the row through various kinds of alterations. Active listening is, of course, widely associated with trained musicians and classical music—conversely, popular music is presumed to be treated as entertainment and, therefore, without the same focus. But this is a terrible overgeneralization of even the most divisive of models.

3. Neither "production" nor "consumption" is exactly a precise category. While one may purchase the technologies to make the music possible, whether that is a laptop computer and a program or an iPad and an app, one is not simply consuming, even when listening to clips made by others on YouTube or Vimeo. Similarly, most of us are not "producing," at least not in the sense of a commercial transaction, when we make things with these digital tools. This is, of course, the discussion that underlies such terms as "prosumer" or "pro-am," and it is a significant debate.

References

Bibliography

Baade, Christina. 2014. Lean back: Songza, ubiquitous listening, and Internet radio for the masses. Paper presented at Culture, Value, and Attention at Home, AHRC Cultural Value Project Expert Workshop, Liverpool Hope University, May 22.

Jarman, Freya. 2013. Relax, feel good, chill out: The affective distribution of classical music. In *Sound, Music, Affect: Theorizing Sonic Experience*, ed. Marie Thompson and Ian Biddle, 183–204. London: Bloomsbury.

Kassabian, Anahid. 2013. *Ubiquitous Listening: Affect, Attention, and Distributed Subjectivity*. Berkeley: University of California Press.

Katz, Mark. 2012. *Groove Music: The Art and Culture of the Hip-Hop DJ*. New York: Oxford University Press.

Massumi, Brian. 2002. *Parables for the Virtual*. Durham, NC: Duke University Press.

Meier, Leslie M. 2011. Promotional ubiquitous musics: Recording artists, brands, and "rendering authenticity." *Popular Music and Society* 24 (4): 399–415.

Powers, Devon. 2010. Strange powers: The branded sensorium and the intrigue of musical sound. In *Blowing Up the Brand: Critical Perspectives on Promotional Culture*, ed. Melissa Aronczyk and Devon Powers, 285–306. New York: Peter Lang.

Sterne, Jonathan. 2012. *MP3: The Meaning of a Format*. Durham, NC: Duke University Press.

Thrift, Nigel. 2004. Intensities of feeling: Towards a spatial politics of affect. *Geografiska Annaler* 86B (1): 57–78.

Sound and Media

Danger Mouse. 2004. *The Grey Album*. CD, DM1. Self-released.

Various Artists. 2013. *Fitness Beats (The Running Mix 2014)*. CD, Rhino, B00GZDKTKW.

Various Artists. 2004. *... For a Rainy Day*. CD, Universal Classics, B00005PJBM.

Various Artists. 2012. *Keep Calm and Relax*. CD, Sony CMG, B00736H4W6.

18   Enhanced Bass: On 1970s Disco Culture's Listening Devices

Jens Gerrit Papenburg

"Life is Not All Highs ... Be prepared for the experience that goes beyond listening" (Anonymous n.d.). At the beginning of 1978, the US-American manufacturer of audio equipment dbx used slogans such as these to promote an optically unobtrusive piece of technology: a small black box with a wooden shelf on each side. While John Badham's 1977 box office smash *Saturday Night Fever* made disco visible worldwide at the cinemas, dbx's box made disco boom sonically at dance clubs by targeting a sophisticated (sub-)bass design that resonated strongly with disco's sound aesthetics and sensory practices. Dbx's engineers ensured their box's operability by installing an operate/bypass switch as well as two other controllers (see figure 18.1). The smaller controller is connected to a rather conventional low frequency equalizer, and the larger one enables the regulation of a parameter called "sub harmonic level." The small box is tagged as the "sub harmonic synthesizer" or as the "boom box" (not to be confused with the portable boom boxes of hip hop culture). According to its manual, it functions as follows: "The Boom Box works by sensing the mid-bass frequencies that remain in the program, and by using them as a guide to recreate (synthesize) corresponding amounts of ultra-low bass" (dbx n.d., 5). Generally speaking, dbx's small box is a so-called enhancer, a device for signal processing that is supposed to optimize the sound of recorded music by a kind of fine-tuning.[1] According to the writings of music journalist Kodwo Eshun, enhancers function as a form of "phonoglutamate" (Eshun 1998, 88) or rather—and in case of the boom box more precisely—as a form of "low end glutamate" (ibid., 151).

In this chapter, I use the boom box and its enhanced bass as a starting point to analyze what I call *active listening devices*. Instead of being mere playback technologies, such devices are in fact assemblages of auditory practices, discourses, and technologies that (co-)constitute, on the one hand, the sound of the heard music and, on the other, the materialities and practices of listening. Coming from a musicological perspective that is based on a particular concept of music that conceives of sound, corporality, and technology as inherent and constitutive parts, I will elaborate on how listening devices enhance the sound of the heard and how that

Figure 18.1
dbx Boom Box (photo courtesy of Jens Gerrit Papenburg).

enhancement correlates with new listening practices and materialities, with a historically and culturally specific organization of the "rich physicality" of listening (Erlmann 2010, 17).

The boom box's enhanced bass is a part of a listening device. By synthesizing the
lower realm of the subharmonic, this sound is involved in the organization of a listening at the margin of the audible, at the boundary between audibility and tactility.
The boom box thus (co-)constitutes a domain that the anthropologist David Howes
calls "intersensoriality" (2005, 7–12). Hence, as a part of an active listening device,
the boom box mediates between a history of listening and a history of nonlistening,
between a history of sound and a history of "unsound" (Goodman 2010). Rather than
being a subject of the study of record production, listening devices are a subject of
the cultural analysis of record postproduction. Rather than being articulated to record
producers, listening devices are articulated to the music listener. Rather than being
placed in the professional site of the recording studio, listening devices are placed in
popular spaces, such as discos and clubs. At such places, the sound of recorded music
is optimized in relation to specific auditory practices while listening to it. Or, to put
it in other words: the boom box did not boom everywhere, but did so especially in
the late 1970s disco culture, where a sonic experience "that goes beyond listening"
was generated.

Designing Bass: Enhancement, Subharmonics, and Listening

The dbx boom box organizes sound enhancement as a part of music listening. It integrates the processes of sonic fine-tuning into the listening devices themselves. These devices thus take over functions of postproduction processes, such as mastering (the final sound-tuning stage in postproduction that mediates between the production of music and the release of music as record or soundfile). The adoption of mastering functions by listening technologies starts as an *avant la lettre* enhancement with the application of "tone control" and "loudness" buttons in hi-fi systems in the 1940s and 1950s—which made Glenn Gould, for instance, dream of what he called the "participational possibilities" of the listener (Gould [1966] 1984, 347).[2] This kind of sonic activation of listening technologies continues up to the present day—for example, in the form of the automatic volume control in music apps such as iTunes. That control refers to a specific listening practice that organizes music listening in a particular way as playlist listening (see Papenburg 2013). But instead of being about loudness, tone, or volume control, the boom box's sound enhancement is explicitly about bass and the cultivation of subharmonic frequencies as a part of music.

The first sonic technologies that were described as "exciters" and "enhancers" were used in recording studios. In 1975, the company Aphex Systems introduced its "aural exciter," which, in order to keep its status as a black box, was only able to be leased (see Aphex Systems Ltd. n.d., 1-1).[3] That exciter added additional overtones by distorting a studio production's high frequency range spectrum. This was meant to result in an "improvement" of recorded sound, especially in terms of the understandability of speech and the sensation of sonic direction. The boom box reverses this principle of the aural exciter: instead of additional overtones, it adds additional undertones to a signal. In doing so, the boom box is synthesizing the realm of the subharmonic—by inverting the harmonic scale. Moreover, dbx's boom box is explicitly a listening technology and not a production technology: "*We do not recommend recording the output of the Boom Box; use the Boom Box during playback*" (dbx n.d., 22; emphasis in original). Unlike other exciters and enhancers that found their place in recording studios, dbx's boom box is one of the technologies through which recorded music is listened to. Unlike other enhancers, the boom box specifically enhances the music's bass spectrum.

The exploration of the realm of the subharmonic through musical practices and sound technologies started long before the boom box. For example, subharmonics are produced in vocal techniques such as throat singing in the Altai mountain regions in central Asia and Tibet (see van Tongeren 2004, 26–29)—as well as by rather uncommon synthesizers that were never successfully brought to the market, such as the Mixtur-Trautonium (see Sala 1950) or the Subharchord (see Steinke 2008, 150–151). Moreover, one of the key figures of the institutionalization of musicology is involved in the prehistory of the boom box. In the late nineteenth century, the music theorist Hugo Riemann claimed that he was able to hear the subharmonics—or as Riemann called it, the undertones—of his grand piano. His perception of these undertones (generally considered unhearable) provided the "empirical" basis for his dualistic music theory (see Rehding 2003 and Hui 2013).

Unlike those auditory hallucinations, which were, at most, only "real" for Riemann, the boom box's subharmonics produce material and corporeal affects. Moreover, these affects also differentiate the box's enhancement of the low end from another technique used by listening technologies to enhance recorded music's bass, namely, from the psychoacoustic effect of the "residual tone" or "missing fundamental."[4] Through this effect, bass frequencies are created in the ear of the listener. This form of bass enhancement is used in the 1960s releases of the record label Motown (see Wicke 2011, 81), as well as by current software bass enhancers such as the MaxxBass plugin by the Israeli software company Waves. The residual tones in Motown's releases and the MaxxBass plugins are an interior psychoacoustic effect produced by the listening subject and are not a part of the outer world. In contrast, the dbx boom box produces corporeal affects that resonate with the sound aesthetics and the listening practices in specific forms of popular music, such as disco and, later, house, techno, and dubstep. Instead of merely being a metaphor, disco culture's resonant bodies do indeed correlate with a material dimension of sound that is culturally and historically specific—sound from the artificial domain of the subharmonic.

Disco's Sound Discourse: "Pure" or "Enhanced"?

At the end of the 1970s, enhanced bass also became a topic in the discourse of sound (re-)production that coincided with those sonic events produced by the enhancing technologies in the booming disco culture. The U.S.-American trade magazine *Billboard* started to organize annual "disco forums" in 1976. In doing so, the magazine was exploring and exploiting the disco boom in the United States. These forums were also places for discussions about "enhanced" sound. *Billboard* arranged a panel that was later summed up by the magazine under the heading "Pure or 'Enhanced' Sound Spurs Spark" (see Anonymous 1979; also Kopp 1980). Audio engineers who designed sound systems for discos—a profession that evolved in parallel to the genre's success—participated in that session, such as Alex Rosner and Ed King. The engineers discussed the normative question of whether disco's sound systems should—only—reproduce sound or if such systems should also actively modulate and enhance sound beyond the manual practices of the disco DJ. Should a sound system just be a form of playback technology, or should it also produce new sounds?

On the *Billboard* panel, Alex Rosner—who was already involved in building the sound system for the first proto-disco, David Mancuso's legendary Loft in Manhattan (see Lawrence 2003, 88–90), and who ran the company Rosner Custom Sound Inc.— bemoaned the "growing use of such sound enhancing aids as dynamic range expanders and boom boxes" (Anonymous 1979, 40). Rosner argued against "enhanced sound" and was for "pure sound," referring to the intention of the "artist and recording engineer" (ibid.). However, an advocate for the intention of the artist might be a surprising figure in the economy of 1970s disco culture. For, what Rosner does not take into account in his reflections about the need to safeguard a particular musician's intentions is that there was no autonomous and intentional subject who expressed him/herself in his/her music in the case of the production of disco music, but rather an entire network of human and nonhuman actors who were collaboratively involved in the creation of a disco production. Therefore, this image of the individual star figure itself only exists in the context of disco's marketing. Furthermore, Rosner's statement also indicates that disco's sound systems are not reducible to the status of playback technologies anymore—that they have been transformed into active listening devices. As the latter, these systems were not following the sonic ideals of high fidelity, such as "transparency" or "harmonic balance." Instead of a well-balanced sound, disco systems hold excessive moments of booming bass; instead of high fidelity, they were more interested in moments of ultra high fidelity and sonic hyperreality. The concepts of playback and sound reproduction are also problematized at a discursive level in relation to boom boxes and disco's sound systems.

Becoming a Resonator: Disco Culture's Technologization of the Materialities of Listening

The boom box's enhanced bass resonates with the listening practices of 1970s disco culture. Disco dancers trained themselves in practices in which they were not only immersed in a dancing crowd but also in the sound of music. They trained themselves in practices in which the sound of music not only affected their ears but also intensely resonated with other body parts at the level of materiality. Such practices corresponded with loud, tactile, low, and corporeal sounds.

Even if dbx's boom box was not primarily targeted at disco sound systems, the company obviously had disco installations in mind as one possible application (see dbx n.d., 13). As Dough Shannon puts it in his manual on the disco as an economic enterprise, by synthesizing sub-bass frequencies that were more felt than heard, the boom box could create "a totally pulsating atmosphere … that causes the customers to experience a physical sensation" (Shannon 1982, 159). This atmospheric and corporeal listening by those eager disco revelers—transformed into the "customers" in Shannon's business model—was crucial to the aesthetics of the disco genre. Dbx could incorporate its boom box into the designing of just such a listening experience—one in which the listener becomes a corporeal resonator in a very concrete sense: "The room crawls with low frequency energy, and you, in effect, become a resonator. Sounds become 'Super-eal' (*sic*) and the sensation is unlike anything you have ever experienced from recorded music" (Anonymous n.d.).

Besides the bass enhancer, disco's cultivation of the subharmonic and the bass domain integrated at least two other technologies for the constitution, exploration, and exploitation of the lower end of music: the twelve-inch single, which allows for a better resolution of bass frequencies than other record formats, and the sub-bass speaker (cf. Papenburg 2011). Richard Long, one of the crucial designers of New York's disco sound systems, based his installations on a specific sub-bass speaker: the so-called Levan Horn, named after the famous DJ Larry Levan (see Fierstein and Long 1980).[5] Long established a sophisticated bass design in New York's discos by separating the bass and the sub-bass spectrum. The Levan Horns produce sub-bass frequencies at the threshold of audibility. Long piled a stack of speakers on a Levan Horn in each corner of the dance floor. In this new "geography of listening" (Lawrence 2003, 90), recorded sound constituted an auditory space that went beyond mono or stereo. In this space, directions of sound were determined by frequency: high frequencies came from above—Long hung tweeter arrays on the ceiling above the dance floor (see Fierstein and Long 1980). Mid frequencies came in from the sides, and the sub-bass frequencies were nearly unhearable but tended to address other senses instead: compared to using one Levan Horn, the four Levan Horns of Long's system for New York's club Paradise Garage primarily produced more sound power, creating a more haptic and tactile quality. Long's four-way-system speakers made the dance floor into an immersive sound space.[6] They were not arranged in one direction, as in the case of a stage or a concert hall.[7] Instead, Long's sound system was attempting to create an immersive sonic event, something that is not an inherent quality of sound in general, but which is artificially produced. Moreover, the system's geography of listening enabled the listener to be as mobile as a dancer, moved by disco's rhythms and sonic textures.

However, Long's four-way system has one apparent blind spot: At the time, only a few records included sub-bass frequencies. One exception was the 12-inch single "Dance, Dance, Dance" (released on the record label Buddah, USA, 1977) from Bernie Edwards and Nile Rodgers's glamorous disco project Chic: "It was the first record to feature sub-bass. Before 'Dance, Dance, Dance,' bass tones below a frequency of 60 Hz were taken out in the process of mastering a record, but Rodgers and Edwards realized the effect that these shuddering tones had in a club and insisted that they be kept in" (Shapiro 2005, 162–163). However, the Paradise Garage sound system's Levan Horns did not have to remain silent while all the other records were spinning: Long built a boom box into his sound system to provide a "blend of 25-50 Hz bass synthesized from 50-100 Hz information present on the recording" (Fierstein and Long 1980).

By exploring the realm of the subharmonic, dbx's boom box boomed. It produced enhanced bass, which correlated with 1970s disco culture's listening practices and

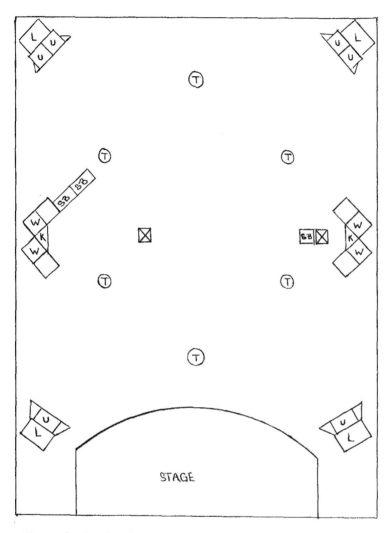

Figure 4- Speaker layout at the Paradise Garage.
         L= Levan Sub-Bass Horn
         W= Waldorf Bass Speaker
         U= Ultima Full-Range Speaker
         K= Pseudo-Klipsch Horn
         SB= Small Sub-Bass Horn
         T= Tweeter Array (hung from ceiling)

Figure 18.2

Speaker layout at New York's Paradise Garage (Fierstein and Long 1980) (diagram courtesy of Alan Fierstein).

materialities. However, the study of active listening devices' enhanced sound opens up a field of research beyond disco's sound systems. First, a delineation of the numerous listening devices of popular music and popular culture can elucidate how listening technology is involved in the constitution and sonic enhancement of the heard object and the listening subject. Or, as the musicologist Peter Wicke puts it:

The same song—listened to through headphones at home, or as part of a 90 minute-long stage performance, or as material used for dancing in a club—is the same song in name only. A differently structured form emerges when the song is made accessible for dancing by the bassline, than, for example, in the case of the subject-centered aesthetic perception through headphones which follows the relationship between text and sound. (Wicke 2003, 118)

A study of the constitutive function of listening devices enables one to focus explicitly on sound in the analysis of the relationship between listening and sound technology, which has become a pertinent subject for sound studies in general. Second, the study of enhanced bass and subharmonics also corresponds with new forms of sound and music. In this regard, the cultural analysis of an enhanced sound system can contribute to a critique of a "fetishization of midrange frequencies" in the "futurist, avant-gardist or rockist legacy of (white) noise music and its contemporary disciples" (Goodman 2010, 28; see also Henriques 2011). Moreover, its dynamic with current smartphone-driven "treble cultures" (Marshall 2014) needs further examination.

Notes

1. Dbx produced a whole series of products for sound enhancement. The company started with the production of noise-reduction technologies—which finally became Dolby systems' unsuccessful competitors—and added technologies such as dynamic range expanders.

2. In the 1950s, sound enhancement was a concept in mastering as well as in the marketing of music. In the context of the introduction of stereo in the late 1950s, RCA Victor remastered a part of its mono back catalog and released it as "enhanced stereo" discs (see Milner 2009, 147).

3. The understanding of the process of music recording as a process of music enhancing dates back to the early 1930s. Conductor Leopold Stokowski, for example, was interested in "enhanced music," which was made possible by the electrified recording process in the early 1930s (see McGinn 1983, 64).

4. On a cultural history as signal analysis of missing fundamentals, see Siegert 2013.

5. Long designed systems for discos and clubs that gained a legendary status, such as the Paradise Garage and Studio 54 in New York or the Warehouse in Chicago.

6. This kind of sonic immersion differs from the immersion created by quadraphonic systems, which were unsuccessfully brought to market around 1970. The latter did not use frequency to create a sonic space.

7. The sound of amplified music should come from the stage in most concert situations—see Thompson 2002, for example, on the extreme directional "modern" sound of the Radio City Music Hall in 1930s New York City. Even the famous and massive sound system of the Grateful Dead, a band that aimed to create a psychedelic experience (see Wolfe 1968, 263), produced directional sound that came from the stage. However, there are sound systems that do not follow that practice: Pink Floyd used the so-called Azimuth Coordinator to control a quadraphonic sound system for the first time at a concert in 1967. By using two joysticks, the coordinator could make sounds fly through the room from one corner to another (see Mason 2005, 314).

References

Bibliography

Anonymous. 1979. Pure or "enhanced" sound spurs sparks. *Billboard* 91 (32): 40, 53.

Anonymous. n.d. Introducing the Boom Box Subharmonic Synthesizer. *Advertisement.*

Aphex Systems Ltd. n.d. Aural Exciter Type III Model 250: Operating Guide and Service Manual. Hollywood, CA: Aphex Systems Ltd.

dbx. n.d. *dbx Model 100 "Boom Box" Sub Harmonic Synthesizer: Instruction manual.*

Erlmann, Veit. 2010. *Reason and Resonance: A History of Modern Aurality.* New York: Zone Books.

Eshun, Kodwo. 1998. *More Brilliant Than the Sun: Adventures in Sonic Fiction.* London: Quartet Books.

Fierstein, Alan, and Richard Long. 1980. State-of-the-art discotheque sound systems—System design and acoustical measurement. Paper presented at the Audio Engineering Society (AES), 67th Convention, New York, October 31–November 3. New York:

Goodman, Steve. 2010. *Sonic Warfare: Sound, Affect, and the Ecology of Fear.* Cambridge, MA: MIT Press.

Gould, Glenn. (1966) 1984. The prospects of recording. In *The Glenn Gould Reader,* ed. Tim Page, 331–351. New York: Alfred A. Knopf.

Henriques, Julian. 2011. *Sonic Bodies: Reggae Sound Systems, Performance Techniques, and Ways of Knowing.* New York: Continuum.

Howes, David. 2005. Introduction: Empire of the senses. In *Empire of the Senses: The Sensual Culture Reader,* ed. David Howes, 1–17. Oxford: Berg.

Hui, Alexandra. 2013. *The Psychophysical Ear: Musical Experiments, Experimental Sounds, 1840–1910.* Cambridge, MA: MIT Press.

Kopp, George. 1980. Sound systems: Sound experts sound off. *Billboard* 92 (33): 53.

Lawrence, Tim. 2003. *Love Saves the Day. A History of American Dance Music Culture, 1970–1979.* Durham, NC: Duke University Press.

Marshall, Wayne. 2014. Treble culture. In *The Oxford Handbook of Mobile Music Studies*, vol. 2, ed. Sumanth Gopinath and Jason Stanyek, 43–76. New York: Oxford University Press.

Mason, Nick. 2005. *Inside Out: A Personal History of Pink Floyd*. London: Phoenix.

McGinn, Robert E. 1983. Stokowski and the Bell Telephone Laboratories: Collaboration in the development of high-fidelity sound reproduction. *Technology and Culture* 24 (1): 38–75.

Milner, Greg. 2009. *Perfecting Sound Forever: The Story of Recorded Music*. London: Granta.

Papenburg, Jens Gerrit. 2011. Hörgeräte: Technisierung der Wahrnehmung durch Rock- und Popmusik. PhD diss., Humboldt University Berlin.

Papenburg, Jens Gerrit. 2013. Soundfile: Kultur und Ästhetik einer Hörtechnologie. *Pop Kultur und Kritik* 2 (1): 140–155.

Rehding, Alexander. 2003. *Hugo Riemann and the Birth of Modern Musical Thought*. Cambridge: Cambridge University Press.

Sala, Oskar. 1950. Das Mixtur-Trautonium. *Melos. Zeitschrift für Neue Musik* 17 (9): 247–251.

Shannon, Doug. 1982. *Off the Record: The Disco Concept*. Cleveland, OH: Pacesetter Publishing.

Shapiro, Peter. 2005. *Turn the Beat Around: The Secret History of Disco*. New York: Faber & Faber.

Siegert, Bernhard. 2013. Mineral sound or missing fundamental: Cultural history as signal analysis. *Osiris* 28 (1): 105–118.

Steinke, Gerhard. 2008. Subharchord II. In *Zauberhafte Klangmaschinen: Von der Sprechmaschine bis zur Soundkarte*, ed. Institut für Medienarchäologie, 149–151. Mainz: Schott.

Thompson, Emily. 2002. *The Soundscape of Modernity: Architectural Acoustics and the Culture of Listening in America, 1900–1933*. Cambridge, MA: MIT Press.

van Tongeren, Mark. 2004. *Overtone Singing: Physics and Metaphysics of Harmonics in East and West*. Amsterdam: Fuscia.

Wicke, Peter. 2003. Popmusik in der Analyse. *Acta Musicologica* 75 (1): 107–126.

Wicke, Peter. 2011. *Rock und Pop: Von Elvis Presely bis Lady Gaga*. Munich: Beck.

Wolfe, Tom. 1968. *The Electric Kool-Aid Acid Test*. New York: Farrar, Straus & Giroux.

Sound and Media

Chic. 1977. *Dance, Dance, Dance (Yowsah, Yowsah, Yowsah)*. Twelve-inch single, Buddah, DSC 121.

Saturday Night Fever. 1977. Director: John Badham. Feature Film, Robert Stigwood Organization/ Paramount (RSO).

19 Baby Monitor: Parental Listening and the Organization of Domestic Space

Andrea Mihm

Going On Air

In countless nuclear families, the baby monitor is part of the basic household equipment that helps organize current everyday life with infants. Its ritualistic handling during the children's bedtime ceremony makes it a firm component of the daily and nightly duties that accompany contemporary concepts and practices of Western parenthood. Just as in the case of disposable diapers, baby bottle sterilizers, or child car seats, the baby monitor fits into the trend of the mechanization and increasing digitalization of child care begun in the 1960s. Through an analysis of this technical device from a cultural-historical perspective, I will point out some of the crucial sonic aspects of sociocultural transformations and challenges in family life and domestic space.[1]

The widely known English appellation *baby monitor* refers to the idea of an automatic control system whose main task consists of observing the child and possibly warning the child's guardian. In contrast, the commonly used German term *babyphon* (thus containing the Greek word *phon*) points more specifically to sound, and a specific one—the crying of a baby.

Although there are many enhancements—from video control to particular apps—I will focus on the more classical and frequently utilized device of plastic-covered sender and receiver. Its principle of operation is quickly explained: in the child's room, the sender's microphone "listens" to whether or not the baby is making sounds. If the infant raises its voice, these sounds (as well as others) are transmitted to the receiver, which is usually located in the parents' immediate vicinity. Hence, a baby monitor can be described as a technical device that is interposed between infant and guardian. Via radio communication, it enables a one-way sonic connection between the different locations of the parents and child.

Accordingly, the application of a baby monitor only proves to be useful based on a particular geographical distance between sender and listener—a situation that emerged with the widespread institution of separate rooms for children and parents. Along with

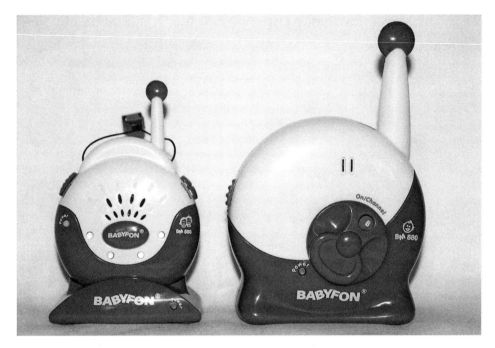

Figure 19.1
Standard baby monitor, 2014 (photo courtesy of Andrea Mihm).

increased economic possibilities, the notion that children need their own private spaces has gained recognition in the broader public since the 1970s (Bellwald 1998, 383–384).

Transmission Target: The Baby's Cry

A second basic premise is the perception that a baby's cry is a momentous signal that needs to be noticed immediately in order to respond promptly to it. Thus, the transmission target is the specific acoustic of filial crying—the primal sonic expression of a newborn and humans' "first language," as Rousseau suggested (Renggli 1992, 205). However, it is certainly not a language of words, but appears more as an eruptive and somehow unregulated amorphous phonation. Though as prelinguistic enunciation, it is also a sign of differentness (Christensen 2003, 130). Crying can require translation and interpretation, a matter that was imputed to the mothers in the first place and one that pediatrics, psychologists, as well as editors of guidebooks and trade journals address. In parallel, long-running debates show that the notion of the direct reply to a baby's cry follows a whole slew of arguments and analyses about the baby's reasons and intentions.

Basically, there are two threads: the first originates from the "Christian theoretical assumptions about the stubbornly selfish nature of all infants who, as carriers of the bad seed of original sin, required rigorous discipline" (Schwartz 2011, 750). Pursuant to this belief, experts in the 1950s and 1960s recommended "letting the noisy brat bawl" (Häussler 1968, 53). On no account should parents—and first and foremost mothers— pick up the crying baby at nighttime because they feel sorry for him or her. In order not to be woken up needlessly by the yelling of the "petty tyrant," a considerable selection of strict advisers even suggested plugging "wax or cotton into the ears" (Anonymous 1966, 10).

In the 1970s, the opposing argument gained acceptance. Among other things, developmental psychologist Mary Ainsworth's research on social interactions between children and caregivers led to the insight that crying is not merely "a problem … of temperament or discipline but of interpretation and mutual reassurance" (Schwartz 2011, 750). According to Ainsworth, the babies who were immediately soothed by their mothers when they started to cry altogether cried significantly less abidingly and intensely than those not appeased by their mothers. In the trade journals of that time, Ainsworth's findings were initially carefully formulated, but slowly took root and became a fundamental part of the daily lives of women with children. Now the mothers' behavior was "right" when they affectionately responded to their crying child and quickly took care of her or him (Mihm 2008, 32–34).

The baby monitor effectively underscored this trend reversal and even exploited it. Not without reason, communication systems for primarily industrial purposes were first advertised in the 1970s as also being useful in the context of child care (ibid., 16) for the sonic surveillance of children's rooms and "to alert parents and sitters to more episodes of crying than they would otherwise have heard" (Schwartz 2011, 753).[2] Although, up until then, so-called intercoms had been used for communication between secretaries and bosses, a shift in function took place: from speech communication (actively undertaken in daily work life through the pressing of a button) to the private, permanent monitoring of a child's sleep. This technological linking component developed from a means of two-way communication into a monitoring instrument—a transition that underscores the changed perception of the child and its vocalizations. In the private sphere of the home, the baby is granted the role of the boss whose full-throated demands are given top priority.

In the beginning of the twenty-first century, the all-embracing self-abandonment of modern mothers in favor of their little ones is no longer considered the ideal breeding behavior. In fact, liberal parents' needs regain importance (Hungerland 2003, 157).[3] The conditions were therefore right for the widespread introduction of the baby monitor. Parents were made to believe that both sides of the issue were reconcilable—not only could the child be monitored, but the parents could also savor a bit of freedom.[4]

Acoustic Surveillance and the Organization of Family Life

Considering the multitude of duties that mothers and fathers have to take on nowadays (child care worker, job holder [if possible], and, last but not least, manager of the nuclear family business), liberties are quite constricted, and the reclamation of leisure time owing to the use of the baby monitor is, above all, wishful thinking. As one mother asserts, "those who have some image of the woman where the mother cozily rests on the sofa during the baby's bedtime are wide off the mark." According to her, "a sleeping baby is a glaring signal" to do the chores (field reports n.d.).[5] Apparently, it is not only the baby's cry but also its silent breathing—and therefore the absence of noise—that signifies a call for action.

As many users report, parental activities during that period of gained time range from household chores like ironing, cleaning, cooking, or gardening to job-related computer work. While the baby monitor is on, the parents first and foremost try to efficiently use their offspring's sleeping time in order to get their work done. At the same time, they are all ears, listening attentively to the transmitted sounds and consciously or unconsciously capturing the quietness.

In this regard, the acoustic device not only constitutes instrumentality for homemaking, but also typifies a particular conception of time management. Owing to the sonic bridging, to some degree it synchronizes the duties of childcare with the organization and maintenance of domestic space or the requirements of the job, as one mother happily affirms: "An utterly useful device that allows me to monitor the child and to do other things at the same time, sometimes even pleasurable things, sitting in the garden, for example" (field reports n.d.). Other parents declare that the baby monitor permits them to go for a meal in a nearby restaurant, or that they use it when visiting their neighbors. Last but not least, some parents entrust the receiver (and thereby their child) to their immediate neighbors, so they can react to the baby's needs while mommy and daddy go out. Used in such ways, the acoustic device does indeed provide parents with a little independence and helps them to keep in touch with others, as well as with each other.

According to the parents, the considerable advantages of this monitoring allow for the combination of flexibility with the feeling of safety—in the handling of the child as well as in the shaping of the parent's life. For example, one parent reports that "due to the Philips baby monitor, I can concentrate on work, and be certain that I will be at my daughter's side whenever she needs me" (field reports n.d.).

As the above statements convey, the baby monitor also satisfies the modern parents' need to keep their offspring under continuous surveillance.[6] And yet, it comes with the territory that the configuration of the baby monitor and the way it is utilized generates an imbalance in the relationship between the guardians and the child, since the primarily one-way transmission allows parents to monitor their child, but not the other

Figure 19.2
"We left our baby at home. But we have a baby monitor with us. It has a range of 5,000 km [3,100 miles]. So we'll hear her right away if she cries." (Cartoon courtesy of Til Mette.)

way around. The following statement made by a Swiss baby monitor manufacturer spells this out quite clearly: "Your offspring won't sense a thing, but you will be able to hear and control everything" (Bébétel n.d.). And while the caregiver has the power to switch the baby monitor on and off, the child, on the other hand, is obligated to transmit personal sounds without being given the choice—and some parents are quite aware of that fact. Tongue in cheek, they talk about their "bugging operations" and "electronic eavesdropping" (field reports n.d.). They report how they "spy on" (ibid.) their child. One parent even compares the monitoring habit with the observation practices of the secret police in former East Germany (ibid.).

Despite such confessions, modern parents generally intend to act in their child's best interests. To them, this acoustic surveillance is a precautionary act. In this regard, the baby monitor is understood much less as a sonic observation device and more as a practical tool in the daily care of children. If the device failed to function in accordance with that purpose, it would constitute a major breach of trust between the manufacturers and the consumers. Whereas producers and vendors consistently assure the public that their devices are fail-safe and interference-free, parents as users often argue the opposite.

Sonic Breaches and the Intrusion into Privacy

Looking at the side effects of this monitoring practice, one can see that the transmitted acoustics are by no means limited to the sounds of a baby's cry or gentle snore. If the field reports of the users are to be believed, undesirable intermediate tones and sonic misroutings are daily occurrences. Not only do guardians sometimes unexpectedly hear the voice of another child, they may also receive other messages, such as sounds from the radios of truck drivers and police, a fact that (especially when initially debuted) has provoked confusion and consternation (field reports n.d.).[7]

Basically, since the transmission uses radio signals and is no longer hardwired, radio frequency interferences occur. Reception interferences happen when there are other devices close by that work on the same wavelength. The acoustics that these attentively listening users have to deal with range from business calls by taxi drivers, hotel staff, paramedics, or firefighters, to radio transmissions and the neighbor's answering machine (ibid.). According to the account of a woman who lives near a hospital, she even picks up the "beeps of the intensive care medical equipment" (ibid.).

Nevertheless, the large majority of interference comes from other baby monitors. This kind of sonic confusion occurs particularly in neighborhoods that have several families with small children, where a number of devices are used in a confined amount of space. Users describe such incidents time and again in their reports: "Right away, on the first day that I used the device—my son was about eight weeks old—all of a sudden a loud 'mama' came out of the baby monitor. With an iron in my hand, I became rooted to the spot. It took me a long time to understand what was going on" (ibid.). Another user reports on their neighbors who take care of their eighty-year-old grandmother living in the same house. For this purpose, they use a similar device as the person reporting,[8] "so, it was quite frequently the snoring of that grandma that came out of the receiver, instead of the sounds of our baby! That was really annoying, especially as we did not know that in the beginning and every time thought 'Oh my god, what's wrong with our baby?' Scared out of our wits, we rushed to our child's room, even though it was just the grandma" (ibid.).

The way in which the users describe those events demonstrates how it feels to accidentally and unexpectedly traverse a sonic line of demarcation. The acoustic frontiers prove to be sociospatial as well, and this unintentional crossing of the audible frontier turns out to be an invasion of privacy. However, it is remarkable that, although the parents are quite bothered by the sounds they receive from other households, very few of them realize that they themselves are potentially targets of a neighborly bugging operation—a fact that is obviously suppressed or rarely taken into consideration.

Besides this correct or incorrect transmission of the baby's crying, some devices also make various continuous noises, such as cracking or beeping, which can be caused by age or malfunction. However, beeps are also deliberately programmed, audible warning signals and may go off either when the batteries have run out or in the case of transmission failure between units. According to one instruction manual, "if radio contact breaks down completely, for example because the baby device is out of range, or because the batteries in the baby device are empty, after a short time the parents' device will emit an alarm tone" (Babyfon 2007, 12).

The beep is thus there for the parents to be able to supervise. In the case of malfunctions, the device sets off an alarm and prompts them to adjust their handling of the device. Thanks to the signal, parents can either change the batteries or reestablish a connection between the units. Therefore, the beep helps the guardians monitor the device and check for proper operation. But the same applies vice versa! With the assistance of the alarm, the baby monitor also controls the guardian. The beep admonishes them to correct their position and to bring the devices (and thereby themselves) into range. As a result, the users have to adapt their own behavior to the technical device. They are prompted not to leave certain areas or, rather, to return to those areas as soon as they leave the designated, permitted space.

Switched Off

In summary, the baby monitor conveys far more than just the cry of a baby. The small, mostly inconspicuous device, its applications, and its aural practices reflect and even construct specific concepts of family, proper upbringing, and parental responsibility. It is involved in the organization of our concept of home—as "a physical space, a social space, an information space" (Stewart 2003, 12) and, most notably, as an acoustic space.

In its short history, the baby monitor has been promoted as an important technology for modern child care and homemaking. Its acoustic bridging partly enables guardians to synchronize the manifold requirements of home and caretaking, and thus it clearly promises and allows them an increased scope of action. However, it has also become apparent that this conceptualization of the monitor falls short and excludes many other aspects—for example, the baby monitor as a control device to which the babies have no choice but to surrender, a device that uses sounds and hearing as mediums of observation. As the parents are also monitored, it is clear that the baby monitor's transmission technology is highly ambivalent. On the one hand, it allows modern caretakers more time for various activities; on the other hand, it forces them to stay within its defined area and to by no means traverse its transmission boundaries.

Moreover, in the end, the baby monitor also raises privacy concerns owing to radio frequency interference. Quite often, the common sonic frontiers between public and private events are unexpectedly crossed.

Notes

1. As a member of the cultural scientific research committee "Kulturwissenschaftliche Technikforschung" at the University of Hamburg, I examined the baby monitor in its various manifestations and everyday applications, primarily in the German-speaking domain. With the help of an empirical approach that included interviews, the analysis of historical parenting guides, journals, advertisements, and 254 user reports, I investigated its emblematic ascriptions and constitutive meanings in the context of family life and child care (Mihm 2008).

2. Further developments concerning the technical functionality and outside appearance—as well as the possible interpretations of those aspects—are described in Mihm 2008, 14–27.

3. At the same time, present-day parents have to deal with the requirement that they provide an "optimal" nurturing environment for their offspring. Correspondingly, the amount of time and money that parents spend on their children has increased exponentially (Maihofer 2001, 34).

4. Analyzing baby monitor manufacturers' advertising, one can see that their arguments are based precisely on this idea. By showing a laid-back mother—no fathers are shown—reading a newspaper, drinking coffee with a friend, or doing some gardening, marketing experts take advantage of the parent's desire to both enjoy some individual freedom and be an attentive mother. The baby monitor is always with the mother—on the table, fastened at the belt, or hanging around her neck like a pendant.

5. All field reports were taken from the web portal http://www.ciao.de, accessed August 2006 (archive Andrea Mihm). The web reports are written by consumers for consumers. Therein the authors describe their experiences with certain products to help others with their buying decisions. The reports contain records similar to diaries and everyday life narrations; they are personal reviews of the product and sometimes feature characteristics of advice literature.

6. On this note, the device is inserted into a comprehensive arsenal including many different protection and control systems. As an example, one could mention the manifold practices throughout the range of medical care. Here, children in particular are subjected to close supervision, even before birth. Via cardiotocograph, the fetal heart sounds are transmitted and monitored during periodic examinations—and so the physical (and mental) development of a child gets compared to specified norms before and after birth (see Hungerland 2003, 143–147; Mihm 2008, 50–76).

7. In a 2005 investigation by Stiftung Warentest (one of the largest German consumer organizations involved in examining and comparing goods and services), the testers also arrived at the conclusion that most devices turned out to be fairly susceptible to radio interference (Anonymous 2005, 69).

8. Thus, the device does not exclusively serve as a baby monitor. As in the cited example, people sometimes use it in quite unconventional ways. Just as the personal computer at first was bought for the purpose of education and was subsequently used as a game machine or has "been relegated to tops of wardrobes or backs of cupboards" (Silverstone, Hirsch, and Morley 1992, 24), the same applies to the functions of the baby monitor, leading to its sonic targets and ascriptions changing during its lifecycle for some families—for instance, when it serves as a doorbell amplifier, when it is used to monitor the basement of a house, or when it acts as an internal telephone. Users thereby convert it to be utilized in idiosyncratic and creative ways. Unhinged from its actual (and from the manufacturer's earmarked) scope of duties, the device is provided with completely new meanings. It can therefore also be converted from a monitoring instrument into a communication medium.

References

Anonymous. 1966. Ihr Baby schläft am besten allein. *Junge Mutti* 4:10.

Anonymous. 2005. Nächtliche Aufpasser—Babyfone—Worauf Sie achten müssen: Ton- und Bildqualität, Reichweite, Störanfälligkeit, Elektrosmog. *Stiftung Warentest* 11:68–71.

Babyfon. 2007. *Operating Manual Babyfon*. BM 880 ECO.

Bébétel. n.d. *Bébétel: Das Babyphone per Telefon! Product information by Leitronic AG*. http://www.babyjoe.ch/Sicherheit/Babyphones/BEBETEL-Babyphone-Bluetooth.html (accessed May 15, 2014).

Bellwald, Waltraud. 1998. Das Kinderzimmer. In *Kind sein in der Schweiz: Eine Kulturgeschichte der frühen Jahre*, ed. Paul Hugger, 383–390. Zürich: Offizin.

Field reports. n.d. Reports taken from web portal http://www.ciao.de (archive Andrea Mihm) (accessed August 15, 2006).

Häussler, Siegfried. 1968. *Ärztlicher Ratgeber für die werdende und junge Mutter*. Baierbrunn: Wort und Bild.

Haudrup Christensen, Pia. 2003. Kindheit und die kulturelle Konstitution verletzlicher Körper. In *Kinder, Körper, Identitäten: Theoretische und empirische Annäherungen an kulturelle Praxis und sozialen Wandel*, ed. Heinz Hengst, and Helga Kelle, 115–138. Munich: Juventa.

Hungerland, Beatrice. 2003. Und so gedeiht das Baby! Altersgerechte Entwicklung und Gesundheit als gesellschaftliche Norm und Leistung. In *Kinder, Körper, Identitäten: Theoretische und empirische Annäherungen an kulturelle Praxis und sozialen Wandel*, ed. Heinz Hengst and Helga Kelle, 139–160. Munich: Juventa.

Maihofer, Andrea, Tomke Böhnisch, and Anne Wolf. 2001. *Wandel der Familie*. Ed. Hans-Böckler-Stiftung, Arbeitspapier 48. Düsseldorf: Zukunft der Gesellschaft.

Mihm, Andrea. 2008. *Babyphon: Auf einer Wellenlänge mit dem Kind: Eine kleine Kulturgeschichte*. Marburg: Jonas.

Renggli, Franz. 1992. *Selbstzerstörung aus Verlassenheit: Die Pest als Ausbruch einer Massenpsychose im Mittelalter; Zur Geschichte der frühen Mutter-Kind-Beziehung.* Hamburg: Rasch und Röhrig.

Schwartz, Hillel. 2011. *Making Noise: From Babel to the Big Bang and Beyond.* New York: Zone Books.

Silverstone, Roger, Eric Hirsch, and David Morley. 1992. Information and communication technologies and the moral economy of the household. In *Consuming Technologies: Media and Information in Domestic Spaces*, ed. Roger Silverstone and Eric Hirsch, 15–31. London: Routledge.

Stewart, James. 2003. The social consumption of information and communication technologies (ICTs): Insights from research on the appropriation and consumption of new ICTs in the domestic environment. *Cognition Technology and Work* 5:4–14.

20 Over-Hearing: Techniques of Popular Listening

Jacob Smith

Many scholars in the field of sound studies have sought to understand the ways in which listening practices have responded to the sound technologies of the past century. R. Murray Schafer ([1977] 1994) describes this dynamic in terms of "schizophonia"; Steven Feld (1994) responds with a discussion of "schismogenesis"; Emily Thompson (2002) and Karin Bijsterveld (2008) explore the subject of noise in the modern soundscape and acoustic design; Susan Douglas catalogs what she refers to as radio's "repertoire of listening" (2005); David Goodman (2010) focuses on the "distracted" radio listener; and Kate Lacey (2013) makes a distinction between the "listening out" and "listening in" that has been fostered by broadcasting (for an overview of "modes of listening," see Pinch and Bijsterveld 2012). Jonathan Sterne's notion of "audile technique" has been particularly generative. Sterne gives an account of how telegraph operators and the users of stethoscopes embodied an audile technique that was characterized by "logic, analytic thought, industry, professionalism, capitalism, individualism, and mastery" (2003, 95). Sterne's account of the emergence of forms of rationalized listening is offered as a counternarrative to arguments that posit "sight as the sense of intellect and hearing as the sense of affect" (ibid., 99). In this essay, I am interested in listening practices that run counter to Sterne's counternarrative: techniques of popular listening that are irrational and even occult.

I use the term *over-hearing* to describe this category of listening, but not as it is understood in its common usage as being the act of listening to a speaker without their knowledge. Instead, I refer to the episodes in the cultural history of listening in which hearing is discursively framed as overactive, overindulgent, or excessive—when consumers of sound are thought to be listening too much or to the wrong sonic objects. In the following, I will provide several examples of over-hearing that rely on the manipulation of sound technologies to play sounds at slow speeds or in reverse, and I will conclude by suggesting some avenues for further research.

Backmasking

Thomas Edison's invention of the phonograph allowed sounds to be captured and preserved, with one unexpected result being that sound could be experienced in novel ways. During public exhibitions of Edison's tinfoil phonograph in 1877–1878, showmen did more than demonstrate the device's ability to accurately reproduce a performance—they also displayed how it could produce a variety of sonic effects (Feaster 2007, 76). For example, exhibitors rotated the phonograph mechanism at different speeds to change the pitch of the recording, and they turned the cylinder backward to play sounds in reverse. Edison and his colleagues were known to have recorded the words "mad dog" so that, when played backward, the result was "God damn." In some cases, backward sound was offered as a demonstration of how the machine could generate an "infinite variety" of new music from preexisting musical compositions, an effect that was sometimes described by exhibitors and reporters as a "musical kaleidoscope" (ibid., 104–106).

By the 1910s and 1920s, public exhibitors shepherded interest in the phonograph toward a model of refined entertainment in the home, with the phonograph presented as a musical instrument, not a musical kaleidoscope. During the same period, the dominant experience of the medium changed from two-way functionality—both recording and playback—to one-way playback only—a shift encouraged by the gradual industry-wide adoption of nonrecordable discs (on the rise of the disc format, see Millard 2005; Thompson 1995). The history of phonograph exhibition tracks a change from two-way to one-way use in another sense—from the display of forward and reverse sounds to forward only. As I have discussed elsewhere, a cohort of conservative Christian pundits and lecturers brought reversed sound back into public consciousness in the early 1980s by claiming that subliminal messages were hidden in the products of the phonograph industry (Smith 2011). *Backmasking* was the name given to the process whereby messages were thought to be placed in popular recordings such that their full meaning could only be discovered when the record was played in reverse. In April 1982, the CBS Evening News with Dan Rather aired a segment that exposed the backmasking supposedly found in records by Led Zeppelin, Electric Light Orchestra, and Styx (Bivens 2008, 90–94).

The theory of backmasking was investigated in countless homes by curious teenagers, including myself. Listeners who carefully rotated their long-playing phonograph records (LPs) in reverse, straining to find hidden messages in the garble of backward sound, were over-hearing—listening too closely to the wrong sounds and applying their muscles of perception to purposes that many considered to be dubious or frivolous. Nonetheless, backmasking was an example of what William Boddy calls a "vernacular theory of electronic communication," and was less influential as a critique of

Satanic recording artists than as a prompt for vernacular practices of occult listening (Boddy 2004, 3; see also Acland 2012, 33).[1]

Slow-Motion Sound

Turntables could be played in reverse, but they could also be slowed down. The great advocate of slow-motion sound was the French film director and theorist Jean Epstein. In his 1955 essay "The Close-up of Sound," Epstein wrote that "the great natural sounds" such as wind, the sea, thunderstorms, and fire were comprised of "complex compressions of infinitely particular and numerous tones" emitted by different sources. These "tangled wave packets" could only be fully appreciated, he claimed, by slowing them down (Epstein [1955] 2012a, 360–370). In "The Delirium of a Machine" (1955), Epstein claimed that slow motion increased the "discriminating powers of hearing," providing listeners with "a capacity for analyzing more finely a sonic set, to distinguish in it otherwise imperceptible components, and to discover timbres until then truly unheard of" ([1955] 2012b, 376–377). Epstein put his theory into practice in films like *Le Tempestaire* (1947), which features stunning sequences in which sounds of the ocean are reproduced at various speeds.

The same year that Epstein wrote these essays, an American named Jim Fassett released the first of several LPs that demonstrated Epstein's claims about how slow-motion sound could increase the discriminating powers of the ear. Fassett began his professional life as a music critic in Boston and then became a musical director for CBS Radio, as well as the host of New York Philharmonic Symphony Orchestra broadcasts from Carnegie Hall. He was fascinated by tape recording technology and collaborated with CBS technician Mortimer Goldberg on several idiosyncratic LPs. The first of these, *Strange to Your Ears* (Columbia, 1955), consists of common household sounds that are made strange by slowing them down or playing them in reverse. Fassett shared Epstein's desire to hear natural sounds in a new way, and he had a particular interest in bird calls: "I thought there must be hidden beauty in them that I couldn't hear," Fassett told an interviewer (McPartlin 1956, 25). During one sequence of *Strange to Your Ears*, Fassett announces that he will reveal the "fantastic intricacies of sound" that are hidden in "the sweet, familiar song" of the canary bird. Fassett slows down the trills of a canary until they are transformed into what he calls "weird, unearthly" sounds that he compares to a "screeching bird of the marshes," human laughter, and a baying dog. Fassett uses deceleration to defamiliarize a common household pet, with one result being that the canary's song is made legible as aesthetic expression—he even stages a duet between his manipulated bird calls and a symphonic flute-player.

Fassett continued his investigation of birdsong on the LP *Symphony of the Birds* (Fassett 1956). The first half of this album consists of a musical work composed entirely of bird recordings that have been manipulated in various ways. The second half of the LP begins with chirping sounds, over which Fassett speaks: "Imagine you have suddenly come upon an open, sun-drenched clearing deep in the forest ... And all around you, you hear a myriad of birdsong. No single voice distinct, all blending in a constant but changing tonal patchwork." The chirps fade out, to be replaced by a low-pitched cooing sound. "Now imagine another spot," Fassett tells the listener, "where you find only glimmers of light that can penetrate the dense foliage of the trees above and around you" and where "the indiscriminate blur coming from the throats of hundreds of varieties of birds starts to take on a pattern of sound." The rest of the record alternates between these two imaginary spaces, one in which bird species are heard "as in nature," and the other a "semi-real darkness of a world transformed" by doubling or tripling the duration of the recordings.

Fassett's records, like Epstein's films, mobilize slow-motion sound, revealing the hidden patterns to be found in natural sounds. To the extent that they used sound technologies to scrutinize mundane, everyday sounds, Epstein and Fassett based their creative work on practices of over-hearing. Notably, both men rationalized their artistic pursuits by comparing them to scientific investigation. In *The Delirium of a Machine*, Epstein compares the "micro-audition" of slow-motion sound to the images provided by telescopy, microscopy, and radioscopy (Epstein [1955] 2012b, 376–77; Turvey 2008, 3).[2] Similarly, Fassett frames his tape manipulation in terms of the revelation of "hidden patterns" in nature. In Sterne's terms, both men enlist over-hearing in the service of audile technique. By the mid-1950s, most home stereos were equipped with variable speed selectors—the fallout of the postwar "War of the Speeds"—and slow-motion sound was no longer the sole domain of network engineers and film editors (see Morton 2004; Millard 2005, 207). In fact, playing a disc that was intended for 78 rotations per minute (rpm) on the 33 or 45 rpm setting became a common mistake. Slow-motion sound thus became available for vernacular investigation, and popular practices did not always share Epstein and Fassett's scientism.

800% Slower

Teenagers in the mid-1960s began playing the Kingsmen's record *Louie Louie* (1963) at slow speed, inspired by an urban legend that its slurred vocals were sexually explicit when decelerated. In this case, over-hearing was motivated by titillation rather than by the desire to reveal hidden patterns in the natural world (see Marsh 1993).[3] More recently, digital audio tools have fostered vernacular practices of slow-motion sound. In 2010, the programmer Paul Nasca created the digital application PaulStretch, which has the capability of stretching audio to many times its length while maintaining its original pitch. PaulStretch gained international renown when the DJ Nick Pittsinger used it to transform Justin Bieber's track "U Smile" from a three-minute pop song to a thirty-five minute ambient epic. Pittsinger's creation spawned an Internet meme, with posts often including the phrase "800% slower" in their title. Where Epstein and Fassett had placed ocean waves and birdsong under the sonic microscope, these amateur remixes make popular music, film and television theme songs, and computer start-up sonicons "strange to our ears" (Jarvis 2010; Richards 2011). With "800% slower" mixes, we see slow-motion sound's potential to foster what might be called "fractal listening," in which a brief recording is made to produce a seemingly infinite variety of sound. However, fractal listening of this kind is not new to the digital era; in some regards, it marks a return to the pleasures of the "musical kaleidoscope" of a century before.

To conclude, I submit that over-hearing can be understood as audile technique's distorted echo in the era of popular sound media (see Gunning 2007).[4] The category of over-hearing reminds students of sound to attend to the multiplicity of coexisting listening practices and the cultural hierarchies that are constructed within them. Media historians might seek out public debates concerning vernacular or subcultural listening practices associated with the consumption of sound media. Future research on over-hearing might also attend to practices of repeated listening, building upon Theodore Gracyk's theories about timbre and memory, Barbara Klinger's study of the repeated viewing of movies in the era of home video, and Tia DeNora's discussion of the everyday uses of recorded music (Gracyk 1996; Klinger 2006; DeNora 2000). Practices of over-hearing might concentrate on the noises of sound technologies rather than their intended signals; research in this area can help to heed Kate Lacey's call to put noise back into the history of media. Jeffrey Sconce's study of the otherworldly associations given to radio static and Douglas Kahn's work on "natural radio" would be excellent models for such a project (Lacey 2013, 91; Sconce 2000; Kahn 2013).

Finally, the "over" in over-hearing might be understood as an overflowing of the banks of the aural sense into other sensory modalities. Once sound became a material commodity, it could be slowed down and reversed, but it could also be touched and handled, appreciated for its visual form, or consumed in tandem with carefully selected images. The domain of over-hearing includes modes of haptic listening, in which sonic texts are felt as much as heard, or situations in which sound design intersects with product design or interior design. As such, this line of investigation offers an opportunity to explore what Steven Connor calls the "fertility of the relations" between the senses (Connor 2004, 154; Howes 2006, 383; Auslander 2004; on haptic hearing, see Marks 2000, 183). There is much that can be learned, then, from the study of vernacular listening practices, even—and maybe especially—those that, on first hearing, sound rather strange to our ears.

Notes

1. Also see Charles Acland on "vernacular cultural critique" (Acland 2012, 33). The backmasking debate has much in common with Acland's discussion of "subliminal media," which Acland understands as "a popular explanation for the operations of media, mind, technology, and representation" (27–28).

2. Epstein engages in what Malcolm Turvey calls a "revelationist" film theory, finding cinema's most significant property in "its ability to uncover features of reality invisible to human vision" (Turvey 2008, 3).

3. It is not surprising, then, that these listening practices prompted adult anxiety and even an FBI investigation.

4. Tom Gunning claims that "the heritage of natural magic follows scientific thought and practice for centuries like a shadow" (2007, 101).

References

Bibliography

Acland, Charles R. 2012. *Swift Viewing*. Durham, NC: Duke University Press.

Auslander, Philip. 2004. Looking at records. In *Aural Cultures*, ed. Jim Drobnick, 150–157. Toronto: YYZ Books.

Bijsterveld, Karin. 2008. *Mechanical Sound*. Cambridge, MA: MIT Press.

Bivens, Jason C. 2008. *Religion of Fear*. Oxford: Oxford University Press.

Boddy, William. 2004. *New Media and Popular Imagination*. Oxford: Oxford University Press.

Connor, Steven. 2004. Edison's teeth: Touching hearing. In *Hearing Cultures*, ed. Veit Erlmann, 153–172. Oxford: Berg.

DeNora, Tia. 2000. *Music in Everyday Life*. Cambridge: Cambridge University Press.

Douglas, Susan. 2005. *Listening In: Radio and the American Imagination*. Minneapolis: University of Minnesota Press.

Epstein, Jean. (1955) 2012a. The close-up of sound. In *Critical Essays and New Translations*, ed. Sarah Keller and Jason N. Paul, 365–372. Amsterdam: Amsterdam University Press.

Epstein, Jean. (1955) 2012b. The delirium of a machine. In *Critical Essays and New Translations*, ed. Sarah Keller and Jason N. Paul, 372–379. Amsterdam: Amsterdam University Press.

Feaster, Patrick. 2007. The following record. PhD diss., Indiana University, Bloomington.

Feld, Steven. 1994. From schizophonia to schismogenesis. In *Music Grooves*, ed. Charles Keil and Steven Feld, 257–289. Chicago: University of Chicago Press.

Goodman, David. 2010. Distracted listening: On not making sound choices in the 1930s. In *Sound in the Age of Mechanical Reproduction*, ed. David Suisman and Susan Strasser, 15–46. Philadelphia: University of Pennsylvania Press.

Gracyk, Theodore. 1996. *Rhythm and Noise*. Durham, NC: Duke University Press.

Gunning, Tom. 2007. To scan a ghost. *Grey Room* 26 (winter): 94–127.

Howes, David. 2006. Cross-talk between the senses. *Senses and Society* 1 (3): 381–390.

Jarvis, Mat. 2010. Microscopics. http://www.microscopics.co.uk/blog/2010/paulstretch-an -interview-with-paul-nasca/.

Kahn, Douglas. 2013. *Earth Sound Earth Signal: Energies and Earth Magnitude in the Arts*. Berkeley: University of California Press.

Klinger, Barbara. 2006. *Beyond the Multiplex*. Berkeley: University of California Press.

Lacey, Kate. 2013. *Listening Publics*. Cambridge: Polity Press.

Marks, Laura U. 2000. *The Skin of the Film: Intercultural Cinema, Embodiment, and the Senses*. Durham, NC: Duke University Press.

Marsh, Dave. 1993. *Louie Louie*. New York: Hyperion.

McPartlin, Joan. 1956. Symphony of the Birds. *Daily Boston Globe*, October 7, 25.

Millard, Andre. 2005. *America on Record*. Cambridge: Cambridge University Press.

Morton, David L. 2004. *Sound Recording*. Baltimore, MD: The Johns Hopkins University Press.

Pinch, Trevor, and Karin Bijsterveld. 2012. New keys to the world of sound. In *The Oxford Handbook of Sound Studies*, ed. Trevor Pinch and Karin Bijsterveld, 3–35. Oxford: Oxford University Press.

Richards, Chris. 2011. A 44-minute disc that takes 4 months to play? It's a stretch. *Washington Post,* December 31, C01.

Schafer, R. Murray. (1977) 1994. *The Soundscape*. Rochester, VT: Destiny Books.

Sconce, Jeffrey. 2000. *Haunted Media: Electronic Presence from Telegraphy to Television*. Durham, NC: Duke University Press.

Smith, Jacob. 2011. Turn me on, dead media: A backwards look at the re-enchantment of an old medium. *Television and New Media* 12 (6): 531–551.

Sterne, Jonathan. 2003. *The Audible Past*. Durham, NC: Duke University Press.

Thompson, Emily. 1995. Machines, music, and the quest for fidelity: Marketing the Edison Phonograph in America, 1877–1925. *Musical Quarterly* 79 (1): 131–171.

Thompson, Emily. 2002. *The Soundscape of Modernity*. Cambridge, MA: MIT Press.

Turvey, Malcolm. 2008. *Doubting Vision*. Oxford: Oxford University Press.

Sound and Media

DJ Nick Pittsinger. 2010. *U Smile*. Soundfile, white label.

Epstein, Jean. 1947. *Le Tempestaire*. Feature Film, France Illustration.

Fassett, Jim. 1955. *Strange to Your Ears*. LP, Columbia, ML-4938.

Fassett, Jim. 1956. *Symphony of the Birds*. LP, Ficker, C1002.

Kingsmen. 1963. *Louie Louie*. LP, Wand Records, 143.

21   Technological Sensory Training

Michael Bull

I recall walking along a deserted beach in Greece in the 1980s and observing a lone sun worshipper who was looking out to sea while wearing a Walkman. I was struck as to why anybody would want to enhance or change what was, for me, an idyllic and deserted landscape by listening to music through headphones. (Remember, this was a time before the ubiquitous mobile phone had become a part of our everyday life.) Over the subsequent fifteen years, this observation led me to conduct ethnographic research into the everyday use of Walkmans (Bull 2000) and then iPods (Bull 2007). What interested me was the technological mediation of the Walkman into everyday user experience. In the act of closing off the world through the use of the Walkman, the user was creating his or her very own sound world—difficult to study because of the very privatizing of the experience. Yet while the technological intervention was largely an auditory one, the sensory ramifications were likely to be broader—such as the person I observed looking out to sea—so the landscape was also being mediated through sound. William Gibson, the science fiction writer, perfectly encapsulated the transformative effect of the Walkman: "The Sony Walkman has done more to change human perception than any virtual reality gadget. I can't remember any technological experience since that was quite so wonderful as being able to take music and move it through landscapes and architecture" (Gibson 1993, 49).

Listening as a Discriminating and Distinctive Experience

Historically, the ears had been interpreted as both passive and democratic: passive in so much as the ears were open to all sounds; it was this very openness that constituted their democratic nature. The use of Walkmans and then iPods broke this sensory mold. Their use was neither passive nor democratic, but rather discriminating and distinctive. The use of headphones demonstrated that the passivity of the ears was merely an historical effect, now technologically superseded. Technology came to the aid of the ears through the invention of headphones (or, even more intimately, earphones); headphones empowered the self, enabling people to hear what they wanted to hear and thereby screen out the world, creating a private auditory universe through the use of mobile technologies such as the iPod. In doing so, people transformed their relationship to the social world in which they lived, to others around them, and to themselves—their cognitive processes through which they channel all experience. With the rise of mobile technologies, users were able to join together the disparate parts of experience through music—users were getting used to media being ubiquitous (Kassabian 2013). They were also getting used to being able to control both their environments and their cognitive states sonically. The following user quotes come from *Sounding Out the City*:

I hate being on a train or bus without having it if I'm by myself. I get really bored. I just like something to listen to. (Bull 2000, 66)

It's like looking through a one-way mirror. I'm looking at them but they can't see me. (ibid., 76)

I think it creates a kind of aura. Sort of like. Even though it's directly in your ears you feel like it's all around your head because you're coming. Because you're really aware it's just you. Only you can hear it. I'm really aware of my personal space. My own space anyway ... I find it quite weird watching things that you normally associate certain sounds with. Like the sounds of walking up and down stairs or tubes coming in or out. All of those things that you hear—like when you've got a Walkman on you don't hear any of those. You've got your own soundtrack. You see them and it looks like they're moving differently because you've got a rhythm in your head. (Walkman user, ibid., 22)

Media Technologies as "Second Nature"

Users increasingly occupy the world through the habitual, mediated use of media technologies, whereby the use of these technologies becomes "second nature" to them. Walkman and iPod users have therefore gone through a very particular form of sensory training. Sterne has situated the origins of such usage in the cultural training and expectations located in the early communication technologies of the West—the telephone, phonograph, and radio. These technologies problematized our understandings of the relationship between interior and exterior sound worlds. Sterne demonstrated that

audile technique was rooted in a practice of individuation: listeners could own their own acoustic spaces through owning the material component of a technique of producing that auditory space—the "medium" that now stands for a whole set of framed practices. The space of the auditory field became a form of private property, a space for the individual to inhabit alone. (Sterne 2003, 160)

Sterne's is a compelling tale of auditory evolution in which the privatization of sound is located in the ears of the bourgeoisie—an impulse and desire progressively democratized throughout the twentieth century with the mass embracing of radios, telephones, and automobiles. The use of technologies such as the iPod, from this viewpoint, becomes the latest phase of a particular historical trajectory.

Walter Benjamin first coined the phrase "technological sensory training" when he argued that technologies had "subjected the human sensorium to a complex kind of training" (1973, 165). From this perspective, a whole range of cultural, historical, and technological factors mediate the way we see, hear, touch, and taste. In the techno-logically mediated world of the twentieth and twenty-first centuries, this training has involved the simultaneous enhancement and restriction of our sensory experience of the world—enhancing the senses of sight, hearing, and, most recently, touch to the exclusion or diminishment of taste, smell, and embodiment itself.

The Abolition of Distance?

For Crary, the history of media use has produced significant changes to our under-standing of space and our cognitive relation to it, "scrambling the relations between these two poles" (2013, 29, 33, 81). Crary is not the first to point to the dramatic transformative power of media technologies to potentially transform our horizon of experience, our moral compasses, and our experience of place and space—as well as its ability to foreground the visual nature of this transformation. Virilio, for example, pointed to the role of the media in creating a transformed "logistic of perception" (1989, 27), and Crary, in an earlier work, discussed a "restructuring of perceptual experience" (1999, 14).

Martin Heidegger, as in the case of Walter Benjamin, understood the dramatic multisensory impact of media technologies on our understanding of space, arguing that

the frantic abolition of all distances brings no nearness; for nearness does not consist in shortness of distance. What is least remote from us in point of distance, by virtue of its picture on film or its sound on the radio, can remain far from us. What is incalculably far from us in point of distance can be near to us. (Heidegger 1978, 165)

Writing in the late 1940s, Heidegger highlights the multisensory abolition of mediated distance embodied in media use. Later, Peters would note the specific auditory nature of this transformation of the real and virtual, "the succession from the 'singing wire' (telegraph), through the microphone, telephone, and phonograph to radio and allied technologies of sound marks perhaps the most radical of all sensory reorganizations in modernity" (Peters 1999, 160). Benjamin, in parallel to this, had previously discussed the revolutionary impact of the moving image upon sight:

By close ups of the things around us, by focusing on hidden details of familiar objects, by ex-ploring commonplace milieus under the ingenious guidance of the camera, the film, on the one hand, extends our comprehension of the necessities which rule our lives; on the other hand, it manages to assure us of an immense and unexpected field of action ...With the close up, space expands, with slow motion, movement is extended. The enlargement of a snapshot does not simply render more precise what in any case was visible, though unclear: it reveals entirely new structural formations of the subject's impulses. (Benjamin 1973, 230)

Like Heidegger, Peters and Benjamin emphasize the transformed sensory and cognitive relation between proximity, distance, and, importantly, presence. Biocca and Levy celebrate this sensory transformation in relation to virtual reality technologies, stating that "we are building transportation systems for the senses ... the remarkable promise that we can be in another place or space without moving our bodies into that space" (Biocca and Levy 1995, 23). If the twentieth and early twenty-first centuries are described as ones in which a "hallucination of presence" becomes ever more pervasive, then the very relationship between the traditionally held materiality concerning the relationship or distinction between subject and object becomes problematic (Crary 2013). This is what Virilio is referring to when he mourns "the loss of the phenomenological dimension that privileges lived experience" (Virilio 1989, 63).

To War with the Senses

Explanations of sensory training move from notions of a Foucauldian disciplining and monitoring of the self that appears to be embodied in much of the mundane and everyday use of a range of media technologies—from the television to the smartphone—to those notions proposing a general militarization of the self embodied in video game playing and the like. This view echoes Theodor Adorno's earlier critique of the culture industry, namely that "amusement under late capitalism is the prolongation of work" (Horkheimer and Adorno 1979, 137), thus tying the culture industry to the industrial complex of which the military complex formed an integral part.

In 1983, while on a visit to Walt Disney World in Florida to talk to children from the International Youth Initiative, the American president Ronald Reagan had this to say about sensory training and the media:

Many of you already understand better than my generation ever will the possibilities of computers. In some of your homes, the computer is as available as the television set. And I recently learned something quite interesting about video games. Many young people have developed incredible hand, eye, and brain coordination in playing these games. The Air Force believes these kids will be outstanding pilots should they fly our jets. The computerized radar screen in the cockpit is not unlike the computerized video screen. Watch a twelve-year-old take evasive action and score multiple hits while playing Space Invaders, and you will appreciate the skills of tomorrow's pilot. (Reagan 1983)

The militarization of the senses described by Reagan has become a prominent concern in the early years of the twenty-first century. Walter Benjamin's analysis of the sensory transformation of sight, distance, and proximity through film and photography can be extended to video game playing, which likewise involves intense visual training. It has been found, for example, that video game playing helps users to "improve their spatial resolution, meaning their ability to clearly see small, closely packed together objects, such as letters … these games push the human visual system to the limits and the brain adapts to it" (Green and Bevalier 2007, 91).

The synergetic relationship between gaming and warring alluded to by Reagan has become widely recognized. For Friedrich Kittler, "media technologies discipline, mutate, and preempt the affective sensorium. Entertainment itself becomes part of this training" (cited in Goodman 2010, 34). The rise of virtuality, the moral ambiguity of sensory distance, and the militarization of the senses are exemplified by the use of drones. "Drone" is the nickname for an unmanned aerial vehicle (UAV). These planes, equipped with sensors, both color and black and white TV cameras, image intensifiers, infrared imaging for low light conditions, and lasers for targeting, are usually directed from a base camp thousands of miles away from their target. Operatives use the visual data proffered by the cameras on the drone to locate targets and decide whether to attack. Drawing out the comparison with gaming, one operative said, "It's like playing a single game everyday but always sticking on the same level" (Darwent 2013, 52).

Even the virtual terrains in which war game players immerse themselves are not so dissimilar to the real theater of war encountered in the Middle East, for example:

The construction of Arab cities as targets for US military firepower now sustains a large industry of computer gaming and simulation. Video games such as America's Army and the US Marine's equivalent, Full Spectrum Warrior, have been developed by their respective forces, with help from the corporate entertainment industries, as training ads, recruitment aids and powerful public relations exercises. Both games—which were amongst the world's most popular video games franchises in 2005—center overwhelmingly on the military challenges allegedly involved in occupying, and pacifying, Orientalized Arab cities. (Graham 2006, 265)

Such games manifest a deep concern for visual and aural fidelity:

For added realism, footsteps, bullet impacts, particle effects, grenades, and shell casings are accorded texture specific impact noises. A flying shell casing clinks differently on concrete, wood, or metal, for instance, and the distinction is clearly heard in the game. Likewise, footsteps on dirt, mud, wood, concrete, glass, and metal are sounded correctly. (Zyda et al. 2003, quoted in Bogost 2007, 78)

The "real" is thus rendered as a three-dimensional copy on the screens of both the consumers of games and military personnel until the copy becomes virtually indistinguishable from the real for the viewer. There is thus a convergence between the mundane everyday media experience, as experienced by our iPod users and, now, our smartphone users, and the exceptional as evidenced by drone technology reflected in video game use. As De Souza e Silva and Frith note in relation to the use of global positioning systems (GPS) on smartphones: "With an increasing popularization of location aware technologies, the relationship between mobile technologies and places requires a new perspective because these technologies strongly influence how people interact with their surrounding space and how they understand location" (2012, 165). The usage of these technologies continues to form an important part of the sensory reorganization of spaces by users (Goggin 2011) whereby users monitor themselves and distant others in a Foucauldian network of permanent observation.

References

Benjamin, Walter. 1973. *Illuminations*. Trans. Harry Zohn. London: Penguin.

Biocca, Frank, and Mark Levy, eds. 1995. *Communication in the Age of Virtual Reality*. Hillsdale, NJ: Erlbaum.

Bogost, Ian. 2007. *Persuasive Games: The Expressive Power of Videogames*. Cambridge, MA: MIT Press.

Bull, Michael. 2000. *Sounding Out the City: Personal Stereos and the Management of Everyday Life*. Oxford: Berg.

Bull, Michael. 2007. *Sound Moves: iPod Culture and Urban Experience*. London: Routledge.

Crary, Jonathan. 1999. *Suspension of Perception: Attention, Spectacle, and Modern Culture*. Cambridge, MA: MIT Press.

Crary, Jonathan. 2013. *24/7: Late Capitalism and the End of Sleep*. London: Verso.

Darwent, Charles. 2013. How truth and fiction become blurred. *Independent on Sunday*, August 25, 52.

De Souza e Silva, Adriana, and Jordan Frith. 2012. *Mobile Interfaces in Public Spaces: Locational Privacy, Control, and Urban Sociability*. New York: Routledge.

Gibson, William. 1993. Time out. *October* 6:49.

Goggin, Gerard. 2011. *Global Mobile Media*. London: Routledge.

Goodman, Steve. 2010. *Sonic Warfare: Sound, Affect, and the Ecology of Fear*. Cambridge, MA: MIT Press.

Graham, Stephen. 2006. Cities and the "War on Terror." *International Journal of Urban and Regional Research* 30 (2): 255–276.

Green, C. Shawn, and Daphne Bevalier. 2007. Action-video game experience alters the spatial resolution of vision. *Psychological Science* 18 (1): 88–94.

Heidegger, Martin. 1978. *Basic Writings*. Trans. David Farrell Krell. London: Routledge.

Horkheimer, Max, and Theodor W. Adorno. 1979. *Dialectic of Enlightenment: Philosophical Fragments*. Trans. John Cumming. London: Verso.

Kassabian, Anahid. 2013. *Ubiquitous Listening. Affect, Attention, and Distributed Subjectivity*. Berkeley: University of California Press.

Kittler, Friedrich. (1986) 1999. *Gramophone, Film, Typewriter*. Trans. Geoffrey Winthrop-Young and Michael Wutz. Stanford: Stanford University Press.

Peters, John D. 1999. *Speaking to the Air: A History of Communication*. Chicago: Chicago University Press.

Reagan, Ronald. 1983. Remarks during a visit to Walt Disney World's EPCOT Center near Orlando, Florida, March 8. www.presidency.ucsb.edu/ws/?pid=41022.

Sterne, Jonathan. 2003. *The Audible Past: Cultural Origins of Sound Reproduction*. Durham, NC: Duke University Press.

Virilio, Paolo. 1989. *War and Cinema: The Logistics of Perception*. Trans. Patrick Camiller. London: Verso.

Zyda, Michael, Alex Mayberry, E. Casey Wardysski, Mike Capps, Brian Russell Shilling, Martin Robaszewski, and Margaret Davis. 2003. Entertainment R & D for Defence. IEEE Comp Graphics and Applications, January–February 2003, 28–36.

Concepts of Listening

22   Listening as Gesture and Movement

Paul Théberge

Introduction: The Look of Sound

I remember watching the Who perform, many years ago, in the documentary film *Woodstock* (Michael Wadleigh, 1970): onstage, the Who were renowned for putting on a dynamic show, and while vocalist Roger Daltrey may have been in the spotlight much of the time, it was guitarist Pete Townshend, swinging his arm in broad circular movements and leaping into the air, who truly commanded the stage. In an era before the use of onstage, large-format video screens, Townshend's exaggerated movements were, no doubt, necessary to capture the attention of the audience at a distance. But there was also something compelling about the wild, windmill attack on the strings of his guitar: it was as if Townshend had found a gesture that was in some way commensurate with the massive sounds that emanated from his amplifier and loudspeakers. Certainly, given the electrical power behind those speakers, a flick of the wrist would have been sufficient to create a big sound. In music performance, however, sounds need to appear to be produced by gestures: Townshend understood this, and in reaching for a gesture appropriate to the volume of sound, he gave visual impetus to the sounds of his amplified guitar (and no doubt inspired a generation of fans to imitate him, on air guitar, when they listened to his recordings).

In another movie theater, years later, I recall watching *The Bourne Supremacy* (dir. Peter Greengrass, 2004). In one sequence, Jason Bourne had just taken a number of body punches in a fight with an adversary, and I too was feeling a bit pummeled by the low-frequency sounds punching out from the theater's subwoofers; later, Bourne placed a clip into a semiautomatic pistol, and although I do not have any direct experience with this kind of weapon, the gesture and the sound effect were striking—I had a strange impression that, somehow, I knew what this might feel like. I wondered how it was that within the body there exists some resource through which this kind of resonance can be invoked, and how it is that a sound *effect* can become a sound *affect*.

These two very different experiences suggest that when we listen to sounds we do not use only our ears: gestures and bodily movements, the visual and the tactile, also play a role in our perception of sound. That these experiences were mediated through the technologies of sound and image reproduction is of no small consequence: a large part of the postproduction process in cinema is devoted to putting sounds together with images, taking sounds that have been separated from the maker of the sound (Schafer 1977) and putting them into new relationships with objects, events, actions, movements, and gestures. Moreover, the uncommon intensity with which media are able to represent visual and sonic events in the "real" world allows them to connect with our everyday experiences in both subtle and powerful ways.

In this essay, I want to explore some of these relationships between listening, gesture, movement, and technology, and to do so I will draw on literature in the cognitive sciences. What interests me here, however, is not the science of the brain per se, but the character of listening in an environment saturated by mediated sounds and images. So, in conjunction with this literature, I will also explore a small number of examples in popular music, film, and games. While these media are, in many ways, quite different from one another, there may in fact be similarities in how they produce affective responses in individuals.

Gesture and Popular Music

Pete Townshend's windmill guitar strumming reminds us that—even in an age dominated by studio recording and electronic reproduction—many genres of music have retained a close connection with the art of performance. And this is as true today as it was in the late 1960s: although fans probably *listen* to more music via headphones and loudspeakers than ever before, they still flock to live performances (or watch YouTube videos) to *see* their favorite bands. Of course, "live" performance is also a highly mediated experience (Auslander 1999), and there are genres of popular music, such as electronic dance music (EDM), that do not rely heavily on conventional modes of performance. In the latter case, DJs play a central role in producing the effect of a performance and, more importantly, dancers embody the music through their own movements. Taken together, these contexts seem to raise the question of whether there is a fundamental connection between our experience of musical sounds, gestures, and movement, a connection that escapes sound reproduction only to be filled in by various media or by the listener? Is there more to music than meets the ear?

Over the past decade, researchers in music and cognitive studies have developed a renewed interest in musical gesture: the turn to music has largely been regarded as an extension of previous research in cognitive studies dealing with movement, embodied cognition, gesture, and speech. To some extent, this renewed interest has also been facilitated by the development of new technologies of motion capture and analysis. Many of these studies have used classical music, musicians, and audiences as research subjects, but their findings are perhaps equally applicable to popular music. For example, a number of researchers (e.g., Vines et al. 2011) have studied expressive movements employed by musicians and how these might have an impact on a listener's perception of emotion in music. Their findings suggest that seeing musicians perform enhances one's perception of the emotional nuances in music in a way that simply listening to a recording does not—our experience of music is thus intensified through multimodal forms of perception. Such work is highly suggestive of the potential response of audiences to a range of gestures employed by musicians (especially rock musicians such as Townshend) and the importance of these gestures in the experience of musical sound.

Other researchers have been more emphatic in their claim for the connections between music and the body: "We conceive of musical gestures as an expression of a profound engagement with music, and as an expression of a fundamental connection that exists between music and movement" (Godøy and Leman 2010, 4). For these researchers, musical gesture is understood as including not only the range of movements made by performers in producing sounds and in conveying emotion to an audience, but also movements made as part of the act of listening: tapping one's foot to the beat, dancing, or playing air guitar (ibid., 5). Both music production and reception are thus understood within an "action-based" frame of reference (Leman 2007).

Despite this evidence, it could still be argued that our experience of listening to disembodied sounds via audio technologies, or via the playful recombination of sounds and images in music videos, has weakened our sense of the fundamental connections between sound and gesture. However, there is also evidence—even within popular culture itself—that suggests otherwise. In his analysis of bass drum sounds in EDM, Hans T. Zeiner-Henriksen (2010) has argued that dancers do not only move in time to the beat of the music (through a simple process of "entrainment"), but, rather, their movements are more complex, often responding to strong melodic movements in the bass or to other dynamic, timbral, and pitch-related interplays that suggest corresponding bodily gestures. In these cases, the issue is not so much a direct, causal relationship between gesture and the production of sound (the sounds are produced electronically) but the correspondences that can be drawn between movements in musical sound and the gestures dancers employ in response to them.

To return to the example of the electric guitar, the popularity of digital games such as *Guitar Hero* and *Rock Band* suggests that the imitative pleasures of playing air guitar are perhaps more central to rock fandom than previously thought. Games such as *Guitar Hero* demand a more precise knowledge and a more rigorous form of interaction with rock music than does the simple imitation of the large-scale gestures produced by Pete Townshend. Indeed, to follow along in time with the guitar solos central to gameplay, gamers must be able to anticipate and reproduce, in precise detail, a series of movements that is nearly as intricate as the production of the solo on a real guitar. In her ethnography of music-game culture, Kiri Miller (2009) argues against critics who dismiss video games as a mere substitute for "real" musical practice: Miller understands games such as *Guitar Hero* as confounding our notions of the "virtual" and the "real" and suggests that they encourage both a conceptual and an affective engagement with music that is significant in its own right. Most important for my argument here, Miller compares playing these games to dancing based on the way in which both facilitate a form of "heightened listening"—a form of listening that is generated through gesture, immersive engagement, and bodily movement (ibid., 410–411).

Bodily Schema and Foley as Sonic Gesture and Movement

The idea that listening and bodily movement are linked is not unique to music, and I would argue that there are ways in which we experience forms of heightened listening through gesture and movement in our everyday lives (although these experiences are perhaps as subtle as they are ubiquitous, and thus less self-conscious than experiences resulting from our engagement with the production or reception of music). For example, our ability to meaningfully interpret the difference between hearing the sound of relaxed breathing and the more rhythmic, labored breathing of a runner is, in part, the result of our own embodied experiences. More important, embodied experiences of this kind become one of the bases upon which we create aesthetic representations (in speech, music, and visual art) and derive affective meaning from them. Drawing on work in phenomenology, cognitive science, and pragmatist philosophy, Mark Johnson (2007) has argued for an embodied sense of knowledge, where movement and spatial experience form what he calls "image schema" in the mind. For Johnson, "An image schema is a dynamic, recurring pattern of organism-environment interactions" (2007, 136). The most fundamental schemas are those associated with our perception of space, movement and balance, force and intensity, and boundaries and containment, among others. Furthermore, these schemas are not limited to any one sensory form of perception: from an early age, our sensorimotor experiences are essentially cross-modal and affective in character (ibid., 137–144).

Although Johnson is interested in how image schema relate to fundamental issues of cognition, mind–body relations, and aesthetic form, my concern here is in how schemas give rise to our sense of meaning through bodily movement and how this might relate to our affective engagement with sound in media. We feel this clearly, I think, in films like James Cameron's *Avatar* (2009): in a series of opening sequences, the main character, Jake, is confined to a wheelchair where we see him (in intimate close-ups) gripping the wheels of the chair, pushing himself forward, and later lifting the weight of his body out of the chair and into the avatar transfer pod. In sharp contrast to these images of bodily confinement, when Jake first emerges from the transfer pod in the form of the avatar, we see (and hear) him stumble about, gain his balance, and then run outside with the sheer joy of unrestrained physical movement. The key point in these sequences is not that we, as audience members, simply "see" actions on the screen or "identify" with Jake's character, but that we intuitively *feel* the "affective contour" of his actions (see Johnson 2007, 144). In this way, the film draws heavily on image schemas, resonating with our own sensorimotor experiences to produce meaning and affective engagement—a form of engagement that is perhaps essential for this film because the audience must be made to *feel* that the transformation of Jake's "real" body into that of the avatar is genuine and credible.

Sonically, nowhere is this play of cinematic gesture, bodily movement, and schema more evident than in Foley—the art of producing human gestural sounds for film and other media. Unlike most sound effects produced, edited, and synchronized in film postproduction, Foley sounds are unique in that they are essentially performed in real time—synchronized with character movements as they appear on screen. Foley sounds are thus closely linked to movements as they occur in the temporal domain, and Foley artists develop a finely tuned sensitivity to the rhythm and dynamics of motion (the sound and gait of a character's footsteps, the rustle of their clothing, or the sound of their hands grasping or manipulating objects). Even though the sound objects used in Foley studios often bear little resemblance to the bodies or objects we see onscreen—no one gets punched in Foley sessions but lots of other objects do—the sounds are nevertheless capable of evoking image schema within us because they are produced by gestures and movements that are analogous to those that we see. It is the dynamic contours of the movement, and the uncommon precision and detail with which the sounds are picked up through techniques of close miking, that allow the studio-produced sounds to both stand in for the "real" ones (a kind of semiotic substitution) and to find resonance with the schema of audience experience. The semiotic, cognitive, and psychological role played by Foley and image schema in film aesthetics is brilliantly depicted in Peter Strickland's film *Berberian Sound Studio* (2012). The main character, Gilderoy, is a meticulous British sound editor hired by a producer of Italian horror films to work on his latest project. As the postproduction process unfolds, Gilderoy begins to show signs of becoming psychologically involved with— and increasingly unhinged by—his work. In particular, being required to supply Foley sounds for violent torture scenes appears to affect him deeply: even though he only rips the stems off radishes and stabs cabbages in the studio, we see the psychological effect of these gestures on his face as he watches the horrific scenes onscreen (interestingly, the audience never sees what Gilderoy sees, we only witness his reactions). In this way, *Berberian Sound Studio* manages to reveal the artifice of postproduction sound while it simultaneously explores the visceral effect that sounds can have on us by virtue of their links to schema derived from our experiences of gesture, physical force, and movement.

Conclusion

Sound studies scholars have, of necessity perhaps, attempted to separate sound experience from the dominant narratives of visual culture. In validating this separation of the senses, however, we may have downplayed some of the essential, multimodal links between our experiences of sound and image, gesture and movement. Indeed, R. Murray Schafer's (1977) notion of "schizophonia," which posits the separation of sound from the maker of sound in the phonographic era, appears to apply to the listener as well: nowhere in Schafer's work do we have a fully articulated notion of an embodied, listening subject.

But as I have attempted to demonstrate here, the aesthetic reception of both music and the everyday sounds represented in dramatic media such as cinema cannot be fully appreciated unless we understand them as linked to processes of gesture, movement, and embodied experience. The fact that contemporary modes of reception take place within the logics of sound and image reproduction does not negate embodied experience—it reinforces it: the detailed, highly amplified sounds of live concerts and cinema exhibition turn sound into a form of tactile experience (see Henriques 2003). Understood in this way, listening is not simply a function of the ear but an activity that requires a subject who is affectively engaged in an experience that is at once sonic, visual, and tactile—an experience that is articulated to the body through sound, gesture, and movement.

References

Bibliography

Auslander, Philip. 1999. *Liveness: Performance in a Mediated Culture*. New York: Routledge.

Godøy, Rolf Inge, and Marc Leman, eds. 2010. *Musical Gestures: Sound, Movement, and Meaning*. New York: Routledge.

Henriques, Julian. 2003. Sonic dominance and the reggae sound system session. In *Auditory Culture Reader*, ed. Michael Bull, and Les Back, 451–480. Oxford: Berg.

Johnson, Mark. 2007. *The Meaning of the Body: Aesthetics of Human Understanding*. Chicago: University of Chicago Press.

Leman, Marc. 2007. *Embodied Music Cognition and Mediation Technology*. Cambridge, MA: MIT Press.

Miller, Kiri. 2009. Schizophonic performance: Guitar Hero, Rock Band, and virtual virtuosity. *Journal of the Society for American Music* 3 (4): 395–429.

Schafer, R. Murray. 1977. *The Tuning of the World*. New York: Knopf.

Vines, Bradly W., Carol L. Krumhansl, Marcelo M. Wanderley, Ioana M. Dalca, and Daniel J. Levitin. 2011. Music to my eyes: Cross-modal interactions in the perception of emotions in musical performance. *Cognition* 118:157–170.

Zeiner-Henriksen, Hans T. 2010. Moved by the groove: Bass drum sounds and body movements in electronic dance music. In *Musical Rhythm in the Age of Digital Reproduction*, ed. Anne Danielsen, 121–140. Burlington, VT: Ashgate.

Sound and Media

Avatar. 2009. Director: James Cameron. Feature Film, Twentieth Century Fox.

Berberian Sound Studio. 2012. Director: Peter Strickland. Feature Film, MPI Home Video.

The Bourne Supremacy. 2004. Director: Paul Greengrass. Feature Film, Universal Studios.

Woodstock. 1970. Director: Michael Wadleigh. Documentary, Warner Home Video.

23   Concepts of Fidelity

Franco Fabbri

A Very Short History

The term *high fidelity* was invented "by Harold A. Hartley, an engineer working at English Electric in about 1926, as a description of the improved radios and electric gramophones" (Dearling and Dearling 1984, 86). According to the same source, "the first widespread use of the words 'high fidelity' occurred around 1934, used by radio stations in America, in particular WQXR in New York" (ibid.). Based on industrial advances that took place in the 1930s and 1940s, the so-called "golden age of high fidelity" spans from the 1950s to the mid-1960s; audiophiles define the late 1960s (when transistors were substituted for tubes in hi-fi amplifiers) as a "near-death experience for high fidelity," followed by "the rise from the ashes" (1970s), then by a decade when "the fashion-magazine gatekeepers take over" (1980s), and then by another decade of innovation, the 1990s (Olson 2005). At the end of the 1990s, the hi-fi concept was definitely still popular enough to be used as a reference for the inventors of Wi-Fi, a concept now familiar to all users of laptops, tablets, smartphones, and so forth, while hi-fi seems to be relegated to an elite of old-fashioned audiophiles.

Fidelity to the Concert Hall

Although the high fidelity concept emerged in the age of electric recording and has since been associated with technological advances and a growing obsession for measures (such as frequency range, signal-to-noise ratio, third harmonic distortion, and the like), fidelity was already a key concept in the mechanical age, before any measuring technology for audio signals was introduced. Edison's tone tests were aimed at convincing prospective buyers that they would not be able to distinguish the sound of a recording from a live source (a singer or an instrumentalist performing the same material at the same time): both the record player and the performers were hidden behind curtains, and attendees were asked to tell if the live source was active or not.

The trick was that the record player was never stopped, so the noise produced by the scratching of the needle on the record's surface was always present. It can be said that the test itself was based on the acceptance of an audile technique (Sterne 2003, 98–99) consisting in the "unconscious" cancellation of the system's inherent noise—a technique that any listener to phonograms such as shellac or vinyl records or cassettes remembers.

For decades, high fidelity has been primarily associated with the idea of recreating the sound of a concert hall in the listener's living room. The coincidence of the golden age of high fidelity with the advent of stereophonic sound (first with open-reel two-track tape recorders in 1955–1956 and with Westrex stereophonic records after 1957) is proof of the importance of the listener's sensation (even on the basis of an imperfect, illusory technology) in comparison with any technical measurement (Perlman 2004, 789). Early demo records offered live recordings, often commented on by a radio-styled presenter, with the aim of directing the listener's attention to the spatial "realism" of the recorded sound.

The stereophonic effect is properly created if the listener is located at the vertex of an equilateral triangle, with the system's two loudspeakers located at the two other vertices. The bulkiness and high cost of early stereophonic systems were congruent with the living room of the so-called average listener; hi-fi was for upper-middle-class families, and the recorded repertoire was geared to their tastes: classical music, jazz, crooners, refined orchestral "light music"—no rock 'n' roll. It may just have been chance, but the decline of rock 'n' roll in the late 1950s in the United States coincides precisely with the rise of stereophonic records. Pop and rock records, at that time, were aimed at different listening contexts: jukeboxes, FM radios, small monophonic record players used by teenagers in dorms or in their own rooms, where collective listening sessions with peers often took place.

Fidelity to the Recorded Master

It is known that the Beatles recorded on two tracks until the second half of 1963, and their stereo releases (especially made for the U.S. market) simply presented the vocal track on one channel and the basic instrumental track on the other; but it was not just because of such technical limits that producer George Martin was not in favor of stereo mixes: he thought that the band's audience was not interested in stereo, and most probably would not own a stereophonic player. Multitracking was more common in the United States, but most pop records were mixed down to mono anyway. A notable exception is represented by the "techno" bands of the early 1960s, which were later dubbed "surf bands," such as the Ventures in the United States and the Shadows in Britain: in particular, US bands were supported by middle-class students living on the West Coast, who loved the technological sound and look of electric guitars and

Figure 23.1
Albums by instrumental bands (later dubbed "surf bands") were among the first records aimed at the youth market to be released in stereo.

amplifiers (Hoskyns 2003, 58). So, starting in 1960–1961, the albums of the Ventures and the Shadows were mixed in stereo.

The advent of transistors in professional and hi-fi audio equipment design brought about many changes: although it worsened the quality of amplifiers according to audiophiles (Olson 2005), it also made hi-fi equipment cheaper and more compact, making car stereos possible. It was also at the foundation of rapid improvements in professional audio equipment, allowing the development of multitrack recorders and multichannel mixing desks, which in turn enabled producers and musicians to use the recording studio as a compositional tool. Only seven years separate "I Want to Hold Your Hand" (the first Beatles' recording on four track) from Pink Floyd's *Atom Heart Mother*, or from Miles Davis's *Bitches Brew*, two out of a multitude of albums where the idea of "a concert hall in your living room" cannot be applied, not only because they are based on studio recordings, but mainly because—due to overdubbing and editing—an actual live performance of the material presented in those records never took place anywhere.

The experience of a stereophonic recording became more and more widespread; in 1967, stereo headphones were made available to the general public for the first time. Only one year earlier, eight-track stereo had been made available for car usage, and by the end of the decade, car radios mounted stereo cassette players (Philips). Although the loudspeaker spatial configuration in a car is far from the ideal triangle, and the aerodynamic noises, along with the engine, make the overall signal-to-noise ratio problematic for hi-fi listening, the combination of proximity and loudness nonetheless makes the car listening experience similar to that allowed by headphones. The car became one of the main places for the individual and collective listening of recorded music, second only to listening at home or in college settings.

Recording studios soon adopted alternative speaker systems in order to allow engineers, producers, and musicians to listen to the actual end result of their work: near field monitors, simulating the typical sound of domestic stereos and shelf loudspeakers, and car stereo speakers. Far from trying to recreate the conditions of a live event in a concert hall, musicians, producers, and engineers listened to their recordings to check if the result corresponded to their intentions, their artistic and/or commercial ambitions, or simply to their liking.

The recording was not a medium: it was an artifact. It has been so forever, in spite of any ideology of transparency. But the mass diffusion of hi-fi in the late 1960s and 1970s introduced a new concept: that of the fidelity *to the recorded master*. And the recorded master became the paradigm, even for live performances. During the same period, public address (PA) systems evolved from the primitive configurations of early Beatles' performances to the spaceship-looking PAs (sci-fi as well as hi-fi) of Pink Floyd in 1969–1970. By the early 1970s, lying down silently on the grounds of sports arenas or at open-air festivals, thousands of young people attended performances by groups such as Pink Floyd, Genesis, or others—performances that were often based on the enlarged dynamic range and stereophonic effects made available by big hi-fi PA systems—performances that were aimed at recreating, in those real places, the imaginary soundscapes originally conceived for headphones, students' rooms, or cars. It was the original concept of high fidelity, just reversed.

Diverging Concepts: "Walking" Away from Audiophilia

As international popular music in the 1970s exploded into a multitude of competing genres, scenes, and functions, a similar diversification took place in playback apparatuses and listening modes: cassette decks (which entered the hi-fi market thanks to the adoption of Dolby noise reduction), ghetto blasters, sound systems, and discotheque PAs. In a burst of confidence based on the phenomenal rise in stereo equipment sales, the audio industry ventured into one of the most catastrophic projects in its history: quadrophony. Competing standards, high prices, and unsatisfactory technical results affected the products, and it became clear that many users of stereo equipment did not place loudspeakers correctly, never listened to recordings sitting in the canonical vertex, and used music in their homes for background listening or as a form of decoration.

In the golden age of stereo—and even later for audio enthusiasts—"do-it-yourself" was a way to aim for perfection in sound reproduction; at the end of the 1970s and in the early 1980s, do-it-yourself became a way to contradict and manipulate industrial standards and to oppose the ideology of high fidelity itself. During that same time period, Sony launched the Walkman.

To some extent, the Walkman was an answer to the listeners' desire for freedom from being obliged to sit in front of a couple of loudspeakers (or, with quadrophony, at the very center of a square) or to stay close to the stereo system with headphones plugged into it. The name of the product, and the images related to it, suggest the idea of walking and seem to imply an outdoor environment, but it could simply mean that one could "walk away" from the stereo and from the awkward postures that stereo listening sometimes implied.

The Walkman also had the effect of obliging users to listen through headphones (earlier) or earphones (later): it made headphone listening a mass phenomenon and favored private, individual listening instead of collective listening. And, in an age of technological advances, it was an immensely successful product based on the downgrading of existing technical possibilities: a cassette player that could not make recordings without connectors to plug it into a stereo preamplifier, a small, pleasurable object whose design was more important than its audio specifications. Was it hi-fi?

It is impossible to answer that question only in terms of measurements. One could say that the copy of a newly bought album on a chrome dioxide cassette, made with a "good" cassette deck from a "good" record player, played on a "good" Walkman and listened to through a "good" pair of headphones, could offer a better listening experience than that of a commercial cassette recording of the same material listened to on an average hi-fi system—especially if one considers that the former could be listened to anywhere, including in environmental conditions masking possible defects (such as tape hiss) in the recording. It goes without saying that mobility was the key to the Walkman's success, but it can be argued that the Walkman's place in media history is also owing to its—so to speak—relativistic attitude toward fidelity. Up to that point (and with the possible exception of other mobile systems such as car stereos), fidelity and hi-fi had been based on absolute values: the sound in a concert hall, the closest similarity of recorded sound to live voices or instruments, technical measurements approaching ideal specifications, the standardized geometry and acoustics of the listening space. The Walkman was the first step in another direction—that of the best fidelity for a specific listening function, mode, and/or environment.

Absolutist and Relativistic Approaches to Fidelity

When Sony released the Walkman, their engineers were already working on the CD, which benefited from the disaster of quadrophony, prompting Sony and Philips to agree on a common standard rather than competing with different systems. But the CD project was based on the old, absolutist ideology of hi-fi: it was, in many respects, that ideology's final accomplishment. It must be noted that no effective mobile CD player was produced until almost a decade after the introduction of the product in 1982–1983. Even at the beginning of the 1990s, CDs could not be used in car stereos if the road's surface was not completely smooth, and portable CD players were bulky, heavy, and suffered from any occasional bump—you could not really walk with an early CD Walkman or Discman. The problem was only solved in 1993, when a buffer memory was added to mobile CD players, allowing them to compensate for any interruptions caused by the laser losing its place. Record industry commentators seem to forget that, up until the mid-1990s, consumers (even those who did not own a CD player, but who had friends who owned one) listened to cassettes that were recorded copies of commercial CDs. They were listened to on Walkmans, personal stereos, and car stereos—and those cassettes were of higher audio quality than commercial cassettes, of course, but also than commercial vinyl records. Although the signal-to-noise ratio of a Dolby cassette is inferior to that of a CD, and the crosstalk is larger, those parameters are better than those of vinyl records, which, until just a few years earlier, had been considered the pinnacle of high fidelity. The culture of home taping (originated by the boom of Dolby tape decks in the 1970s) was largely stimulated by the lack of adequate mobile listening systems for CDs. Nonetheless, the record industry (faithful to the old high fidelity standards) was more worried by the appearance of the digital audio tape (DAT) recorder, fearing that consumers could make bit-by-bit copies at 44.1 kHz of the industry's precious digital masters. At a time when recording studios were investing millions of dollars in purchasing multitrack digital recorders and pursuing the perfect sound advertised by the record industry, CBS proposed the implementation of the so-called copycode, which was based on a circuit (to be included in all DAT recorders) that would stop the recording function when it detected a "hole" in the frequency range of commercial CDs. Studio owners were supposed to purchase expensive new digital equipment and then erase an entire (albeit thin) frequency range from recordings in order to make them copy-protected. When Sony bought CBS, putting the copycode project to an end, those studio owners applauded.

Compression, Listening Environments, and Just-Right-Fi

Alternative mobile systems were launched even before Discman personal stereos became more functional after the buffer memory was added. The Digital Compact Cassette (DCC), launched by Philips in 1992, was a complete failure. Sony's MiniDisc, also launched in 1992, enjoyed moderate success, especially as a portable recorder. Both were based on compressed audio formats—precision adaptive subband coding (MPEG-1 Audio Layer I) and Adaptive Transform Acoustic Coding (ATRAC), respectively. Although audiophiles scorned them for not conforming to the highest audio standards, they sounded better than analog cassettes and vinyl records for the average

Figure 23.2
An ad trying to suggest to customers that the Digital Compact Cassette would become as successful as the cassette or CD. It did not!

user. For a niche of consumers, the MiniDisc was an introduction to compressed digital formats, and discussions about any audible difference between noncompressed and compressed digital audio started then, about three years before the MP3 format's final specifications were issued.

The advent of downloading and MP3 players was seen by hi-fi advocates as the final demise of the concept. And it is true that hi-fi per se is not popular anymore: there is still a niche of audiophiles who buy extremely expensive "high end" or esoteric audio systems, but the hi-fi tag is not even used for selling audio equipment in shopping malls or department stores. Moreover, equipment that used to be considered hi-fi in the 1970s (such as Dolby cassette decks or record players) is now very difficult to find. Many music listeners behave as if they were not at all interested in the audio quality of the music they are listening to: couples that share one earphone each, for example, or kids playing ringtones at high volume with their mobile phones.

It is also true that record producers in the past years have overcharged new and rereleased masters with extreme compression, in order to make them sound louder in noisy environments when listened to with poor earphones (the so-called loudness war). However, when I hear someone say that we live in an age of low fidelity, I consider that to be an unfair judgment. Both owing to the sound aesthetic of late 1970s punk and experimental music, and through Murray Schafer's writings on modern soundscapes (1977), the term received a considerable boost. Urban soundscapes are lo-fi (though they seem to be less and less so, at least in some parts of the world), and high-end audio equipment such as PCs and Macs with sophisticated sound processing software can be used to produce music with a lo-fi feel to it. However, in general, it can be said that now more people than ever listen to an unprecedented quantity of music with an average quality that largely surpasses that of preceding decades. Listeners seem to be adapting to a "just-right-fi" concept: a degree of audio quality that works with the environmental conditions, modes of listening, and music functions. In addition, it must be noted that lossy compressed audio formats such as MP3 are based on the same effect and principle—masking—that lets a system's degree of audio quality depend on the sonic conditions of the environment. If the iPod can be seen as a radical development of the Walkman concept, then just-right-fi is the development of the relativistic attitude toward fidelity established by the Walkman's designers, under popular demand.

References

Dearling, Robert J., and Celia Dearling. 1984. *The Guinness Book of Recorded Sound*. London: Guinness Superlatives.

Hoskyns, Barney. 2003. *Waiting for the Sun: Strange Days, Weird Scenes, and the Sound of Los Angeles*. London: Bloomsbury.

Olson, Lynn. 1996. *The Soul of Sound*. Oregon: Nutshell High Fidelity.

Olson, Lynn. 2005. *A Tiny History of High Fidelity*. http://www.nutshellhifi.com/library/tinyhistory1.html (adapted from Olson 1996).

Perlman, Marc. 2004. Golden ears and meter readers: The contest for epistemic authority in audiophilia. *Social Studies of Science* 34 (5): 783–807.

Schafer, R. Murray. 1977. *The Tuning of the World*. Bancroft: Arcana Editions.

Sterne, Jonathan. 2003. *The Audible Past: Cultural Origins of Sound Reproduction*. Durham, NC: Duke University Press.

24   Loudness Cultures: Practices, Conflicts, Discourses

Susanne Binas-Preisendörfer

During the 2010 World Cup, the tones emitted by the vuvuzelas created the game's soundscape—and also motivated the curious, the irritated, and the critics to spring into action concerning the instruments' volume and omnipresence. While the critics brought up the possibility of hearing damage in the debate, others pointed out that, with their usage of the vuvuzelas, the South African audience was reclaiming its status as an actor in the game (Bonz 2010). It was anticipated that I would tend toward the cultural studies position of "the Other" rather than that of those highlighting a possible health hazard. The often relentless debates around loudness are affected by various discourses. Before those are delineated, however, a few definitional references toward a differentiation of volume and loudness will be explained, as will their dependency.

Loudness, as the volume perceived and/or felt by the listener, is comprehended differently from the physical dimensions of an auditory event—that is, its acoustic pressure, acoustic frequency, or length. In other words, it is not possible to understand loudness based on any physical methods of measurement. Loudness as the psychophysical "dimension" is not unambiguously definable. In audio technology, one starts from the premise that the energy contained in one signal is determinative for the perceived volume of an acoustically emitted signal. The conversion of the physical dimension of measure (acoustic intensity) into the magnitude of psychophysical sensation represents a fundamental question for psychophysics. The loudness term commonly used today was established within the framework of psychophysics (with a loudness unit called *sone*). Above all, traditional models of loudness assume that the perception of loudness is dependent on spectral and time-dependent events. High volume levels are analyzed particularly in noise effect research and in occupational medicine—in the case of professional musicians, damaging effects on human hearing perception have been confirmed by such studies (see Anonymous 2014), but the effect of loud music as so-called self-controlled noise doses by young adults has either been overestimated or could not be explicitly substantiated (Hellbrück 2008).

Legitimate Listening Practices, or The Loudness War Has a Long History

When it comes to popular music, the discussions on loudness are dominated by the so-called loudness war: a war or a battle in the music industry markets for the listener's attention. Today, the loudness war refers to the technological trend in sound studios and radio stations of repeatedly raising the average level on the storage medium without exceeding the technologically determined maximal level. In so doing, producers would like for their recordings to be perceived as the loudest on the radio or on sound carriers. When compared, listeners perceive the louder signal as "better" or value it as the most appealing one. However, since the maximal amplitude is technologically limited, the average level is raised through compression. This compression leads to changes in the sound—a fact that is often negatively valued, especially by audiophile listeners.

This makes it seem as though the loudness war were the result of the digital technology introduced broadly in sound studios and on the radio in the mid-1980s. In effect, digital technology does make it possible to raise the medial sound level easily (on the mixer console). Against this background, the loudness war is a primary topic for studio producers and engineers. It extends far into the public discussion and has led to recommendations from the European Broadcasting Union on "loudness level control, normalization, and reliable maximum level audio signals" (EBU 2014).

Based on specific source studies (Devine 2013), the British pop music and media researcher Kyle Devine demonstrated that the so-called loudness war was in no way a debate that first began in the 1980s: "In fact, loudness has been a source of pleasure, a target of criticism, and an engine of technological change since the very earliest days of commercial sound reproduction" (ibid., 159). According to Devine, loudness was initially an issue for the recording, as well as for the playing, of sounds or music that had to generate a minimum amount of sound pressure in order to mechanically "imprint" itself in wax, tin, or, later, shellac. Although, based on today's standards, the reproductions of sounds at the turn of the twentieth century could in no way approach the original sound at the time of its recording, the most faithful rendering of natural sounds or the most faithful reproduction of an artistic work became the leading category for evaluating technologically reproduced music. Up until today, that still has its discursive application in the evaluation of fidelity or high fidelity, respectively.

As maintained by Devine, the contradiction between listening practices and sound production ideals is responsible for the different positions on the topic of loudness (ibid., 160). Both positions or practices are guided by different interests. Practices of listening can be so variegated, as is how one interacts with music and/or popular music (listening casually, absent-mindedly, attentively, multisensually, dancingly, etc.). Behind the practices of different "listenings" and the ideals of sound reproduction and production are different and contradictory concepts of authenticity, respectively. Whether or not listening experiences are seen as legitimate is decided by the discourse in which they are situated.

In his contribution to the long loudness war history, Devine points to discourse motifs that are strongly reminiscent of cultural-critical argumentations that are still effective today: "Complaints about the general decline of morality and society, about alienation and rationalization, about the sinister dominance of money, of technology, or of the media" (Bollenbeck 2007, 7). Cultural practices, technological developments, and their incorporation into larger society and economic valorization in the media industry are regularly devalued in the framework of cultural-critical observations and contrasted with education ideals for forms of knowledge oriented toward mental objectification (the arts and the sciences).

Musicians and audio technicians criticize the possibilities of digital sound editing and situate the sound experience in opposition to the negatively valued "loudness insanity." They feel pressured by product managers and their arguments—the client wants it that way—and they criticize the music industry's obsession with success. In their opinion, a lot of the productions have nothing to do with a faithful recording, editing, or reproduction of the works created by the artist. Specific listening situations are not addressed by them, which affects practices of collective listening or dancing to music in particular. The berated product manager must always have a radio or dance mix produced as well.

Topoi and Narratives of Cultural Criticism

The parallels to German-language aesthetic (Hanáček 2008) arguments (end of the eighteenth century to the middle of the twentieth century) are obvious, especially in terms of the concepts of autonomous art on the one hand and the everyday aesthetic on the other—two concepts that are considered to tend toward incongruity. Beauty is to be looked for above all in form or in spirit, and not in the material or in the physiological effects of music. For Eduard Hanslick, the sonic took on particularly threatening proportions. For music to be idealized as an art form and raised to the highest level, the sonic had to be formulated based on the theories of aesthetics in order to make music objectively valuable: "However, he [Hanslick] was quite aware of the fact that the majority of listeners demonstrated great affinity for the sonic and felt music more than heard it, in the narrower sense" (Hanáček 2008).

Becoming deeply moved by music or sound was denounced as pathological behavior in the aesthetic traditions, whereas the contemplative is perceived as noble, as artistic listening. Adorno, who wrote that the "showy and pompous" dominates the internal compositional structure (1968, 200), criticized the narrowed dynamic bandwidth for the radio transmissions of his time. In his opinion, the compression of the dynamics threatened the form of the piece, "for only when it can fully develop from pianissimo to fortissimo can the individual motif truly function as the nucleus that already contains the totality" (Hanáček 2008, referring to Adorno's writings about the radio). In his critique of cookie-cutter compositions and standardized stimuli, denunciations of the loss of integrity, the negative effects of technological developments, and, above all, the criticism of mass culture and the culture industry (see Hecken 2014, 4) are broadly elaborated and have led to the formation of a tradition in academic thought. As a preventative measure toward the struggle against noise and overly loud music, the psychologist August Schick recommends "convincing media corporations to get a different taste in listening, or to invent a new composition of youth culture." In the process, he references Adorno, who as a "critical sociologist ... also considered overloud music as a commodity" (Schick 2001, 9).

A central aspect of discursive debate is addressed with the suggestion that the majority of listeners feel rather than hear—whether it be the sound of a symphony or a pop concert. Loudness will become a problem as long as listening practices are not named and/or remaining silent is required in the tradition of aesthetics.

In the German language, there is a wide selection of words or, rather, prefixes that differentiate the verb *hören* (to hear). In the process, they point to a variety of hearing practices—in other words, differing situations of the perception and appropriation processes of sound events and music: *aufhören* (to listen up/pay attention), *mithören* (to listen in/overhear), *überhören* (to not hear/to ignore), *weghören* (to not listen/to ignore), and *zuhören* (to listen to). However, the majority of these words focus on aspects of language comprehension. Thus, the meaning of sound phenomena and a large amount of music—which distinguishes itself "radically from language as the other form of human communication based on sound" (Shepherd 1991, 103)—is still not adequately understood. Listening concentratedly *to* something, that is, *zu*hören, was accepted as the only legitimate (or highly valued) listening practice for a long time in musicology. The introduction of an attribute (e.g., absent-minded or casual listening) indicates that it is not just sounds or music that are at the focus of attention; other aspects can play a role as well, such as when music is listened to on the car radio or while dancing. "The appeal of music is primarily and originally somatic and corporeal, not cerebral and cognitive" (ibid., 104). Perceptions of loudness, therefore, should be analyzed in the context of the situational experience of sound in specific spaces of experience.

The reasons for which high levels of loudness are supported—for example, in the context of music as a medium for community formation (sound recording productions and popular music concerts from diverse genres, disco and club patronization)—and for which they are rejected—for instance, in the context of aesthetic discourses as well as in discussions concerning (public) health, hearing damage, and the acoustic ecology movement (Schafer 1977)—are incomprehensible without looking at the respective

temporal relations and aesthetic as well as social concepts (Schulze 2012) that are negotiated concerning loudness.

1960s rock music was one of the forms of noise pollution most criticized by Murray
Schafer. He justified this by pointing to the fact that its sound level had risen to over
one hundred decibels, which was (supposedly) higher than any music that had ever
been heard on the planet (Breitsameter 2010). Popular music phenomena ended up
in an argumentation context alongside the acoustic phenomena, or rather effects, of
industrialization and the densification of traffic infrastructure. On the other hand, one
could claim that the cultural-critical argument narrative is being applied here. Popular
music is put on a level with a noise-polluted environment, in particular owing to its
intrinsic mechanistic amplification through loudspeakers (schizophonia!) in order for
the music to be experienced in the different listening situations. The German philolo-
gist and cultural studies scholar Georg Bollenbeck understands cultural-critical motives
as "a normatively charged mode of contemplation." Cultural-critical objections originated during processes of social modernization—that is, primarily during processes
of industrialization. In their evaluation of these social transformations, they should
"sensitize to the impositions of modernity" (Bollenbeck 2007, 10–11).

Loudness as Transgression and as Code for Cultural Self-Assertion

Almost at the same time, British cultural studies were considering questions regarding the creation of style in the subcultures active at the time. "Clothed in chaos,
they produced Noise [sic] in the calmly orchestrated Crisis [sic] of everyday life in the
late 1970s—a noise which made (no)sense in exactly the same way and to exactly
the same extent as a piece of *avant-garde* music" (Hebdige (1979) 2002, 114–115).
For early British punks most of all, loudness acted as a kind of code that permeated
through all of their created practices' individual elements—from clothing, to movement patterns while dancing, to jargon, to music—like a recurrent theme (homol-
ogy). Loudness can be understood here as a threateningly staged symbol of group
solidarity and group identity for outsiders. At the same time, loudness signifies the
sonic self-assurance of a group of actors, as in the vuvuzela example. Similar phenomena can be reported out of the world of various metal stylistics. For that matter, loudness is never restricted to the level of sound, for example in death metal
(double bass, tremendous tempo, lower-tuned strings, growling), but instead reaches
far into the reservoir of symbols and symbolic behavior (see Chaker 2011). Loudness
is deeply rooted in the value system of all kinds of genres and stylistics of popular music. If your ears aren't ringing, it wasn't good. "This amp goes to 11!" says
a character in *Spinal Tap*, a film about a half-fictitious metal band; the trailer for a
documentary film about rock music and its bands of the second half of the 1980s in
East Germany (*Flüstern und Schreien*, 1988) contains the directive: "This film must be

listened to loudly."[1] "Aspects of group life … made to reflect, express and resonate" different style elements and their codes (Clark et al. (1975) 2006, 43). The usage of the feedback effect toward a perception of the guitar sound as distorted (Jimi Hendrix at Woodstock) became the topos of a successful stage performance. In deciding about whether to attend a concert, the possibility of experiencing higher sound levels may play an important role. Performers call out from the stage that the audience is not loud enough. Call-and-response orchestrates a communal experience that generates a hypersocial space. In the process, loudness is only one moment in a club or live event that is evaluated as successful in which an aesthetic "surplus production" takes place by means of light, sound, rhythm, temperature, smells, and performance. You could really exhaust yourself: physis or the illusion of physical expenditure plays an important role and can be interpreted as a sonic strategy of cultural and social self-assertion.

Figure 24.1
Subheading for *Flüstern und Schreien: Ein Rockreport* (DDR 1988, Dieter Schumann), © DEFA-Stiftung.

Figure 24.2
Slam dancing to the *Feeling B* song "We always want to be well behaved"—*Flüstern und Schreien: Ein Rockreport* (DDR 1988, Dieter Schumann), © DEFA-Stiftung.

In recent years, clubs were shut down in some German cities because it was impossible to comply with the maximal values set by the Federal Control of Pollution Act. "Finding a space that meets the requirements of the performance and of the building regulations for noise protection is extremely rare in practice. After 10 pm, the maximal noise level that is permitted at the next row of houses is 55 decibels: that is about as loud as table talk" (Schölermann 2012, 1). For that reason, joint associations of clubs and concert promoters demanded a separate approach to noise pollution for cultural venues. In their opinion, clubs are cultural spaces and not freeways (Herzog 2012). Some of the clubs affected by the law have existed for decades. The building boom in Berlin led to substantial densification and overcrowding in the inner-city area. The singer of the Berlin band Knorkator is indignant: "'That is a crime against culture' … for more than 30 years he has been coming to this club—and now he is too loud for the neighbor in the newly erected residential building next door"(Heine 2010). SO36,

Figure 24.3
Punks talking with the director—*Flüstern und Schreien: Ein Rockreport* (DDR 1988, Dieter Schumann),
© DEFA-Stiftung.

another club threatened with closure, was one of the most important self-organized meeting places for the punk and new wave scene in the late 1970s and early 1980s. Bands such as Die Toten Hosen, Einstürzende Neubauten, Die Tödliche Doris, and Dead Kennedys performed there.

Conclusion and Research Perspectives

The discourses on loudness in and about popular music delineated here can be summarized in the following way:

Music Scene Actors
Some musicians, and especially sound engineers, are using developing technological methods as an opportunity to criticize the—according to them—diminishing

recording quality for music. In relation to listening practices, their sound quality requirements provide the starting point for this criticism. Musicians who lean toward particular genres, or who stand for genres such as rock, punk, industrial, metal, and so forth, express a lack of understanding for the desire to close up clubs or for enforced noise-reduction measures. Similarly to their fans, they articulate and stipulate the importance of loudness for a successful performance and experience of music.

As part of the music economy, club operators are attempting to challenge noise-reduction measures, as such measures tend to take away the foundation of their business. Since clubs and concert venues constitute the infrastructure of specific scenes, musicians are also a part of this discourse—it would be short-sighted, therefore, to assume that musicians are being instrumentalized by clubs and journalists.

Academic Discourses

When dealing with questions of loudness, one is confronted with the point, declared as fact, that the environment is polluted by noise. Sound studies—which was expressly developed as a consequence of the acoustic ecology movement—is based, in the German-speaking realm above all, on the paradigm that people had unlearned listening as a result of this noise-polluted environment, and that listening (what kind?) must be accorded new weight and academic attention. Research on popular music thus required multiple forms of legitimization, including legitimization of physically high volumes, the mechanistic practices of their production, communication, and reproduction, as well as their integration into commercial contexts such as "listening practices." As always, it holds in empirical psychology that youth should either not listen to popular music at all—especially not over loudspeakers—or only in a controlled fashion.

In contrast, cultural studies-oriented popular music studies focus on the functions of loudness as code and as aesthetic experience spaces for group distinction and for community-building. In so doing, they are aiming for an understanding of different listening practices.

In the music scene and in an academic space structured by disciplines, one comes across areas of conflict concerning the interpretational sovereignty of legitimate actions and positions. Ethnographically oriented studies try to understand culturally active subjects' actions that are related to each other. Since the experience of music (i.e., listening practices) cannot be reconstructed in the laboratory and can at most be compared, what must be focused on in the future, in my opinion, are participatory methods and the situation of listening or, rather, interpretive listening. To that end, an intensive discussion of methodological, theoretical, and practical questions of qualitative research is necessary, as they have thus far barely been developed in relation to the medium or, rather, the material of the medium of sound.

In the tradition of cultural studies and de/constructivist oriented theories, the field of popular music studies undermines the concept of, or desire for, an existing supra-individual aesthetic and its corresponding norms. This essay has attempted to delineate how deeply rooted cultural-critical topoi are, in the sciences, oriented toward the cultural and the social, and it makes the point—one which should be taken seriously—that they are always a reaction to the gains and losses of modernity.

Translated by Jessica Ring

Note

1. Thanks to Dietmar Elflein for the reference to these interesting "details" in his talk, "This Amp Goes to 11!—(Hard) Rock, Heavy Metal und Lautheit" as a part of the lecture series LAUTHEIT at the Institut für Musik der Universität Oldenburg during the winter semester 2011–12.

References

Bibliography

Anonymous. 2014. *Occupational and Environmental Medicine.* http://oem.bmj.com/content/suppl/2014/05/01/oemed-2014-102172.DC1/oemed-2013-102172_press_release.pdf.

Adorno, Theodor W. 1968. Orpheus in der Unterwelt. *Der Spiegel*, November 11, 200.

Bollenbeck, Georg. 2007. *Eine Geschichte der Kulturkritik: Von Rousseau bis Günther Anders.* Munich: C. H. Beck.

Bonz, Jochen. 2010. *Vuvuzela.* http://www.jochenbonz.de/2010/06/.

Breitsameter, Sabine. 2010. Hörgestalt und Denkfigur. In *Die Ordnung der Klänge: Eine Kulturgeschichte des Hörens*, ed. R. Murray Schafer, trans. and re-ed. Sabine Breitsameter, 7–29. Mainz: Schott.

Chaker, Sarah. 2011. Kill your mother/Rape your dog: Zur Rolle von Musik in transgressiven Jugendkulturen und Szenen. In *Medien, Ethik, Gewalt: Neue Perspektiven* (Schriftenreihe Medienethik, vol. 10), ed. Petra Grimm, and Heinrich Badura, 205–34. Stuttgart: Franz Steiner.

Clark, John, Stuart Hall, Tony Jefferson, and Brian Roberts. (1975) 2006. Subcultures, cultures and class. In *Resistance Through Rituals: Youth Subcultures in Post-War Britain*, ed. Stuart Hall and Tony Jefferson, 39–131. London: Routledge.

Devine, Kyle. 2013. Imperfect sounds forever: Loudness wars, listening formations, and the history of sound reproduction. *Popular Music* 32 (2): 159–176.

EBU Operating Eurovision and Euroradio. 2014. R128: Loudness normalisation and permitted maximum level of audio signals. Recommendation. Geneva. https://tech.ebu.ch/docs/r/r128.pdf.

Elflein, Dietmar. 2011. This amp goes to 11! (Hard) Rock, Heavy Metal und Lautheit. Paper presented at lecture series LAUTHEIT at the Institut für Musik, University Oldenburg, winter semester 2011–12.

Hanáček, Maria. 2008. Das Sonische als Gegenstand der Ästhetik. *PopScriptum* 16 (10). http://www2.hu-berlin.de/fpm/popscrip/themen/pst10/pst10_hanacek.htm.

Hebdige, Dick. (1979) 2002. *Subculture: The Meaning of Style*. London: Routledge.

Hecken, Thomas. 2014. Kulturkritik, Pop, Poptheorie. Unpublished transcript presented at symposium Kulturkritik und das Populäre in der Musik, Fachhochschule Düsseldorf, March 21, 2014.

Heine, Hannes. 2010. Stumpen will Stammgast bleiben. *Der Tagesspiegel*, June 21. http://www.tagesspiegel.de/berlin/stumpen-will-stammgast-bleiben/1867750.html.

Hellbrück, Jürgen. 2008, Das Hören in der Umwelt des Menschen. In Musikpsychologie: Das neue Handbuch, ed. Herbert Bruhn, Reinhard Kopiez, and Andreas C. Lehmann, 17–36. Reinbek near Hamburg: Rowohlt.

Herzog, Marco. 2012. ClubConsult: Erkennen, Planen, Handeln. http://www.clubcommission.de/themen/clubconsult (accessed November 13, 2014).

Schafer, R. Murray. 1977. *The Soundscape: Our Sonic Environment and the Tuning of the World*. Rochester, VT: Destiny Books.

Schick, August. 2001. Lärm und überlaute Musik: Psychologische Analyse und Präventive Maßnahmen. Univ. Oldenburg. *Berichte aus dem Institut zur Erforschung von Mensch-Umwelt-Beziehungen* 38:12.

Schölermann, Karsten. 2012. Liste des Grauens: Eine kleine Musikclubpolemik. Paper presented at conference PLAN! POP12, Initiative Musik gGmbH, Alteglofsmein, May 2, 2012.

Schulze, Holger. 2012. The body of sound: Sounding out the history of science. *SoundEffects* 2:197–209.

Shepherd, John. 1991. Music and the last intellectuals. Special issue: Philosophy of Music and Music Education. *Journal of Aesthetic Education* 25 (3): 95–114.

Sound and Media

Flüstern und Schreien: Ein Rockreport. 1988. Director: Dieter Schumann. Documentary, DEFA, GDR.

25   Existential Orientation: The Sound Knowledge of Fans

Diedrich Diederichsen

How is a connection drawn between the technologically recorded sounds of pop music, the specific voices, the specific sounds of effect pedals, types of amplifiers, synthesizer models, guitar designs, and the creation of countercultures, subcultures, urban tribes, and other communities of identification? Is it possible to speak of knowledge in this context, of a particular kind of knowledge, of translation, of language? What is the relationship between individual hearing—often operating on the basis of desires and projections—and the creation of collectives of youth, minorities, and other outsiders? Is a particular knowledge of how to read and embody music crucial for the creation and maintenance of such collectives?

Listen to Contingent Reasons Instead of Intentions

Similarly to noises in the real world, we learn to understand technologically recorded sounds as caused. I use the word *caused* here in contradistinction to *intended* or *planned*. These opposites are not absolutely antagonistic to one another: that which is caused can also be intended (even by the same source, if it is capable of intentional acts). It is more a matter of a contrasting perspective. When we listen to classical music, we ascribe the sounds with various intentions: the self-expression of the composer, the interpretation, the expression of the musician, and so forth; the material and physical frameworks, the historical properties of the instruments, the acoustic dimensions of the space, emerge at best as forms of resistance that must be overcome or controlled, or as atmosphere, as passive component. When we listen to pop music, in addition to the music we expect to recognize individuals as mechanical causes that express themselves bodily and unintentionally: a certain physicality of the singer, a sound effect to be attributed to a specific machine or technology—we hear it that way, as if the body of the singer, the grain of the machine, wanted to speak to us. It is in this way that the fans' fetishisms and the artist's mannerisms come into being—and they are of much greater importance for the creation of forms of knowledge than the actual musical notions and intentions for the development of an aesthetic practice.

Conventionalizations of the Unintentional: Pragmatics, "Music," Visuality

However, if, in the case of pop music and other music oriented toward recording, it were only about transmitting live broadcasts made up of young people's physicality and the technology of the instruments, amplifiers, and the studio, it would not be necessary to explain why listening to pop music is such a social phenomenon—one that creates communities, operates with overt inclusions and exclusions, and that has repeatedly aimed to establish, differentiate, and newly consolidate generations, peer groups, sects, and subcultural tribes. It seems crucial that the recipients are prepared to interpret these transmissions in a particular way and to alphabetize them, as it were—beyond sensationalization and fetishization. For the fans and the recipients, the causes are singular and physical and yet still capable of having a meaning, which is of course only logically possible if that which is transmitted can be recognized; in other words, when the singular submits to a certain type classification.

In order to achieve that, pop music's sonic material follows at least three conventions. First, the transmission of the specific, the unique, and the physical is based on often very narrow rules. However, those rules are not musical, but have rather the architecture of symbolic systems similar to those of language. In that, they draw close to phenomena out of the pragmatics of verbal communication and are akin to certain familiar units from it, such as *yell*, *praise*, *preach*, and *whisper*, or certain physical frameworks and "disabilities" in speaking (hiccups, stuttering, hoarseness, nasalization, gender and age-specific voice qualities and their subversion, and a deficient, but overtly employed, vocal spectrum) in order to pair the singularity with an interpretable speech act convention. The second convention is "music." Auditive indications of the singular—made available through technological recordings—are embedded into a specific concept of music, almost always based on simple structures in pop music and often deriving from folk contexts. A third convention comes from the link between pop music and its accompanying images, a link that has become increasingly close over the course of its history: photographs in magazines and on record covers, moving images on television shows and in films depicting concerts and events occurring backstage. I will return to these three conventions and their function as context at a later point.

Therefore, fans' knowledge consists, on the first level, of relatively conventionally listening for linguistically and pragmatically unambiguous and indexically singular sounds and, in turn, linking these to meanings or projecting onto them. This is a very specific knowledge that one does not find in such a marked way in other artistic formats: physical singularity and presence are very much in the foreground in the performative arts—but they do not have to be spelled out, as they either comply with the logic of mastery of already available material (in the case of most so-called allographic arts, such as theater, classical European music, etc.) or they can be attributed to the dramatic intentions of the performer. On the contrary, in pop music it is about treasuring the unintentional for the most part; it is precisely about *not* assigning the singular with some purposeful intention, but instead about what happens to the particular mastered genre and the instrument when it is used in a targeted way.

However, rock 'n' roll's first lyricist Chuck Berry insisted that his hero, the fictitious guitarist Johnny B. Goode, "could play the guitar just like ringing a bell." Whether that is about a church bell, a doorbell, an erotic play on words à la Anita Ward ("Ring My Bell"), or something familiar ("It rings a bell"), it is something that is known and simple; regardless of its requirements of virtuosity, it is no trouble at all for the virtuoso. The unintentional, played from the hip, becomes a symbol that is as clear as the ringing of a doorbell or a church bell, a simple communicative symbol that indicates, without ambiguity and without room for interpretation: there is someone outside who would like to enter, or the church service has begun. That is the secret mechanism: How does a sound generated by the body or by a machine, albeit in combination with the supporting conventions of speech and music, become a social pragmatic element of youth culture (and, more generally, counterculture or subculture) tribalism?

Exclusive Fan Knowledge: Tribalistic Readings

That which Chuck Berry ascribes to the virtuosity of the guitarist appears to have more to do with the contrast between conventionality and singularity (as subsequent pop music cultures made clear) than with a harmony (however achieved) between sound causation and sound mastery, between sound and music. Pop music communication begins in the gap between the two: this gap, this divergence between (musical) convention and (individual, singular, physical) causation initially follows a model that is already established in popular music anyway—and not only in popular music. A well-rehearsed tradition permits individual deviation on the part of the respective performer, which is not only protected by tradition and convention—in fact, they reinforce one another. The intoning of a song, the performing of dance steps and rituals can only survive as a cultural standard when it is believably protected through individually and generationally specific interpretation. However, this dialectic of conformism and empowerment is fundamentally influenced by a model of individuals and community that draws on either the personal acquaintanceships of the actors (in the village) or at least on the awareness and recognition of the stereotypical social functions of those involved (cop, priest, prostitute, clown, etc.).

In pop music, fans and recipients develop a knowledge that works instead with two unknowns: they react to the causes of sound, which they know only in their recorded form, and they understand them as if they were symbols instead of indices (and at the same time still also indices). The smoke does not just tell them that a fire is burning somewhere—that too—but also which most emotionally symbolic text will be transported with the smoke. Of course, in the process, the music provides (mostly vague) emotional temperatures, and of course the text often delivers very conventional, incidental but also at times pointed, unerring stories that aid in the construction of a context. However, the knowledge—a knowledge that is capable of differentiating between a cool and an uncool sound, between an embarrassingly, anxiously generated guitar strumming and one that is casual and seemingly unintentional, between an absurd, overly dramatic power chord and an appropriately grandiose power chord—is only mildly dependent on this contextual information. This knowledge modulates the judgment of a person (of an author) based not on information about their reconstructable intentions and content decisions and not based on their ability, education, or worldview. Instead, this judgment is formed on the basis of the—seemingly—much more reliable information about the physicality of the person concerned.

Alongside pragmatics and music, another context is the specific visuality, the widespread images of not only the performer, but also the illustrative and often associative images which accompany the performances: from psychedelically coded mandalas to road movie landscapes. In addition to symbolic and indexical images, iconic symbols also become important when it comes to channeling pop music fans' sound knowledge and to explaining the improbability of its social consequences. Independent from the various quasi-deflective or metonymic consequences that accompany this form of parallel reception—one identifies oneself with the image of a singer, desires the pianist's body, is disgusted at the simplicity of a refrain, and is strangely drawn to the raw grain of a guitar sound characteristic—something like a socially oriented substrate is generated: one learns that certain things in all of the three symbolic fields within a tribalistic reading are permitted and others not at all.

Naturally, this knowledge is fragile. What is accepted and what is not—perhaps because it has become too common (accepted too greatly by the unwanted mainstream outside of the tribe) or too particular and sophisticated (too artistic and now a purely artistic and no longer communicative-oriented usage of symbols)—is incessantly newly negotiated. Either the boundaries shift—between 1966 and 1981, a particular distorted guitar sound migrated from rebellion rock to soft rock—or the tribes multiply. After 1990, in electronic dance music one can observe that sound knowledge reacts very specifically to certain sonic attractions, either treasuring them or rejecting them, and often with relatively little contextual knowledge originating from the music product, but rather based on concrete experiences with particular sounds in nightlife or at parties. The rejection of or attraction to a sound effect or a rhythmically repeated acoustic color is then closely linked to concrete people and their physicality: the technological sound source is no longer attributed to an artist as subject, but rather to an attendant tribe, with whom one would like to engage or not.

Three Types of Sound Knowledge

Therefore, one can differentiate between three types of sound knowledge: the knowledge of belonging, the knowledge of suitability, and something like a vocabulary, a lexicon. However, in the beginning, there is desire: it originates in the unresolved gaps between the specificity of the sound, especially in its human-individual causation (and, later, mechanical-technological: types of amplifiers, sound effects, rhythm machines), and in its publicness, its supposedly transparent interpretability. It is what love stories are made out of: on the one hand, the desired person is the creature of my fantasies and obsessions; on the other hand, she is real and leads a public life. Desire resides within this mysterious tension. This tension corresponds to the remarkable connective efficiency of the fans and recipients in a bipolar pop music world. One pole is the private and intimate environment of the children's bedrooms in which this desire is formed and, as such, fortified by the fact that the articulation of voice and noises appears to be exclusively directed at me, the room's more-or-less lone inhabitant. The opposite pole is the remarkable experience with public spaces in which these indexical sounds, which I have desired and interpreted exclusively, suddenly mean something to other people as well. In the switching back and forth between both intimate and public poles, the relationship between simply desired signs with relatively open interpretation is on the one hand fixed; the process of fixation is observed from the outside: the fan, moving through the public space, carries symbols of his or her enthusiasm, fashion and style attributes that tell of the relationship between intimate experience and social integration. This slowly stabilizing relationship is experienced as a special knowledge, a better knowledge than conventional knowledge; one that has its own lexicon and makes claims of general validity—this knowledge is quasi self-generated. My childhood bedroom desires and their public ratification through the peer group and subculture first created the meaning, developed a semantic horizon with which desire and affect can get along with a (different, better, younger, incomplete) social world. The mastery of this relationship between affect and social meaning is experienced as an existential knowledge directly related to my body—and is, for that reason or in spite of that reason, a better knowledge. It is better because it is connected to an existential experience of identification and empowerment: my feelings have become an element of a larger social structure. Needless to say, this historical description mainly refers to Western middle class definitions of childhood and adolescence of the post–World War II era.

Later, I (and others like me) rationalize this better knowledge into one of suitability. I declare the desired sound to be a rebellious one or an emotional one, or I construct another relationship to a referable, familiar type of content that allows me to speak about the sound, to interpret it, and to inscribe something similar to intention or expression in parallel to the person who is generating the sound or responsible for it. This does contradict the original perception of unintentionality; I do not, however, interpret this suitability as directly or consciously expressed, but rather as a suitable expression of a symptom or of an almost natural cause. "This drum solo sounds so far-freakin-out. I want to take it with me," says a dancing hippie in a Raymond Pettibon drawing on the cover of the Minutemen album "Double Nickels on the Dime" (1984). This drum solo is the suitable expression for the world that I desire, that I want to have or to which I would like to belong. The dancing hippie then loses balance and dances off the roof of a house down into the abyss. The unintentional causation represents the kind of required suitability when privileged knowledge that cannot be sufficiently explained simply by rationality or even artistic impressiveness is concerned.

Finally, it is remarkable that alongside the key sounds that are to be ascribed to a particular time of a particular community (scratching, 1981–1985: urban, style-conscious male youth; flageolet chords on acoustic guitar, 1966–1973: sensitive, artistically interested female and more nontraditional male middle-class youth), there is a larger lexicon that is less intensely received and cannot be as easily applied for identification or attribution to one tribe. Instead, it forms more the background of significant sounds; in other words, it is the reservoir out of which those key sounds originate. The blast beat drumming style in metal, which has characterized most of the fast post-traditional metal styles since the late-1980s, appears nevertheless in a large variety of metal substyles that are usually mutually exclusive between subcultures and tribes. It does, however, seem important for the drum style to be recognizable as originating from a larger metal universe (and so belonging to metal on a symbolic level), before it specifically and indexically categorizes a particular drummer or band and, thus, a subcultural meaning, one that only refers to a very specific scene—although it is important for its self-conception that it initially emerged out of something larger, possibly in the course of disaffiliation.

That is then the path of movement through which the originally singular sounds (caused by voice or effect) travel: from the simple identification of an empirical body or machine, to the meaning of the differentiation of a specific group originating in convention and context, to a symbolic meaning for a larger group—which then turns into either a rival or even opposing force, or into a background required by the next specific sound image creation.

In conclusion, it would be useful to discuss how (frequency, degree of aggression or emotional pathos, context-relation) the corresponding symbols are provided. To put it simply, it would also be possible here to describe a path or a panorama—that of the displayed indexical effect, the novelty effect and key sound, to those sound causes that also make up the material for musical modulations (e.g., a Stratocaster with wah-wah pedal), to a given tribal-specific sound played in a rhythmic loop on a techno track whose meaning for the tribe functions as something like a synthesis out of the highlighted and incorporated sound symbol: on the one hand, the symbol delivered in four-four time is extremely present and put on display; on the other hand, however, it also constitutes musical material.

Translated by Jessica Ring

References

Sound and Media

Minutemen. 1984. *Double Nickels on the Dime*. LP, SST Records, SST 028.

26   Corporeal Listening

Holger Schulze

Situated Listening in Everyday Life

At any given moment in the early twenty-first century, we are surrounded by more devices that are potentially sound emitting and sound generating than ever before in cultural history: the number of loudspeakers in any location we may enter is at least equal to—and often much greater than—the number of human beings, as everyone will presumably be carrying at least one cell phone, tablet device, or portable game console. This cultural experience has developed into a commonly shared reality for years now. After all, it is the pure physicality and also corporeality of loudspeakers in every space, every location (in our shopping malls, public spaces, cars, and public transport, or in any bar, dance club, or open-air concert festival) that has paved the ground for a reflection of this basic physicality. How does this relationship between situativity and corporeality take form?

Physically speaking, the spatial area in which we exist is never empty anyway. And even with a lack of sound-emitting gadgetry, it is—again, physically speaking—densely filled with materials, gases, objects, particles and dust, air currents and fumes: these are the material carriers of sound waves or, to be more precise, the disturbance of an equilibrium (which a sound wave actually is) is diffused via these materials. Sensing sound is, after all, defined as an activity realized by a medium—thus turning us, the listeners, into a medium of resonance ourselves. Multiple waves of pressure reach our skin, bones, flesh, and our diaphragms, eardrums, vocal chords, and cochlea almost constantly throughout our lives. Individually felt corporeality is, as such, a function of situative listening. Sound studies analyzing individual sonic experiences in a given situation are actually realizing a disciplinary fusion between material culture and sensory studies in the field of the auditory.

Listening and sound practices in popular culture are, to a large extent, practices of everyday life. Following Lefebvre's concept of the everyday, we might state that, in listening, we experience materially as an "intersection of the sector man controls and the sector he does not control" (1947, 40). Thus, the concept of everyday life is also closely related to minor, highly personal, and mostly repetitive situations (in which moments of exaltation, excitement, and ecstasy also find their place). It is this aspect of repetition that brings it close to the recently observed phenomena of ubiquitous listening (Kassabian 2013).

Many listeners—professionals from the field of cultural theory or sound research, as well as fascinated laypeople, amateurs, and passionate fans—have realized that what we listen to is physical, unstable, dynamic, and highly immersive, not to say invasive. We might be affected, in fact we could be transformed by such vibrations. But how could this corporeal experience be described?

Kinesthesia and Vibration

Since the nineteenth century, common and established models of listening (Helmholtz 1863; Fletcher 1929; Beranek 1954) have focused almost exclusively on corporeal aspects of the human ear in specific, mostly seated and static, listening situations. Listening was and is almost exclusively researched in the context of immobile and often synthetically constructed listening situations—usually in *the laboratory* (or, more recently, filling out an online survey). Corporeal reactions beyond those of the double organ of hearing, as well as a more dynamic listening position—let alone one that is in permanent motion—are still not included in the listening models that we inherited from our ancestors in physical acoustics, in signal processing, and room acoustics.

This traditional focus on seated, semiotic, structural, and spiritualized listening (Johnson 1995) was questioned as soon as artistic and popular practices proposed highly and immersively corporeal listening practices such as dancing, strolling, destroying objects, or performing. Between rock 'n' roll, soundwalk, skateboarding, rave and hip hop culture, a field emerged fully populated by popular practices of corporeal listening. These historical concepts of listening began to be opened up and relativized as soon as the historicity and the culturality of these concepts became widely accepted. Concepts of seated, calm, and silent listening needed to be reevaluated as actually highly elaborated, quite exotic, and, therefore, extreme and fascinating forms of listening and auditory *téchniques du corps* (Mauss 1936). But even in the early twenty-first century cultural research on sound, such genuine kinesthetic aspects of corporeal listening were just beginning to be a focus of research or, at least, a consistently disrupting and motivating observation. In particular, in the field of *dance studies* and *artistic research,* the concept of kinesthetic vibration became more and more of a crucial and even defining concept.

The effects of spatial and material vibration, which have been known for over half a century and assiduously reduced through the science of building and room acoustics, have now become a major constituting element in the anthropological experience of sonic environments—as part of a *hearing perspective* (Auinger and Odland 2007).

From this *point of audition* (Chion 1994, 89–92), vibration is nothing to be avoided, excluded, or systematically negated; it is instead a fundamental component of listening and, moreover, of any subtle, intricate, and intensive approach to perceiving a given situation. Thus, vibration is merely a symptom of movements taking place in a given space-time continuum; vibration is only absent where no events whatsoever occur, where no transformations and no causal relations take place materially—where, basically, no human or extra-human life exists. This simple physical and biological fact was summed up in the rather exalted concept of an *Ontology of Vibrational Force* (Goodman 2010, 81–84), a bold proposal by Steve Goodman in which he amalgamated current discourses on bass culture and bass materialism with seventeenth-century traditions of genuine pantheist philosophy by Baruch Spinoza as taken up in the philosophy of sensorial immanence by Gilles Deleuze.

The aspects of kinesthetic listening are closely related to the vibrational effects of movements in the known universe. In *Dance Studies* (Malnig 2008) and also in *Voice Studies* (Ihde [1976] 2007; Sowodniok 2012), the visceral basis of sound events in corporeal resonances is not to be neglected: it is one of the driving and controlling forces in cultural practices of dancing as well as in corporeal practices of vocal exercise and vocal excess. The whole vestibular system and, especially, its otolithic organ is part of the inner ear, and as such it is physically closely related to the cochlea. We orient ourselves, anatomically and sensorially, in almost the same way we try to discern the sounds to which we listen. Sound and movement, resonance and spatial orientation are deeply interlinked and in many situations hardly differentiable. Our humanoid perceptual profile embodies sound in every single movement of our lives.

Given these basic insights, how can we then, after all, undertake a research project on the grounds of such an ontology of vibration while acknowledging the fundamental status of the vestibular system in listening? I will provide three examples from my own recent research on a materialist anthropology of sound in everyday life. These three examples shall provide a basis for my argument as they come from the fields of (a) *functional sounds*, (b) *les sports glissants*, and finally, comprehensively (c) *dancing practices* as listening practices.

Three Corporeal Listening Practices

(a) Functional Sounds: "It's ringing! Sorry, I need to catch this call!" Functional sounds, such as ringtones (Gopinath 2012), computer sounds, car signals (Bijsterveld et al. 2013), navigational signals in public buildings, even the sounds of kitchen appliances or your personal setup of multimedia apparatuses in the living room: all of these sounds in everyday life bear a strong and intrusive corporeality. By means of field research and critical analysis, we might get a deeper insight into what happens when we listen to, and immediately react to, such functional sounds in the way they seem to imply is required—and we might understand how these sounds were conceived and produced in the first place.

In our research on this issue in the Sound Studies Lab,[1] we noted the following elements, which seem to be characteristic of functional sounds: functional sounds do indeed penetrate everyday listening, and as such they tend to become an integral part of our common listening expectations and sound experiences. Our corporeal listening is thus becoming increasingly shaped to fit these expected sounds, generating strong experiences such as the phantom ringtone (Gopinath 2012; Maier, Schneider, and Schulze 2015). This incorporation and embodiment of sonic expectations leads precisely to a rather crucial point when speaking about sounds: How do we understand sounds—and how do we react to them accordingly?

It seems, as far as our field research on designing and applying functional sounds in everyday life goes, that there is a kind of almost *situative* semantic ascription to sounds (Maier 2012). Sonic signals are not to be understood as perfect reproductions based on a perfectly distinct matrix (following structuralist theories of semiotics); instead, our understanding of these signals is constituted mainly by a highly situative network of meanings and a major corporeal intrusiveness of all those functional sounds. This interpretation of their situative corporeality of sounds then extends to other fields of research.

(b) *Les Sport Glissants*/The Sliding Sports: The soundscape of a given bodily activity or—in its most excessive and elaborated form—the soundscape of a sport is, as such, constituted by its sensorial, kinesthetic, and skillful use of and access to the environment in combination with and mediated by a specific piece of sports equipment. Following the research of Gunter Gebauer and others in the field of historical anthropology of sports, we might distinguish more traditional sports of disciplinary and linear excess of the human body from new, less disciplinary and more individualistic and, spatially, almost kinesthetic approaches to sports as in *les sports glissants* (a common notion in French; Gebauer 1992). Whereas both forms of sport do aim for a certain excess, the excess in the latter case is one of highly performative bodily movements, of gestures, of skills, and of kinesthetic and material mastery; it is far less characterized by the genuine, traditional joy of disciplinary practices applied to one's own body while training. Undoubtedly a major transformation, which, on the one hand, gives way to many new hagiographies, rebel narrations, and apotheosis of these new sports, and which, on the other hand, recently even affected the industry's effort to promote those more traditional nineteenth century sports—such as running—by placing it in a framework with similar attributes as the sliding sports through the usage of data mining and social media (see Gopinath and Stanyek, in Born 2013).

The sound events structuring those activities are, as such, closely related to the bodily movements of the actors and their intentions to present themselves through the medium of one or many pieces of sports equipment on a sporting stage. To a large extent, the sounds of sports are also made up of the functional sounds of equipment, gear, clothing, and gadgetry—in addition to the bodily sounds of breathing and the flexing of muscles. The difference in sonic traces is generated here by a differing multitude of materials, of habitus, and—lest we forget—of the sporting environments that provide a resonating space. The act of gliding itself, through an urban or other environment (by skateboard, roller skates, snowboard, parkour, you name it), is framed and accentuated by not only the participants' clothes, looks, haircuts, and body modifications—but also by the intentional, audible sonic traces they lay out before a (willing or unwilling) audience. For instance, sonically speaking, skating is quite a hardcore activity, with harsh initial (acoustic) reflections and massive, but flexible, materials. And there are some quite heavy beats in there as well.

The sound practices presented here (see Maier, "Sound Practices," this volume) are thus again characterized by a situative, corporeal, and—even more so—performative joy of movement. And they generate a sonic profile that is recognizable by expert listeners and practitioners as well as by laypeople in this field.

(c) Dancing as Listening: The moment we enter a space filled with rhythmic vibrations and repercussions on many macro and micro levels, we cannot help engaging in this *vibrational nexus* (Goodman 2010, 91–94). We are enveloped in a performative *rhythm-analysis* (Lefebvre 1992). Engaging in the rhythmic activity called dancing is the basic example of the experience of the *noncochlear sound* (Kim-Cohen 2009)—sound that activates many other organs, sensoriums, zones of our skin, muscles, and bones than just those of the inner ear.

Dance practices in popular culture (Malnig 2008) are basically derived from, and chained to, specific styles of music in their characteristics of rhythm, distortion, novelty sounds, and novelty rhythms and breaks—in other words, to a specific *sound concept* (Wicke 2008). This specific perception of a pop song or a club track is, as such, to be understood as an extension of the musical performance into the realm of a culturally and historically specific corporeality: the breaks and effects in a 1960s twist song trigger and legitimate significantly different dance moves than those in a 2010 dubstep track. Take the example of a highly infectious dance track from 2013: *Happy* by Pharrell Williams (Williams 2013). The infectious quality of this song is exemplified in its extraordinary twenty-four hour music video by We Are From LA and Yoann Lemoine (a.k.a. Woodkid) (Williams 2013): it consists of 360 situationally interlinked performances by ordinarily dressed and performing dancers, actors, and entertainers (e.g., even Jimmy Kimmel and others) and only twelve performances by the singer himself.

All of these performances take place in everyday situations: at a parking lot, in a kitchen, backstage, on a boardwalk close to a street on the outskirts of a city. The thoroughly casual, swingy, and almost effortless dancing demonstrated by these performers represents the many forms of everyday dancing we all might be involved in now and then. Nodding our heads, tapping our feet, swinging with our hands, hips, and heads are the most subtle and rather microscopic forms of dancing we might observe in these Western(ized) capitalist cultures these days. And yet, with these microdances, we are beginning to engage in the previously mentioned rhythmanalysis. We enter the vibrational nexus. We resonate with noncochlear sound.

In this and other cases of casual microdancing, the cultural dispositions in body movements and corporeal expression are transcended: by leaving our day-to-day work routines and goal-oriented ways of walking efficiently, we may open up a realm of corporeal movement and responsivity that allows for a revision, a rewiring, and a renewal of our bodily profile. We enter the realm of sound and music in order to dance. In this interpretation, dancing must indeed be understood as a *form of listening*.

Perception and Corpus

"When we ask someone to speak about what they have heard, their answers are striking for the heterogeneity of levels of hearing to which they refer" (Chion 1994, 25). Corporeal listening encompasses a whole range of individual and often highly idiosyncratic listening practices that go beyond the listening models established in the history of science. As soon as a corporeally extended listening model is applied as the basis for a research project, a whole variety of specific changes must be implemented in the pursuit of that research. Research on corporeal listening practices, and likewise on their listening situations, demands approaches to research that lead the researcher to visit, to experience, and to explore those places, situations, and listening practices that are their subjects. It is, as such, an example of object-driven research as known from ethnographic material culture studies (Miller 1998): *the research follows the objects*. Corporeal listening as such—this has become clear—is intrinsically related to situational listening: the impact of corporeality is a quality of the situativity of listening.

The researcher's subject will, then, as a kind of poetic justice, be part of this listening situation as well: he or she will also be under the influence of a specific sound concept and a material sonic event. As a consequence, the specific sonic materiality (Cox 2011; Cobussen, Meelberg, and Schulze 2013; Schulze 2015) of the given situation translates into a specific kinesthetic and corporeal materiality: a corporeality that constitutes itself via specific intrinsic movements and reactions to extrinsic movements as a form of bodily tension. *Un corps, c'est donc une tension* (Nancy 1992, 126). Research then evades the realm of mere signs; it also becomes a form of *corporeal* research—and as such it could also be presented in forms accessible to corporeal sensibilities. As I am finishing this chapter right now, I sit, rather tense, at my desk as the deadline approaches. I might look out the window, as I just heard an unexpected trombone playing in our backyard. And quickly getting back to writing this chapter, I might state that corporeal research in this form is open to various contingencies, idiosyncrasies, and historicities that drive, shape, and constitute research. And conducting that kind of research in every situation would mean that those crucial forces are not negated, ignored, or repressed—instead, they would simply be accepted as qualities of how humanoids live and work and exist. Research conducted by such creatures should not repress those qualities.

Note

1. http://www.soundstudieslab.org.

References

Bibliography

Auinger, Sam, and Bruce Odland (a.k.a. O+A). 2007. Hearing perspective (Think with your ears). In *Sam Auinger*, catalog, ed. Carsten Seiffarth and Martin Sturm, 17. Vienna: Folio.

Beranek, Leo. 1954. *Acoustics*. New York: McGraw Hill.

Bijsterveld, Karin, Eefje Cleofas, Stefan Krebs, and Gijs Mom. 2013. *Sound and Safe: A History of Listening Behind the Wheel*. New York: Oxford University Press.

Chion, Michel. 1994. *Audio-Vision: A Universal Experience? Sound on Screen*. New York: Columbia University Press.

Cobussen, Marcel, Vincent Meelberg, and Holger Schulze. 2013. Towards new sonic epistemologies: Editorial. *Journal of Sonic Studies* 1. http://journal.sonicstudies.org/vol04/nr01/a01.

Cox, Christopher. 2011. Beyond representation and signification: Toward a sonic materialism. *Journal of Visual Culture* 10 (2): 145–161.

Fletcher, Harvey. 1929. *Speech and Hearing. With an Introduction by H. D. Arnold*. New York: D. Van Nostrand.

Gebauer, Gunter, ed. 1992. *Aspekte einer zukünftigen Anthropologie des Sports*. Clausthal-Zellerfeld: Deutsche Vereinigung für Sportwissenschaft.

Goodman, Steve. 2010. *Sonic Warfare: Sound, Affect, and the Ecology of Fear*. Cambridge, MA: MIT Press.

Gopinath, Sumanth. 2012. *The Ringtone Dialectic: Economy and Cultural Form*. Cambridge, MA: MIT Press.

Gopinath, Sumanth, and Jason Stanyek. 2013. Tuning the human race: Athletic capitalism and the Nike+ Sport Kit. In *Music, Sound, and Space: Transformations of Public and Private Experience*, ed. Georgina Born, 128–148. Cambridge: Cambridge University Press.

Grossmann, Rolf. 2008. Verschlafener Medienwandel. *Positionen—Beiträge zur Neuen Musik* 74:6–9.

Helmholtz, Hermann von. 1863. *Die Lehre von den Tonempfindungen als Physiologische Grundlage für die Theorie der Musik*. Braunschweig: Friedrich Vieweg.

Ihde, Don. (1976) 2007. *Listening and Voice: Phenomenologies of Sound*. Albany: SUNY Press.

Johnson, James H. 1995. *Listening in Paris: A Cultural History*. Berkeley: University of California Press.

Kassabian, Anahid. 2013. *Ubiquitous Listening: Affect, Attention, and Distributed Subjectivity*. Berkeley: University of California Press.

Kim-Cohen, Seth. 2009. *In the Blink of an Ear: Toward a Non-Cochlear Sonic Art*. New York: Bloomsbury.

Kittler, Friedrich. 1985. *Aufschreibesysteme 1800/1900*. Munich: Wilhelm Fink.

Lefebvre, Henri. 1947. *The Critique of Everyday Life*. London: Verso.

Lefebvre, Henri. 1992. *Éléments de rythmanalyse*. Paris: Éditions Syllepse.

Maier, Carla J. (neé Müller-Schulzke). 2012. Driftende Klangzeichen. *Situation und Klang: Zeitschrift für Semiotik* 34 (1–2): 109–123.

Maier, Carla J., Max Schneider, and Holger Schulze. 2015. Situative signals in sonic conflicts: Elements of a sound design theory. In *The Auditory Culture Reader*, ed. Michael Bull and Les Back. New York: Bloomsbury.

Malnig, Julie, ed. 2008. *Ballroom, Boogie, Shimmy Sham, Shake: A Social and Popular Dance Reader*. Champaign: University of Illinois Press.

Mauss, Marcel. 1936. Les téchniques du corps. *Journal für Psychologie* 32 (3–4): 271–293.

Miller, Daniel. 1998. *Material Cultures: Why Some Things Matter*. London: UCL Press.

Nancy, Jean-Luc. 1992. *Corpus*. Paris: Editions Métailié.

Schulze, Holger. 2015. Der Klang und die Sinne: Gegenstände und Methoden eines sonischen Materialismus. In *Materialität: Herausforderungen für die Sozial- und Kulturwissenschaften*, ed. Herbert Kalthoff, Torsten Cress, and Tobias Röhl. Munich: Wilhelm Fink.

Serres, Michel. 1985. *Les Cinq Sens: Philosophie des corps mêlés*. Paris: Grasset. Trans. by Margaret Sankey and Peter Cowley as *The Five Senses: A Philosophy of Mingled Bodies* (New York: Continuum, 2009).

Sowodniok, Ulrike. 2012. *Stimmklang und Freiheit: Zur auditiven Wissenschaft des Körpers*. Bielefeld: transcript.

Wicke, Peter. 2008. Das Sonische in der Musik. *PopScriptum*. 16 (10). https://www2.hu-berlin.de/fpm/popscrip/themen/pst10/pst10_wicke.htm.

Sound and Media

Williams, Pharrell. 2013. *Happy*. Music Video, Back Lot Music/i Am Other/Columbia Records, BLM0251.

III Producing Sonic Artifacts: Hands-On Popular Culture

Material Sound Concepts

27   Records on the Radio

Thomas Schopp

A History of Media Constellations

Since the second half of the twentieth century, commercial radio in the United States and elsewhere has depended on recorded music. The creative search for "unheard" sounds in American recording studios that began in the 1950s and the subsequent emergence of new musical genres strongly influenced radio programming. By broadcasting special selections of records, programmers defined new target audiences. Disc jockeys offered their listeners meaningful experiences of recorded music and helped popularize it. Records and radio, as two distinct sound media, defined a permanent and successful partnership.

While terms such as *intermediality* or *media convergence* are regularly used to describe the cultural reality of interconnected media in the digital age, it is important to point out that media constellations have a history. In recent years, scholars have suggested a move from the traditional study of single media to a study of media constellations (Bolter and Grusin 1999; Thorburn and Jenkins 2004). A history of media constellations has to analyze how one medium reflected or used another medium. Another, yet more difficult analytical approach might try to demonstrate how two media reflected or used each other. This conceptual shift could provide a new understanding of the complexity of technological change in the twentieth century.

My chapter offers a case study of a *history of media constellations*. I would like to take a look at a constellation of radio and records known as *freeform radio*, which was a popular type of music radio in the United States in the late 1960s. At first, I will situate freeform radio in its specific technological, social, and cultural contexts. I will then explore how the sound of freeform radio and the discourse of the deejays reflected aesthetic and cultural aspects of the recorded sound of music. As a theoretical tool, sound can be used to analyze the relations between internal elements (deejay talk) and external ones (records) that constitute the radio program as a whole.

Freeform Radio in the 1960s

Freeform radio emerged in the second half of the 1960s when the American radio system was encouraged to develop new programs for the FM band. Until then, commercial stations had focused on the AM band, which was dominated by derivatives of the Top 40 format. The Top 40 format featured hit singles, fast-talking disc jockeys, many commercial spots, and station identifications. Compared to AM radio, FM radio had the general advantage of better sound quality. Owing to a different technological process, FM signals covered the largest part of the audible frequency range. Moreover, FM signals could be broadcast in stereo.

Freeform radio was produced both by experienced disc jockeys who had grown tired of the rigid programming structures of AM radio and by industry newcomers. While professionals in different parts of the country began to develop alternative forms of broadcasting on the FM band, it is Tom Donahue (1928–1975) who is most often credited for the establishment of a consistent and sustainable programming strategy on the stations KMPX-FM and KSAN-FM in San Francisco. The common denominator between Donahue and his fellow deejays was an interest in the long player as a recording medium. AM radio would play only singles, no album cuts. The freeform deejays demonstrated that FM radio and the LP complemented each other very well. They proved that young adults—who were the ones buying the most albums—were eager to tune into radio stations that played album cuts on the air.

The term freeform referred to the fact that the new radio stations, unlike stations in the AM band, worked without format clocks and general playlists. Music programming was based on individual selections that the deejays themselves prepared on the basis of their tastes and expertise. They played songs from a variety of popular styles such as psychedelic rock, folk, country, blues, jazz, and, eventually, classical or Indian music. However, freeform radio should not be misunderstood as a radio format without commercial strategies and professional attitudes. One reason for its success was that it was able to make the boundaries between the station personnel and its listeners—many of whom belonged to the hippie subculture—less perceptible.

Radio Programming and Recorded Music

As a case study, freeform radio raises general questions about the relationship between records and radio. Inspired by the aesthetics of recorded sound, commercial radio in the United States invented a variety of voice types. In the 1950s, the sonic intensity of rock 'n' roll singles turned the deejay talk into a high-energy performance. It was defined by a fast pace, informal language, regular jokes and exclamations. Although the vocal delivery was meant to sound spontaneous, it was the highly trained style of a professional radio personality. The deejays of the late 1960s developed a different type of radio voice. Freeform disc jockeys would not talk over records on the air like their colleagues on AM radio (Hagen 2005, 344–348). They spoke rather slowly, avoided verbal affectations, and made expressive pauses when processing thoughts. This kind of "mellow" vocal delivery was identified as "cool" and "authentic" by the hippie audience. The disc jockeys presented themselves as experts of popular music who strongly identified with the subcultural attitudes and practices of their listeners.

The image of the deejays as informal music experts not only affected the design of their radio voices but also the discourse they propagated. Unlike their forerunners in the previous decade, freeform deejays provided biographical facts about artists and inside stories about the production of new albums. They constructed a discourse around the idea that popular music in its recorded form had become a kind of phonographic art that was worth reflecting on and discussing. After 1965, singers such as Bob Dylan and bands such as the Beatles, the Grateful Dead, and the Doors discovered the long player as their favorite recording medium. Many artists began to release albums with extended songs and an elaborate sound design. A transcript from a deejay show that Tom Donahue produced in April of 1969 for KSAN-FM offers a perspective on the discursive strategies with which he and other freeform disc jockeys worked. It represents a monologue about the new LP of the band the Charlatans. Donahue self-confidently uses his position as a well-known radio personality to define the cultural status of the band and the aesthetic value of their product. He introduces the Charlatans as the mythical origin of the San Francisco music scene to his audience:

We have the Charlatans up at the station tonight and they brought with them their new Philips LP. And we played a portion of it earlier in the show and I'm gonna play some more of it now. The Charlatans, as I said earlier, are the San Francisco originals. I mean a great part of the San Francisco music scene and other scenes, too, were built around the Charlatans and the Family Dog. The rest of it is outgrowth. You are the trees and the leaves but they are the roots. And in a way I told them I sort of put them down for cutting an LP. Because I dug 'em as a myth. You know and then years from now when people say: "What was the group you really …" "The Charlatans." "I never heard of them." "No man, they never made a record. They're just in memories." And sometimes you establish yourself as a legend better that way than otherwise. Fortunately, the Charlatans have made an LP that's good enough to keep them remembered for a long time. Gonna hear "Ain't Got the Time," "Wabash Cannonball," and "Folsom Prison Blues." (Donahue 1969)

Another dimension that indicates how radio sound reflected the recorded sound of music is the temporal structure of programs. The Top 40 radio of the 1950s featured a fast pace and a dense sound design. The disc jockeys took the sonic intensity of rock 'n' roll records—short songs, technically enhanced instrumental sounds, and an up-tempo beat—as a general paradigm for the construction of their programs. Deejays such as Alan Freed (1921–1965) would play a large number of singles per hour. They talked over intros and occasionally sang along with the singers on the records, thus intensifying the experience of music on the radio. They made sure that the flow of program elements (music, talk, commercial spots, jingles) was seamless. Pauses were classified as "dead air" and banned from the program.

Freeform radio established a contrasting concept of program flow. The deejays would program so-called segues or sets—that is, sequences of songs. Segues were composed according to different categories, such as artists, topics, or moods. Due to the increased duration of album cuts, freeform programs were comprised of fewer elements per hour than Top 40 radio. Deejays used pauses on the air in order to loosen the program flow.

Summary and Further Research

Freeform radio successfully captured and enforced important trends in record production in the late 1960s. Its deejays promoted the idea that records had turned into a kind of phonographic art that deserved to be reflected on and discussed in public. They translated this knowledge into the content and form of their program. The deejay shows on freeform radio provided meaningful experiences of recorded music for listeners with a subcultural background. Deejays like Tom Donahue developed a kind of "mellow" voice that was similar to the dominant style of speaking in the hippie subculture (Willis 2014, 133–138). They created segues of songs that expressed an artistic approach toward the construction of program flows.

This case study has tried to briefly demonstrate what a history of media constellations could look like. A pluralistic view of media history would provide the groundwork for a deeper understanding of the complex and dynamic interrelations between different media in the twentieth century. Recording and broadcasting media have reflected and used each other in various ways since their beginnings. Early radio plays that were produced "live" in studios regularly deployed recorded sound effects. Television shows such as American Bandstand in the late 1950s featured performances of singers and bands lip-syncing their hit singles. Music television in the 1980s and 1990s predominantly broadcast music videos and, following the model of the disc jockey, established video jockeys. A more detailed study of radio and television programs as composites of recorded and broadcast sound should be of interest to scholars from media studies, music studies, and sound studies.

References

Bibliography

Åberg, Carin. 1999. Radio analysis? Sure! But how? In *Radio-Kultur und Hör-Kunst: Zwischen Avantgarde und Popularkultur, 1923–2001*, ed. Andreas Stuhlmann, 83–104. Würzburg: Königshausen & Neumann.

Bolter, Jay D., and Richard Grusin. 1999. *Remediation: Understanding New Media*. Cambridge, MA: MIT Press.

Crisell, Andrew. 1996. *Understanding Radio*, 2nd ed. London: Routledge.

Douglas, Susan J. 2004. *Listening In: Radio and the American Imagination*. Minneapolis: University of Minnesota Press.

Föllmer, Golo. 2013. Theoretisch-methodische Annäherungen an die Ästhetik des Radios: Qualitative Merkmale von Wellenidentitäten. In *Auditive Medienkultur: Techniken des Hörens und Praktiken der Klanggestaltung*, ed. Axel Volmar and Jens Schröter, 321–338. Bielefeld: transcript.

Garner, Ken. 2003. Radio format. In *Continuum Encyclopedia of Popular Music of the World*, vol. 1: *Media, Industry, and Society*, ed. John Shepherd et al., 461–463. London: Continuum.

Hagen, Wolfgang. 2005. *Das Radio: Zur Geschichte und Theorie des Hörfunks; Deutschland/USA*. Munich: Fink.

Hilmes, Michele. 2005. Is there a field called sound culture studies? And does it matter? *American Quarterly* 57 (1): 249–259.

Keith, Michael C. 1997. *Voices in the Purple Haze: Underground Radio and the Sixties*. Westport, CT: Praeger.

Krieger, Susan D. 1976. Cooptation: A history of a radio station. PhD diss., Stanford University.

MacFarland, David T. 1979. The development of the Top 40 radio format. PhD diss. 1972, University of Wisconsin. New York: Arno Press.

Montgomery, Martin. 1986. DJ talk. *Media Culture and Society* 8:421–440.

Sterling, Christopher H., and Michael C. Keith. 2008. *Sounds of Change: A History of FM Broadcasting*. Chapel Hill: University of North Carolina Press.

Sterling, Christopher H., and John M. Kittross. 2002. *Stay Tuned: A History of American Broadcasting*, 3rd ed. Mahwah, NJ: Erlbaum.

Thorburn, David, and Henry Jenkins, eds. 2004. *Rethinking Media Change: The Aesthetics of Transition*. Cambridge, MA: MIT Press.

Wicke, Peter. 2009. Der Tonträger als Medium der Musik. In *Handbuch Musik und Medien*, ed. Holger Schramm, 49–87. Konstanz: UVK.

Willis, Paul. (1978) 2014. *Profane Culture*. Princeton, NJ: Princeton University Press.

Sound and Media

Donahue, Tom. 1969. *Metanomine*. KSAN-FM, San Francisco, April 17. Audio stream, 49:25 min. Online available: Tom Donahue April 17, 1969, http://www.jive95.com (accessed April 4, 2010).

28   From Stationality to Radio Aesthetics: Investigations on Radiophonic Sounds

Golo Föllmer

Conversations with radio professionals and the literature on radio practices quickly reveal that, wherever they may be located, most radio stations invest great efforts to develop and refine a *sound* of their own. Accordingly, in most cases listeners can easily and in almost no time at all distinguish certain stations or station types from others (usually within a few seconds). How is that possible? During those seconds, one might pick up on a discussed topic that might be typical for a certain station, but many topics are valid news items for all stations, so they pop up everywhere. Moreover, as experience has taught us, station identification can work just as well during weather forecasts, phone-ins, music, signature tunes, and other framing sound elements emitted by the radio receiver.

Obviously, many program elements are dealt with in a particular way so that they convey a distinct and coherent *radio identity*, which practitioners grasp with terms such as *stationality*, *Anmutung* (German for *impression, look, and feel*[1]) or *channel identity*. Apparently, practitioners are able to produce, and listeners are able to pick out and decipher, certain *identity markers* of the station's sound.

The idea of a radio station's specific sound can be comprehended more precisely with Paul Théberge's conceptualization of the *sound* term, which he applies to popular music. Théberge points out that "since the rise of the recording and broadcasting industries, the search for the right 'sound'—the sound that would capture the ears and the imagination of the consumer—has been a matter of concern" (Théberge 1989, 99). And he cites musician/producer Brian Eno for an explanation of the term's prevalence: "The sound is the thing that you recognize" (Korner 1986, 76).

Théberge argues that this concept of sound comprises not only those attributes traditionally associated with music—that is, the acoustic characteristics of instruments, vocal, and musical styles. Instead, he describes how multitrack recording as "a particular technology and a particular mode of production" (ibid.) came to exert far-reaching influence on production processes and, thus, extensively shapes the sound of any studio production. Radio programs may be considered a product of a particular technology and a particular mode of production to a similar extent. Just as Théberge claims about popular music production, radio production follows the rationalistic logic "of economic efficiency and technical control" (Théberge 1989, 100).

As in the case of popular music, what is produced in the radio studio is most substantially shaped by audio technologies and the technical requirements of production processes. Like popular music, radio programs target a mass market and thus have to follow standards in order to get recognized most easily. Moreover, popular music is the main content in most radio formats, making it probable that the other "surrounding" sounds in radio are produced in a way that makes them correspond to the music's emphasis on sound and "technicality."

What elements and characteristics comprise sound in the case of radio? A heuristic approach allows for the proposal of four categories:

(a) Characteristic ways of speaking, including rhetorical tropes as well as patterns of pronunciation and prosodic features such as pitch and other types of voice modulation (Bose and Grawunder, forthcoming)

(b) A characteristic arsenal of other, primarily nonvocal sound material, including a choice of music genres (Reese, Gross, and Gross 2002; Schramm 2008) as well as originally produced *packaging elements* like jingles, station IDs, signatures, and so forth (Sangild 2012)

(c) A characteristic way of sequencing and mixing broadcast elements into a *broadcast flow*, resulting in rhythmic patterns, specific densities, and spectral properties (Åberg 2001; Föllmer 2013a)

(d) A characteristic overall spectral sound quality as shaped by typical studio tools for filtering, dynamic compression, aural enhancing, and so forth (Gawlik and Maempel 2009)

Having delimited these four areas of investigation, we can more easily ask about their respective functions, how they are created by radio practitioners, and what are the schemes and categories that permit the processes underlying identification.

However, if we ask these questions in such a general manner, we have to consider an enormous breadth of objects, all of them characterized by something that makes them unique (which is why we can differentiate them in the first place): different institutional forms of radio (private, public service, state, or community radio), different content types (talk radio, news channel, music formats such as adult contemporary, country, contemporary hit radio), different program elements (service elements such as the weather forecast and traffic report, news bulletins, chatter, comedy, etc.), different geographical or geopolitical areas from or to which a station is broadcasting (continents, nations, and regions with their respective cultural and language backgrounds), and, finally, the different eras of radio.

I propose using the term *stationality* and its synonyms *Anmutung* or *channel identity* to refer to the concrete, distinct sounds of individual stations. Apart from that, the term *radio aesthetics* is used to speculate about the general principles responsible for the formation of all kinds of individual stationalities, and, hypothetically, for the rules and schemes that may have been established more generally under the specific technical and infrastructural conditions of a certain time, of a definable phase in radio's evolution. Moreover, the term *identity marker* is used to express the hypothesis that certain sound elements are purposefully used to signal stationality, and thus a radio's identity.

In the following paragraphs, I will sketch three exemplary approaches to questioning individual radiophonic sounds.

Approach I: Historical Sources

The first example will illustrate how we can approach this topic via historical sources *about* radio: the following presents a 1960s witness remembering how radio was made at a certain point in time, in the context of a very special geopolitical situation.

From 1960 to 1972, the German Democratic Republic (GDR) (or East Germany) operated the AM radio station *Deutscher Soldatensender* 935. Meant to indoctrinate West German soldiers as part of the absurdities of the Cold War, it pretended to be broadcast by a West German communist force—and thus made efforts to sound West German. Helga Jacobi, who appeared for years as the voice talent "Viola" in the shows, remembers the orders she got when she first began her career at *Soldatensender*: "It all started with me not ever being allowed to listen to our own [GDR] stations anymore. ... Only western newspapers and western stations! Because you have to adopt the western style, including the timbre, the voice, that diction" (Perl and Kainz, 2000).[2]

Apparently, the station directors considered it necessary to take measures regarding the radio personalities' use of timbre, voice, and diction in order for the station's output to be perceived as authentically West German. Helga Jacobi clearly refers to different forms of media—including newspapers in addition to radio stations. That points to the fact that these measures targeted the voice's sound quality and the way in which words were pronounced, accented, and so forth (i.e., the prosody) and not only the usage of syntax and vocabulary. Interestingly, Helga Jacobi was presumably not provided with any formal training, despite the supposed national importance of this endeavor; instead, she was told to pick up the right diction, timbre, and voice by simply listening to exemplary sources. When talking to radio professionals today, one still hears similar descriptions of how to achieve a particular stationality in one's speech style.

This report leads to two conclusions: first, the process of identifying certain styles is conducted by listeners with such expertise that it is hard to fool them; second, adopting a specific attitude is not a matter of following rules out of the book, but of getting acquainted, getting in tune, so to speak, with the sound qualities characteristic of a specific group of people, in a specific area, at a specific point in time.

Approach II: Listening Modes

What is an appropriate *auditory* approach to radiophonic sound concepts? Let us take three situations with which we might all more or less be acquainted.

(1) As mentioned before, switching on a radio apparatus and flipping through the frequencies in search of our favorite station (while avoiding those we dislike) is a thing we do with great ease. In most cases, it takes us only a couple of seconds, and sometimes only a split second, to put the sounds we hear on a station into a certain category: a radio format, a program type, a journalistic style.

According to Michel Chion (1994), what we do in these moments is *causal listening*: listening with the goal of retrieving information about an audible object or process (25). Based on the years of everyday practice we all have, very detailed information can be collected in this way, and we are able to recognize many patterns quickly. As stated by Chion, the voices of individual persons, certain categories of voices (e.g., male/female, adult/child), and different types of objects (e.g., different automobiles, such as truck/sports car/limousine) as well as more general categories or *groups* of sound (e.g., human, mechanic, outdoor, or watery sounds) can, in many cases, be identified in the blink of an eye. This might even occur without any linking of the sounds to an image or linguistic term, as he proves using the example of a radio announcer we might listen to "every day without having any idea of her name or her physical attributes. Which by no means prevents us from opening a file on this announcer in our memory, where vocal and personal details are noted, and where her name and other traits (hair color, facial features—to which her voice gives us no clue) remain blank for the time being" (ibid., 27).

However, just as in many other socially relevant situations, there is more to decipher in radio. Of course there are musical styles, ways of speaking typical of a specific segment of society, or certain qualities of so-called packaging elements that we like or do not like. Models that can explain our high expertise in deciphering these and in developing preferences regarding them are theories that use the concepts of cultural capital, habitus, and social distinction (Bourdieu 1984; Schulze 1993). According to those theories, we hear a specific cultural capital, a set of values and attitudes (or *habitus*), speaking through all cultural artifacts, including radio. Thus, choosing a specific radio station is, on the one hand, often an expression of the individual desire and social necessity to be part of a specific group and, on the other hand, a way of accumulating further cultural capital and adopting markers signaling a "media habitus" (Kommer 2010) appropriate to the values relevant to the group in which we wish to remain or to which we would like to belong in the future. In short: to achieve our daily distinctive and bonding goals, we apply media-specific forms of socially aware causal listening. Vice versa, radio practitioners use and form specific habitus markers in order to address certain parts of society as their audience. A simple proof for this assumption is the definition of so-called *personas*—that is, exemplary listeners described in detail in terms of their income, tastes, interests, and so forth in most radio stations' stylebooks (a resource defining rules of production and journalistic work).

(2) For the second listening example, let us imagine a heavily formatted radio program where presenters often do little more than use fixed tropes such as the three-element break. This mainstay of media rhetoric usually involves, for instance, stating the current time, the station name (or slogan), and the next music title to be played.

A good portion of radio research has been using what Michel Chion (1994, 28) calls *semantic listening* to examine media products such as the three-element break: taking every utterance of words in radio strictly as *language* signifying the *meaning* inherent in those words. As radio scholars from the cultural studies tradition have pointed out, this analytical approach misses the point. Asking what is in it for the listener, they comprehended the function of those repetitive, content-empty broadcast elements with principles such as foregrounding the interpersonal over the ideational (Montgomery 1986), constructing an "intimacy at a distance" (Tolson 2005, 15) between presenter and individual listener, enhancing "sociability" as a social relationship between broadcaster and a larger audience (Scannell 1996, 23), "domesticating" radio by making it part of the facilities and requirements appropriate to the private home (Lacey 2002), and "secondariness" as radio's primary function—namely, that of being not attentively listened to in many cases, but overheard while doing and focusing on something else (Crisell 1996). Summing up these lines of thought, Carin Åberg argues that very often in radio "the presenter is regarded more as a pleasant voice than a serious reporter" (2001, 83).

How can radio practitioners foreground the interpersonal, construct intimacy, enhance sociability, and exploit domestication and secondariness using schematic rhetoric tropes? Although the wording of three-element breaks is indeed rather similar even across different station types, the way they are delivered differs considerably. The diction, the voice, the way music and packaging elements are mixed in and layered in respect to the presenter's speech: all this conveys much more than meaning inherent in words.

(3) To describe the characteristics of these elements, we finally arrive at the third way of listening described by Chion: *reduced listening* (1994, 29), a term originally coined by *musique concrète* composer Pierre Schaeffer with reference to Edmund Husserl's phenomenological concept of reduction (ibid., 216): "Reduced listening has the enormous advantage of opening up our ears. … The emotional, physical, and aesthetic value of a sound is linked not only to the causal explanation we attribute to it but also to its own qualities of timbre and texture" (ibid., 31). Schaeffer had developed the concept in order to move away from the dominance of causal attributions when listening to music

constructed by concrete sounds—that is, when using recordings of sound emanating from everyday objects. By means of repetition and other forms of distancing this kind of material from its origins, he strived to achieve a mode of listening detached from learned attributions and significations.

In listening to sonorous objects [*objets sonores*] whose instrumental causes are hidden, we are led to forget the latter and to take an interest in these objects for themselves. ... In fact, Pythagoras' curtain is not enough to discourage our curiosity about causes, to which we are instinctively, almost irresistibly drawn. But the repetition of the physical signal, which recording makes possible, assists us here ... by exhausting this curiosity, it gradually brings the sonorous object to the fore as a perception worthy of being observed for itself. (Schaeffer 2004, 77)

As scholars interested in the sounds of particular media, such as the radio, we can adopt this listening strategy in order to start questioning and understanding why specific elements in a medium sound the way they do.

A good example to explain this is the *swoosh*. A swoosh is a synthetic sound element often used right at the beginning of certain types of radio packaging elements, such as transitions, bumpers, or drop-ins. When listening to the swoosh semantically, it is often first related to science fiction (the spaceship *swooshing* by). The interpretation may then be that the swoosh is used to signify modernity, energy, a dynamic situation, and so forth—but since it gets repeated over and over again, day in, day out, it becomes degraded, reduced to a gimmick sound that is awkwardly calling for attention.

However, turning to *reduced listening* enables us to temporarily disregard the learned signification and to pay attention to the sound spectrum of the swoosh. This listening reveals that, due to the extremely wide and quickly changing spectrum of the phasing noise that forms the swoosh, when the swoosh comes in, it is capable of masking any sounds contained in the music being played at that moment. Therefore, it proves to be an ideal element to prevent rhythmic, harmonic, or other "collisions" or disharmonies during the transition between two music titles, to ensure smooth broadcast flow. Apparently, sounds like this are not used in order to convey meaning to the listener, but to make him or her feel safe and comfortable when listening to the program.

Approach III: Phenomenological Variation

However hard to control in day-to-day radio production, rather simple experiments can reveal indicators that can be assumed to form identity markers to which listeners react when picking (and thus categorizing) their radio programs. In a number of experiments, evidence was found for the importance of several indicators of the formation of stationalities.

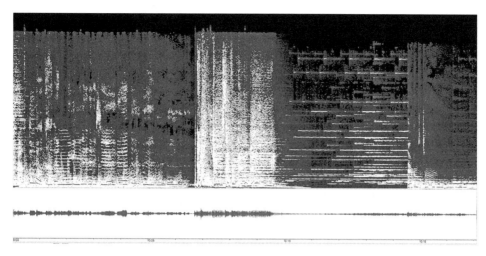

Figure 28.1
Spectrogram of a seventeen-second passage from the Kiss FM morning show in Spain, recorded online on June 13, 2014. At the bottom, the timescale and waveform are shown; the spectrogram in the upper part of this figure shows the hearing range (20–20–000 Hz) with low frequencies at the bottom, brighter areas equaling higher volumes. The passage shows the ending of a music title (left) and the beginning of another (right). The large bright patch in the center shows a packaging element with much higher overall volume level apparently, which opens with a "splashy" swoosh, also called a "stager," which cuts over fading music with a frequency spectrum that covers the full hearing range—thus effectively attracting attention and masking all other existing audio signals. The spectrogram was made with the software EAnalysis.

A presenter's twenty-second moderation segment at *MDR Figaro*, a regional German *Kulturradio* (radio format focused on cultural topics), was modified according to the method of phenomenological variation. Adapting Philip Tagg's description for the application of this method to popular music (Tagg 1982), the original recording was altered on the basis of heuristic assumptions drawn from reduced listening, switching back and forth between a number of *Kulturradios* such as *MDR Figaro* and a number of contemporary hit radio formats aimed at a much younger audience. The assumptions were:

(1) Kulturradio presenters tend to speak with a lower average pitch than presenters of programs aimed at younger audiences.
(2) Kulturradio presenters tend to use a narrower range of frequencies (in music terminology, *ambitus*) than those of youth programs.
(3) Kulturradio presenters tend to speak more slowly than those speaking to younger audiences.

There are many more differences that could be discussed, such as the degree to which vocal extremes (e.g., a vocal harshness or throatiness to express surprise, speaking more softly to express intimacy, or speeding up to express excitement, etc.) are applied. However, the three characteristics first mentioned were chosen because they are of a type that can easily be modified by technical means today.

Using the professional pitch and harmonics correction software Melodyne, the aforementioned example recorded off of a *Kulturradio* station was raised in pitch by a minor third, it was sped up by 10 percent, and its ambitus was raised from a ninth (fourteen halftones) to more than a twelfth (roughly twenty halftones) by manually dragging single phonemes to different pitch levels. Although the original recording would have been categorized in most cases as stemming from a cultural or news channel, listeners tended to categorize the presenter in the modified example as representing a youth radio station. Without changing any element of the content, that is, by modifying only those elements that are part of the sound of a radio station, the perceived stationality was crucially altered (Föllmer 2013b).

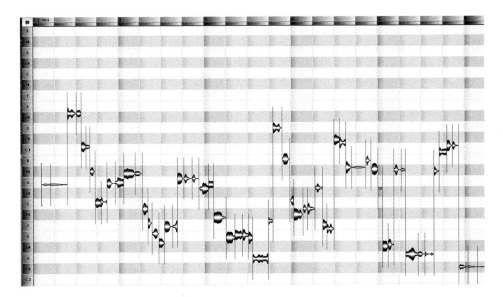

Figure 28.2
Representation of approximately 20 seconds of an original (non-modified) presenter's voice recorded off of the German public service radio MDR Figaro on September 2, 2009, as analyzed with the software Melodyne. The so-called "blobs" represent lengths and pitches (if related to the time scale at the top and the pitch scale on the left) of speech phonemes which are automatically identified and segmented by the software. As the distribution shows, the melodic range is approximately a ninth (fourteen halftones).

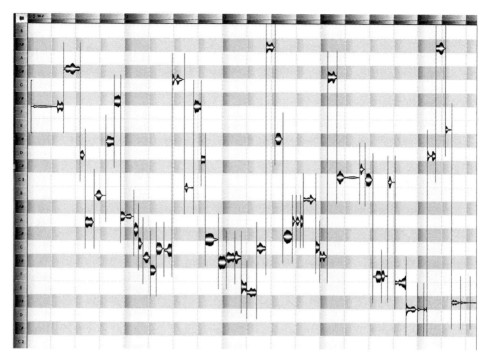

Figure 28.3

Representation of the same voice as in figure 28.2, this time modified using different tools offered by the software Melodyne. Consulting the pitch scale to the left, the distribution of the "blobs" now shows that the average pitch was raised about three halftones and the melodic range was stretched to approximately twenty halftones. Speed was increased about 1 percent, thus the number of measures (time scale at the top) covered by the segment is decreased.

Similar experiments were conducted on the timing, volume leveling, and sound processing of broadcast elements in the mix, in the interest of pinpointing the typical identity markers for certain radio formats. Apparently, the choice of the moment when the presenter speaks over a fading music title, the volume chosen for a music title playing while the presenter is speaking, and the settings used on audio processors in the radio transmission chain play fundamental roles in the formation of stationalities (ibid.).

Conclusion and Outlook

In summary, the approaches discussed in this chapter prove the significance of radiophonic sounds in general, as well as in naming particular identity markers responsible for individual stationalities. It appears that radio producers have been successful in producing broadcast flows that result in continuously coherent radio-phonic sounds or stationalities. The examples presented also revealed individual elements that are of particular importance to the marking of stationality. These identity markers enable producers to communicate stationality repeatedly and at chosen points in time.

Not all of the conceivable and/or undertaken approaches have been discussed here. For instance, the results drawn from the method of phenomenological variation beg for more empirical validation and refinement in a controlled lab environment, which has only taken place to a minor degree up until now. Moreover, although listeners are undoubtedly capable of using identity markers for categorizing purposes, it is unclear which elements they actually use and react to as identity markers, and what the assumptions are that they draw from "identifying" a station, for instance regarding the values and thus the audience it aims to represent. Also, observing reactions to stations from outside an audience's usual reception area might be telling regarding questions about a global radio aesthetics, just as conversations with production experts from different types of radio stations and different nations are likely to help answer the question of whether we can think of radio not only as a collection of the many individual manifestations of what radio can be, but also whether there is a specific, generalizable sound of radio as a whole—an auditory modus operandi distinct from those of other media and unique to radio alone.[3]

Notes

1. This difficult to translate and old-fashioned term comes close to Walter Benjamin's concept of "aura," but without Benjamin's requirement of uniqueness. It is thus close to Gernot Böhme's (1995) concept of "atmospheres," which describes relationships between different sensual qualities of environments—be those rooms, streets, or media—supplying a background to foreground events and perceived subconsciously for the most part.

2. "Angefangen hat es damit, dass ich absolut keinen Sender von uns mehr hören durfte. ... Nur West-Zeitungen und West-Sender! Denn man muss ja den Stil vom Westen raufkriegen, auch ... vom Timbre, von der Stimme, von dieser Diktion." Transcription by the author.

3. At the time of writing, investigations into radiophonic sound have been conducted, for instance, in the context of an interdisciplinary network of European scholars (http://radioaesthetics.org) and as part of the European research project "Transnational Radio Encounters" (2013–2016; http://transnationalradio.org).

References

Bibliography

Åberg, Carin. 2001. Radio analysis? Sure! But how? In *Radio-Kultur und Hör-Kunst: Zwischen Avantgarde und Popularkultur 1923–2001*, ed. Andreas Stuhlmann, 83–104. Würzburg: Königshausen & Neumann.

Böhme, Gernot. 1995. *Atmosphäre: Essays zu einer neuen Ästhetik*. Frankfurt am Main: Suhrkamp.

Bose, Ines, and Sven Grawunder. Forthcoming. How to describe radio voices. In *Sound Bridges, Sound Walls: Broadcasting in the Historical Formation, Mediatization, and Localization of Sound*, ed. Alexander Badenoch, Golo Föllmer, and Hans-Ulrich Wagner. Köln: Herbert von Halem.

Bourdieu, Paul. 1984. *Distinction: A Social Critique of the Judgement of Taste*. Cambridge, MA: Harvard University Press.

Chion, Michel. 1994. *Audio-Vision: Sound on Screen*. New York: Columbia University Press.

Crisell, Andrew. 1996. *Understanding Radio*, 2nd ed. London: Routledge.

Föllmer, Golo. 2013a. Theoretical-methodical approaches to radio aesthetics: Qualitative characteristics of channel-identity. In *Electrified Voices: Medial, Socio-Historical, and Cultural Aspects of Voice Transfer*, ed. Dmitri Zakharine and Nils Meise, 325–341. Göttingen: Vandenhoeck & Rupprecht.

Föllmer, Golo. 2013b. Radio aesthetics—radio identities: A project overview and a methodological study on the experimental variation of radio speech. In *Radio: Community, Challenges, Aesthetics*, ed. Grażyna Stachyra, 225–234. Lublin: Marie Curie-Skłodowska University Press.

Gawlik, Fabian, and Hans-Joachim Maempel. 2009. The Influence of Sound Processing on Listeners' Program Choice in Radio Broadcasting. 126th AES Convention, AES, P24-4, Convention Paper 7785.

Kommer, Sven. 2010. *Kompetenter Medienumgang? Eine qualitative Untersuchung zum medialen Habitus und zur Medienkompetenz von SchülerInnen und Lehramtsstudierenden*. Leverkusen: Budrich UniPress Ltd.

Korner, Anthony. 1986. Aurora musicalis: An interview with Brian Eno. *Artforum* 24 (10): 76–79.

Lacey, Kate. 2002. Radio in the Great Depression: Promotional culture, public service, and propaganda. In *Radio Reader: Essays in the Cultural History of Radio*, ed. Michele Hilmes and Jason Loviglio, 21–40. New York: Routledge.

Montgomery, Martin. 1986. DJ talk. *Media, Culture & Society* 8:421–440.

Reese, David E., Lynne Gross, and Brian Gross. 2002. *Radio Production Worktext: Studio and Equipment*. Boston: Francis & Taylor.

Sangild, Torben. 2012. Radiolab: A radio signature analysis. *Sound Effects* 2 (1): 139–152.

Scannell, Paddy. 1996. *Radio, Television, and Modern Life*. Oxford: Blackwell.

Schaeffer, Pierre. 2004. Acousmatics. In *Audio Culture: Readings in Modern Music*, ed. Christoph Cox and Daniel Warner, 76–81. New York: Continuum.

Schramm, Holger, ed. 2008. *Musik im Radio: Rahmenbedingungen, Konzeption, Gestaltung*. Wiesbaden: VS Verlag.

Schulze, Gerhard. 1993. *Die Erlebnisgesellschaft: Kultursoziologie der Gegenwart*. Frankfurt am Main: Campus.

Tagg, Philipp. 1982. Analysing popular music: Theory, method, and practice. *Popular Music* 2:37–65.

Théberge, Paul. 1989. The "sound" of music: Technological rationalization and the production of popular music. *New Formations* 8:99–111.

Tolson, Andrew. 2005. *Media Talk: Spoken Discourse on TV and Radio*. Edinburgh: Edinburgh University Press.

Sound and Media

Perl, Angelika, and Peter Kainz. 2000. *Der Laubfrosch hat die Farbe gewechselt: Geheimes Radio im Kalten Krieg*. Radio feature, DeutschlandRadio Berlin, November 8, 1 hr., 3 min.

29 World Music 2.0: Updated and Expanded

Thomas Burkhalter

World Music 2.0: The Concept

It is through interviews with musicians in the Arab world, Africa, Asia, and Latin America, and through their tracks and videos on SoundCloud and YouTube, that I came up with the World Music 2.0 concept in 2010 (see online version 2011,[1] and Burkhalter 2012). It is a theoretical category meant as a provocative hint at the fact that there are still many boundaries to cross and stereotypes to break within transnational music networks. The World Music 2.0 concept argues that music from the Middle East, Africa, Asia, and Latin America has the potential of coproducing multisited modernities. Musicians use the possibilities of digitalized music production, distribution, and reception for a more individual and manifold production of music. They construct confident positions and form new multisited avant-gardes of the twenty-first century. On the other hand, these musicians are still caught in the old postcolonial structures and dependencies whenever they aim to reach Euro-American platforms—they need invitations from international funders and festivals (see my article "Sound Studies across Continents," this volume); processes of "othering" are ongoing. The focus on diversity, ethnocentrism, and Orientalism in Euro-American reception platforms still forces musicians to constantly prove their cultural identity, their "Arabness" or "Africanness."

This chapter revisits the World Music 2.0 concept from two main angles: first, new renderings of war in music; second, new renderings of "exotica." It may challenge the World Music 2.0 concept with new sounds, music, and noises; it aims to expand the concept to a model that shows the great variety of artistic and symbolic strategies in the reworking and recontextualizing of acoustic material and references in translocal music production.

SoundCloud: Drinking Coffee to Audio Battles

Today, tracks, compositions, samples, and field recordings from Burkina Faso or Honduras are just a few mouse clicks away.

April 18, 2014. Safely at home in Bern, Switzerland, I sit in front of my laptop, warm coffee in hand, with a fast Internet connection. I browse through the online audio portal SoundCloud—I hope to find tracks that are fresh, powerful, extravagant, surprising, and yet distinct to my ears. I also hope to find tracks that go against the often-overheard idea that contemporary music is a one-way stream, flowing from the United States and Europe to the "world" and never back in the other direction. This mix of interests and motivations led to the fact that the 628 users I follow[2] live across the world—in Africa, the Middle East, Europe, the United States, Switzerland, and Asia. Some upload experimental sound and noise, others club tracks and a few field recordings and radio art.

Today, I grow fascinated and intrigued by a great amount of "war dubs," released by producers from the *grime* networks in the United Kingdom:[3] the Oil Gang mixes gunshots with the typical synth sounds of the grime genre in their track "Novelist." In "State of War," Lemzly Dale offers noises of weapons and electronic string sounds. War sounds hit hard in "Hit Somebody" by Plastician, accompanied by deep bass. Deeco plays with the old fascination of marching band rhythms first in "Fatality"; the track then transforms into a harsh sounding, brutal battle—amazingly powerful through my headphones.

I call grime producer Footsie in London via Skype, and he explains what war dubs are all about: various producers had decided to hold a competition on who could release the most aggressive tracks within a limited time period. Tracks were uploaded to SoundCloud, and the one that reached most listeners would win. In this audio battle, weapons of war mixed with heavy bass, string sounds, and powerful rhythmic patterns seem a convincing path to victory. "War Dubs" created "a lot of noise," Footsie tells me. "All UK media wrote about it."

Beirut: Memories from the 2006 War

My mind starts turning, and memories come back. In 2006, I witnessed sonic booms and air strikes by Israeli fighter jets in Beirut. What now triggers fascination through headphones had once been fear:

Two deep booms came long before the wake-up call. The Israeli air force had bombarded Beirut International Airport, BBC World reported. The next night was even worse. Israeli planes produced one sonic boom after another. It sounded like heavy thunder but louder, and each explosion was accompanied by short cracking noises. It felt as if the sky would crash down on me. Immediately, I became overly cautious. Shall I go out on the balcony to observe what is happening? Definitely not! Shall I record those sounds? No! Don't press your luck! A *boom*, very deep and aggressive, ended the spectacle: The first bomb was fired from a ship into Southern Beirut. The city fell quiet. Birds sang. (Burkhalter 2013, 1)

By spring 2014, over 150,000 people are reported to have lost their lives in the civil war in Syria. Cities such as Aleppo are bombarded daily. Are war dubs from the UK thus ignorant, or morally or ethically wrong? We can ask the same question with these and many other compositions, tracks, and sound collages: in the early part of the twentieth century, composer Luigi Russolo was fascinated by the noises of war, and he describes them in his manifesto *L'Arte Dei Rumori* (2005) in detail. For him and other Futurists, war shows humankind in all of its beauty, passion, and reality (Witt-Stahl 1999). Ludwig van Beethoven "imagined" war in his orchestral work *Wellingtons Sieg* (1813); British electronica artist Matthew Herbert reworked samples from the Libyan Civil War for his album *The End of Silence* (2013). In 2006 in Beirut, trumpet player Mazen Kerbaj went out on his balcony and improvised to the noises of falling bombs. We hear his trumpet together with bombs dropped by Israeli warplanes. In between, we hear car alarms going off due to the reverberation from sonic booms. Kerbaj named his recording "Starry Night" (after a painting by Vincent van Gogh) and uploaded it to his blog.[4] The recording was soon embedded on the website of the UK music magazine the *Wire*. It pushed Kerbaj's international career forward, but led to a lot of criticism:

Lebanese and non-Lebanese argued that I used the situation to become famous. Well, please come and live in Beirut, and try it yourself. To be honest, I preferred to stand on the balcony, to play trumpet and to record those bombs, rather than to stay in the living room and go crazy. When you play, you shift your brain, and you hear those bombs as sounds, and not as killing machines. Just to continue working helped many of us to stay sane. (Kerbaj, interview by author)

Beyond the "using war to gain fame" argument, Lebanese criticism stated that Kerbaj comes from an elite family and lived in a neighborhood that was not being targeted by Israel. Bombs fell a few kilometers away, in Hezbollah-run areas.

YouTube: Drinking Coffee and Watching Exotic Video Clips

In Switzerland, I can afford to take a break from this tough topic, and I pour myself a second coffee. I start surfing on YouTube—perfect for some lighter fare. Again, I receive specific information, through my "friends" on Facebook (to a greater extent than on SoundCloud, situated locally in Switzerland) and—mainly—through the 150 blogs, newspapers, and academic journals that I follow via the Rich Site Summary (RSS) reader feedly.

I have noticed that many musicians have rediscovered the appeal of "exotic" sounds and images. In "Tribal Skank" by Fr3e (2009), white British businessmen and black

Figure 29.1
Mazen Kerbaj and Sharif Sehnaoui, free improvisers from Bern, at Forum Freies Theater, Düsseldorf, 2009 (photo courtesy of Thomas Burkhalter).

British singers and dancers, wearing "bush" costumes and "traditional" tattoos, dance around a fire to an electronic UK Funky tune. Wearing a leopard costume, DJ Ganyani's singer FB sings and dances to South African house music beats in "Xibugu" (2013) in a tourist-style version of an African village. Extravagant video clips come from the Ghanaian rap duo FOKN Bois, comprised of Wanlov the Kubolor and M3nsa. In "Uncle Obama" (2012) by Sister Deborah (sister of Wanlov the Kubolor), the two appear next to monkeys and bananas. In "Sexin Islamic Girls" (2012) they sit in front of an African mud house on a bench with three girls wearing headscarves. The girls also eat bananas (in a rather pornographic way). And in "BRKN LNGWJZ" (Broken Languages), I observe the FOKN Bois walking through a city in "traditional" African costumes and through the countryside in suits. They discuss their multiple identities: "I be" the "mango tree climber," "leaf eater," "seed cracker," "bare-hands orange peeler," "fanti, ga and twi speaker," "condom non-wearer," "polygamist lover," "police briber," "lotto staker," "choir girls fingerer," "macho rapper"—the list continues for 6.03 minutes.

This self-exoticizing stands in contrast to what I kept hearing from musicians and sound artists in the Middle East, in Africa, and South-Eastern Europe: they want to be modern, *in* the world, and recognized within their specific transnational niche music networks. Their aim is to create individual musical identities—separate from self-exoticism, commercialization, and propaganda. With their music, video clips, cover art, posters, and lyrics, they attack the old model of Euro-modernity—or Euro-American modernity. They share the desire for new, multisited modernities with many scholars (e.g., Escobar 2007; Gaonkar 2001; Randeria and Eckert 2009) and individuals worldwide.

On YouTube, however, the body of work dealing with exotica seems to be growing. Is it because of Facebook? For that which is obscure, ironic, and exotic does indeed benefit disproportionately from the avalanche effect of likes and shares.

Ghana: Fighting Pentecostal Churches and Corruption with Parody

In March 2013, I spent two weeks in Accra, Ghana, working on a documentary film about the FOKN Bois. Things became complex. The FOKN Bois do not look toward Europe or the United States exclusively. On the contrary, they fight with and against very local issues and topics—and they do so mostly through parody; their "exotica" disturbs local viewers too. They make fun of corruption in local politics, and they go against the fast-growing influence of Pentecostal churches and rich bishops in Ghana and other African countries. One can watch this at its best in their rap "musicals": *Coz Ov Moni* (2010) and *Coz Ov Moni II (FOKN Revenge)* (2013), both produced exclusively by their local network. In *Ghana is the Future*—the XL teaser to our documentary film— Wanlov the Kubolor explains that their art is a kind of "deformed anger," a way to let their "frustration out."[5] He makes it clear that, while subversive, FOKN Bois are still a patriotic project: "We were never raised or trained or programmed to be patriotic, to care about the country. We're just raised to know that whatever we are learning is for us to get a visa to go somewhere else or to go to heaven" (*Ghana Is the Future*, 2013).

Beirut: Breaking Taboos

In Beirut, Mazen Kerbaj and his colleague Raed Yassin break a variety of taboos when working around the topic of war. Raed Yassin draws heavily on acoustic source material from the Lebanese civil war in many of his sound works. His sound collage "Civil War Tapes" (Burkhalter 2013) mixes and manipulates propaganda songs from various right-wing militias with those of left-wing singers, political speeches from leaders of all the religious groups, news jingles and ads (from Pepsi to Barilla) from around two hundred radio channels, melodies from children's programs, noises from bombs and rifles, the kitschy Arab pop music of singers like Sammy Clark, and much more.

From a Euro-American perspective, Raed Yassin and Mazen Kerbaj generate interest when focusing on the Lebanese civil war—this is what the World Music 2.0 concept argues. From a Lebanese perspective, however, they are breaking a taboo when discussing war in public (Khalaf 2002). From a political and historical perspective, we hear the conjunctions between music, warfare, and the Middle East. From an artistic perspective, we can see them fighting for musical and artistic quality in a region that is full of commercialism and propaganda. And from a biographical perspective, many of the musicians might use music as therapy for dealing with their personal traumatic war memories. Many other perspectives can be added. It is these diverse and contradictory perspectives that I would like to use to expand the World Music 2.0 concept.

Anarchive: Sounds from the Dustbin of History

Similar complexities emerge when we look at the topic of exotica. Many musicians and sound artists enjoy sounds and styles that fall outside the canon of their home countries' official music history. Elites ignored them and referred to them as tasteless, noisy, trash, kitsch, cheap, or Orientalist. In Beirut, Raed Yassin and contemporaries often focus on music and songs from the belly dancing culture of the 1950s and 1960s ("that violated every boundary of authenticity"; see Rasmussen 1992, 69), quickly and cheaply produced Arabic pop of the 1980s (Frishkopf 2010, 18), and new wave dabké and Mahragan (or Electro-Sha'abi) of the 2000s (see Burkhalter 2014).[6] These musicians dig into what Reynolds calls the "anarchive" (2011, 151)—the dustbin of history. "The archive degenerates into the anarchive: a barely navigable disorder of data-debris and memory trash" (ibid.).

Figure 29.2
Raed Yassin performs the anarchive at Zentrum Paul Klee Bern, 2009 (photo courtesy of Thomas Burkhalter).

In ethnomusicologist Veit Erlmann's words, these artists have replaced "pastiche" with "parody" when working around exotica. Erlmann describes World Music 1.0 (what we tend to find in our record stores) with the term *pastiche*, which he defines as a form of parody lacking the polemical or satirical aspect. World Music 1.0 tries to highlight "unspoiled" and "authentic" musical forms and idioms. However, it mixes sounds of a completely commercialized present with the pseudohistorical patina of different places and times (Erlmann 1995, 25). Musicians today joyfully play with the Euro-American fetishizing and exoticizing of the East.

In focusing on the anarchive, the musicians too are also rewriting history. They dig out forgotten artists such as the Lebanese rock band the Seaders, who released an album in the UK in the 1960s, or Egyptian composer Halim El-Dabh. His piece "The Expression of Zaar" (1944), a recording of a ritual in Egypt made with a wire recorder, is today considered the first *musique concrète* piece in history (Burkhalter, Dickinson, and Harbert 2013). In short, the anarchive hints at the fact that modernity in music was always coproduced within networks across the world.

Dreams and Strategies at Play

The more we leave the safe haven of Switzerland, the more hopes, dreams, challenges, possibilities, and strategies come into play. World Music 2.0 is only one possible reading—and one from a Euro-American perspective. For the FOKN Bois, Raed Yassin, and other musicians and sound artists, however, Euro-American platforms seem less important by the year. In their track "Help America," the FOKN Bois collect money for the United States, foreseeing changes in the global economy and joking about it at the same time. They and their fellow artists in Ghana are experimenting with new models to generate income. Their film *Coz Ov Moni I* was partly financed through a small, private goldmine run by Panji Anoff, an art lover and owner of the production company Pidgen Music. Pop stars and DJs in Ghana are becoming ambassadors of mobile phone companies. In a similar way, musicians and sound artists in Beirut and the Arab world are working on a variety of platforms. Along with the other member of the duo, Praed, Raed Yassin tours just as often in Japan as in Europe. Performing in Europe seems to add mainly symbolic value to an artist's name, and this may generate opportunities and income on local and regional platforms in the long term.

Model: Reworking Material and References

If our aim is to analyze audio and video in translocal music making (e.g., around topics such as war or exotica), a few clicks of the mouse are not enough. I may begin to get the feeling that I am an expert of these musics, when in reality I have no sense of the meanings and functions of these musics in other contexts. There are a great variety of artistic and symbolic strategies at play that artists are using to rework and recontextualize material and references (from different places and times) into their music. Asking the following questions might be fruitful:

(1) Music making as a process
   (a) What sounds, images, texts, or topics are being worked with?
   (b) From where and when is this material?
   (c) How is the material transformed (mono- to pop-referential; Burkhalter 2012) and manipulated (e.g., in pitch, speed, running direction; through filters and effects)?
(2) The musician as an actor
   (a) What reasons does the artist give for using the material?
   (b) What position does the new music (and the artist) take in regard to the material (e.g., homage, parody, militant radicalism, total abstraction)?
   (c) Where does the artist/producer live?
   (d) What is the artist's background (e.g., cultural, social, economic, gender, educational)?

(e) How much knowledge does the artist have of the material?

(f) Which "world orientation" becomes evident in his/her music?

(3) Music as a media product

(a) On which reception platforms is the music sold and discussed?

(b) What new meanings get attached to the music on various reception platforms?

(c) What are the discussions/arguments around the music (e.g., legal, ethical, philosophical)?

Soundscape pioneer Murray Schafer told students that they needed to know the names of the frogs they were recording.[7] I do not agree with him. Answering the questions above, however, shows what is at stake—otherwise, sounds remain ghosts of another place. A deeper analysis leads to stories that highlight in detail the possibilities and challenges offered by our increasingly digitized, globalized, and urbanized world.

Notes

1. Weltmusik 2.0, http://norient.com/academic/weltmusik2-0/ (accessed May 1, 2014).

2. See https://soundcloud.com/norient (accessed May 1, 2014).

3. See Set "War Dubs," https://soundcloud.com/norient/sets/war-dub (accessed May 1, 2014).

4. http://mazenkerblog.blogspot.ch/ (accessed April 23, 2014).

5. Watch XL-Teaser: http://norient.com/video/teaser-africa-is-the-future/ (accessed April 23, 2014).

6. http://norient.com/blog/islam-chipsy/ (accessed April 14, 2014).

7. Podcast: http://norient.com/podcasts/murrayschafer/ (accessed April 14, 2014).

References

Bibliography

Burkhalter, Thomas. 2011. World Music 2.0. In *Norient: Network for Local and Global Sounds and Media Culture*. http://norient.com/academic/weltmusik2-0/.

Burkhalter, Thomas. 2012. Weltmusik 2.0: Musikalische Positionen zwischen Spass und Protestkultur. In *Out of the Absurdity of Life: Globale Musik*, ed. Theresa Beyer and Thomas Burkhalter, 28–48. Solothurn: Traversion.

Burkhalter, Thomas. 2013. *Local Music Scenes and Globalization: Transnational Platforms in Beirut*. New York: Routledge.

Burkhalter, Thomas. 2014. The Yamaha PSR OR-700 à la Islam Chipsy. In *Norient: Network for Local and Global Sounds and Media Culture*. http://norient.com/blog/islam-chipsy/.

Burkhalter, Thomas, Kay Dickinson, and Benjamin Harbert, eds. 2013. *The Arab Avant Garde: Musical Innovation in the Middle East.* Middletown, CT: Wesleyan University Press.

Erlmann, Veit. 1995. Ideologie der Differenz: Zur Ästhetik der World Music. *PopScriptum* 3 (3): 6–29.

Escobar, Arturo. 2007. Worlds and knowledges otherwise: The Latin American Modernity/ Coloniality Research Program. *Cultural Studies* 21 (2–3): 179–210.

Frishkopf, Michael. 2010. *Music and Media in the Arab World.* Cairo: American University in Cairo Press.

Gaonkar, Dilip Parameshwar. 2001. *Alternative Modernities.* Durham, NC: Duke University Press.

Khalaf, Samir. 2002. *Civil and Uncivil Violence in Lebanon—A History of the Internationalization of Communal Conflict.* New York: Columbia University Press.

Randeria, Shalini, and Andreas Eckert, eds. 2009. Geteilte Globalisierung. In *Vom Imperialismus zum Empire*, 9–31. Frankfurt am Main: Suhrkamp.

Rasmussen, Anne K. 1992. An evening in the Orient: The Middle Eastern nightclub in America. *Asian Music* 2:63–88.

Reynolds, Simon. 2011. *Retromania: Pop Culture's Addiction to Its Own Past.* London: Faber & Faber.

Russolo, Luigi. 2005. *Die Kunst der Geräusche.* Ed. Johannes Ullmaier, trans. Owig DasGupta. Mainz: Schott.

Witt-Stahl, Susann. 1999. *But His Soul Goes Marching On: Musik zur Ästhetisierung und Inszenierung des Krieges.* Forum Jazz Rock Pop 3. Karben: Coda.

Sound and Media

Burkhalter, Thomas. Norient account on SoundCloud. https://soundcloud.com/norient (accessed December 23, 2014).

Burkhalter, Thomas, and Peter Guyer. 2014. *Ghana Is the Future—XL Teaser.* Documentary, Recycled TV AG and Norient. http://norient.com/video/teaser-africa-is-the-future/ (accessed December 23, 2014).

DJ Ganyani. 2013. *Xibugu.* http://youtu.be/npWF7kvyWc4 (accessed December 23, 2014).

El-Dabh, Halim. 2000. *Crossing into the Electric Magnetic.* CD, Without Fear, WFR003.

FOKN Bois. 2011. *BRKN LNGWJZ (Broken Languages).* https://www.youtube.com/watch?v=wdg -_TRiNkw (accessed December 23, 2014).

FOKN Bois. 2012. *Sexin Islamic Girls.* https://www.youtube.com/watch?v=to4jZDaVHKo (accessed December 23, 2014).

FOKN Bois. 2012. *Coz Ov Moni I.* https://www.youtube.com/watch?v=R_YsQK2Yo3c (accessed December 23, 2014).

FOKN Bois. 2013. *Coz Ov Moni II (FOKN Revenge)*. http://foknbois.bandcamp.com/merch/coz-ov-moni-2-fokn-revenge-dvd (accessed December 23, 2014).

Fr3e. 2009. *Tribal Skank*. https://www.youtube.com/watch?v=qxVXbQPOS1M (accessed December 23, 2014).

Herbert, Matthew. 2013. *The End of Silence*. CD, Accidental Records, AC74CD.

Kerbaj, Mazen. 2006. *Starry Night*. http://mazenkerblog.blogspot.ch/ (accessed December 23, 2014).

Sister Deborah. 2012. *Uncle Obama ft. FOKN Bois*. https://www.youtube.com/watch?v=b2HSo3yywDU (accessed December 23, 2014).

Various Artists. 2014. *War Dub—Grime from the UK, Japan, and …* Bern: Norient. https://soundcloud.com/norient/sets/war-dub (accessed December 23, 2014).

30 Computer Game Sound: From Diegesis to Immersion to Sonic Emotioneering

Mark Grimshaw

Introduction

The cross that any new field of study must bear is that it is defined in terms of one or more already existing and similar fields of study—at least initially. Thus it is with the relatively new field of study of computer game sound: a field of study that defines and examines our perception of, use of, and relationship to sound FX in computer games. In this field, the similar and preexisting field of study that has been most widely used is the study of film sound, particularly as espoused by Michel Chion. Other such fields that have since been used include the study of presence, sonification, cognition, psychology, psychoacoustics, and ecological acoustics. However, once the new field of study reaches a sufficient level of maturity, it is in a position to redefine itself as no longer similar (although perhaps with some overlapping areas), but as a significantly distinct field of study requiring its own terminology and conceptual thinking. In this case, computer game sound has long differentiated itself from film sound through the use of terms such as interaction and immersion, despite making use of a large number of film-related terms and concepts. It is this trajectory that underpins the following brief exegesis of my study of computer game sound, as well as the presentation of the main concepts and debates around the subject and my thinking about the future direction of the field.

At the time of writing in early 2013, it is my contention that the field of computer game sound has reached that level of maturity. At this point, I must make a distinction between sound—that is, sound FX—and music in computer games.

Despite my own background in the study of music composition, then popular music recording and music technology, my interest in computer games is confined to sound FX. This is not merely owing to a personal interest driven by a fascination with sound FX developed during many, many hours of having fun playing computer games; it is because I believe that such game sound, produced and utilized in ways very different from game music, can lead to a better understanding of the player's engagement with the game world than can the study of game music. Moreover, this understanding helps to explain the phenomenon of our perception of the world outside the game and the positioning of our selves within it. Being a motivating factor for the rigorous study of computer game sound, this is a theme that provides support to my contention and one that I develop throughout this essay.

When I began my study of computer game sound ten years ago, only a handful of short articles that directly concerned themselves with the field had been written. I wished to discover what the relationship was between player and sound in the game genre first-person shooter. I chose that genre for my analysis not only because I had been a longtime player of games such as Quake III Arena (id Software, 1999), but also because it seemed to me that such a genre would prove to be the most fertile ground for the study of this relationship. In this genre, the player is visually and conceptually positioned within a three-dimensional game world; I thus sought to discover and assess the role that sound played in immersing the player within that world. There are matters of narrative and plot to contend with, which led to the possibility of debating the function of sound in that respect. In addition, some of these games strive for visual and sonic realism, with others being more fantastical in their premise, so there were also questions of realism, authenticity, and verisimilitude to pursue.

Ultimately, I was able to devise a conceptual framework for the sound of the first-person shooter genre that utilized both existing and new concepts and terminology (Grimshaw 2008a). Thus, terms such as diegetic and acousmatic sound were borrowed from film sound theory (e.g., Chion 1994). In doing so, I was following the work of one of the earliest game sound scholars, Axel Stockburger (2003). However, because of the specific nature of computer games, and in particular first-person shooters, I invented a number of neologisms—some to nuance the concept of diegesis in relation to computer game sound and some to define the ways in which sound FX help construct and allow the perception and navigation of the various visual, spatial, and temporal elements of the game world. I also explored the concept of the first-person shooter as an acoustic ecology, further developing territory first mapped out by Breinbjerg (2005) and basing my thinking on the work of soundscape pioneers R. Murray Schafer (1994) and Barry Truax (2001). I also borrowed from the field of sonification (Kramer et al., n.d.; Ballas 1994) and auditory icons (Blattner, Sumikawa, and Greenberg 1989; Gaver 1986), arguing that the game audio engine is a sonification engine that converts the player's actions, and the player's very presence in the game world, into sound. In all of this, I was concerned with exploring computer game sound's function in the immersion of the player in the first-person shooter game world. To me, this immersion seemed to result from a combination of the player's perceptual response to game sounds and the player's ability to trigger sounds through their actions and presence in the game world. Thus, I also delved into the field of presence studies (e.g., Fencott 1999; Slater 2007).

The Main Debates

Since 2008, an increasing number of articles and several books have been published on the subject of computer game sound; this itself testifies to the distinct nature of the field. Among the books are: *Game Sound* (Collins 2008), which concerns itself primarily with aspects of game sound FX and music production and audio technology; *A Comprehensive Study of Sound in Computer Games* (Jørgensen 2009), which, although limiting itself in the main to sound FX and music in Warcraft III and Hitman Contracts, explores the function and experience of game sound through concepts and means such as diegesis, auditory icons, earcons, and sonification; and *Game Sound Technology and Interaction* (Grimshaw 2011), an anthology spanning a number of approaches to computer game sound, from the technical to the aesthetic and from production to reception. The following exposition is based mainly on the analysis of the first-person shooter genre. Although a number of the concepts may be applicable to a wide range of less-immersive genres, the application of these concepts to particular genres such as music games should be treated with caution.

When compared to that other major sensory modality through which we perceive the game world, namely vision, our hearing affords the opportunity to perceive that extensive part of the game world that exists beyond the narrow confines of the image presented on screen (though it may operate cross-modally with vision in some circumstances). This function and its examination have spurred the development of two main strands of conceptual debate: the first centering on issues of diegesis, and the second centering on immersion.

The conceptualization of sound in the diegesis of computer games derives from film theory (e.g., Chion 1994), from whence other terms such as acousmatic and off-screen sound are also utilized. However, because of a particular property that computer games possess—and that films do not, to any great extent—in the case of games, the theorizing of sound and diegesis has been expanded with new concepts and accompanying neologisms. That property is interaction. In the case of sound, this interaction occurs in two fundamental ways.

First, the player's presence and actions in the game world trigger sounds. At the very least, the presence of the player causes a range of ambient sounds to play. These change as the player moves through the world, and modern first-person shooter games are capable of processing these and other sounds so that their acoustic properties (such as reverberation and filtering) more closely match the illusion of the visual environment. Second, with actions such as walking, jumping, operating radios, and firing guns, the player directly triggers a large variety of sounds. Furthermore, in a multiplayer game, players can hear many of the sounds triggered by other players.

As a consequence of these differences from film, new diegetic terminology has arisen in an attempt to explain and taxonomize game sound in an environment where players themselves contribute to a highly nonlinear soundscape. My own contributions to this debate have included concepts such as *ideodiegetic*, a diegetic sound any one player can hear, and two forms of ideodiegetic sound: *kinediegetic* (those sounds triggered by the player) and *exodiegetic* (those sounds triggered by the actions of other players or bots). In a multiplayer game, each player experiences a different soundscape (a significant point of difference from film and other media)—I have thus suggested the concept *telediegetic* for those sounds heard by one player whose response to such sounds has consequences for another player. The multiple soundscapes of such games may be different and often quite distant from one another geographically, but there is still a relationship between them and their players that affects and even effects the diegesis of the game.

Another contributor to this debate is Jørgensen (2009) who adopts the diegetic/extradiegetic terminology of the film theorists Thompson and Bordwell (2003) and uses the term transdiegetic for those sounds (and music) that transgress the diegetic border and lead to the questioning of the game world's boundaries. Specifically, a transdiegetic sound is an ostensibly extradiegetic sound (of which there are many in games, from the musical score to system sounds) that in fact influences the player's actions and thus affects the diegesis. If extradiegetic sound has no visual source within the game world, Jørgensen suggests that some ambient game sounds are actually transdiegetic because, although their source may not be visible, they have the potential for diegetic effect.

Perhaps the most succinct and thoughtful assessment of the game sound diegetic debate is that provided by Hug, who suggests that the debate itself is obsolete: "If the game is part of the same system as the player, the narrative world and the existential world of the player merge into one" (Hug 2011, 415). Whether this is in fact the case leads us to the next debate, namely that of immersion.

Immersion in the game world, often conflated with presence (e.g., Brenton et al. 2005; Waterworth and Waterworth 2014), is a difficult concept to quantify, precisely because there is as yet no objective means of identifying it. It is related to the notion of engagement with the game but, in some theories, goes further by suggesting that the player is part of the game world and thus is immersed in it. The notion of interaction and its effect on game diegesis forms a large part of the conceptual foundation for thinking about immersion and, as we have already seen, the player is able to sonically interact with and, indeed, co-compose and arrange the soundscape. Because immersion appears to be a perception rather than a sensation, emotions have been suggested as having a function in immersion and, here too, sound has a part to play.

Ekman and Lankoski (2009) suggest that emotion, particularly fear, is a key contributor to immersion. For example, in the game genre called survival horror, fear can be especially attributed to acousmatic sound. Ward (2010) turns to Barthes (1977) to suggest that the perception of immersion derives in part from the embodiment of voices such as those heard in BioShock (2K Games 2007). I suggest that the player's ability to trigger sounds whose hearing has an effect on the game's diegesis means, in effect, that the player is part of an acoustic ecology and thus is immersed in that ecology (Grimshaw 2008b, 2012). This is a particularly interesting thread to follow because it raises many questions about our relationship to sounds and acoustic ecologies outside the world of the game and also provides us with the means to understand that relationship.

Several studies have attempted to provide an objective basis for the theorizing of immersion through sound, but a precise quantification of the concept remains elusive (Grimshaw, Lindley, and Nacke 2008; Nacke, Grimshaw, and Lindley 2010). However, the assessment of player psychophysiology in such studies points the way to an intriguing potential future for game sound that is intimately bound up with diegesis and, particularly, immersion and emotion. This future potential brings me to my concluding remarks.

Conclusion: The Future of Game Sound?

In studying the player's psychophysiology in response to sound in the game world, researchers are pursing two goals (for an overview of such research, see Grimshaw, Tan, and Lipscomb 2013). The first goal is that of quantifying and analyzing the player's psychophysiological response to sound as a means either to study the effects of game sound exposure (and this might often be driven by health issues) or to study the effects of sound on such areas as the perception of immersion or the relationship of sound to image. The latter focus might be for the purposes of improved game design or it might be to make use of the controlled environment of the game world to extrapolate the gained results and insights to the wider world—insights on sound and emotion or the relationship to real world acoustic ecologies, for example.

The second goal is more intriguing. Psychophysiological methods include galvanic skin response, electrocardiography, electromyography, and, importantly, electroencephalography (EEG). The data gained through such methods can be used to make an assessment of the affect state of the player. Not merely levels of boredom or excitement or valence, but also the different emotions the player may be experiencing (although the latter remains difficult to assess). What if we could inform the game engine of the player's psychophysiological state (thus, their boredom/excitement level, their valence, and even their specific emotions at any point in time), and the game engine could then use this information to synthesize or process sounds in real time in order to more closely control the player's emotional experience, to more fully engage and immerse the player, and to guide the player through a more satisfying emotional adventure?

The game engine is already capable of doing this to some extent. As a sonification engine, it sonically responds to the player's presence and actions in the game world, and these actions are likely to be based in large part on the player's affect state. But in the context of sound, biofeedback—or sonic emotioneering—proposes a more fine-grained, subtle, and intimate approach.

Sonic emotioneering is a field that I and colleagues have been working in for some time now, using EEG devices and brain-computer interfaces (Garner, Grimshaw, and Abdel Nabi 2010; Garner and Grimshaw 2011). There are many pieces of the puzzle still to be found. As already mentioned, the accurate quantification of emotions from psychophysiological data remains difficult, although the assessment of the level of boredom/excitement is an easier task. There is also, for example, the question of how to process and synthesize multiple sounds in real time in a practical way, such that it does not use more of the scarce computing resources than necessary to attain the effect (resources that are also managing the graphics and artificial intelligence systems of the game, for instance)—in this respect, the field of procedural audio is promising (e.g., Farnell 2007, 2011). Once emotions can be precisely quantified through psychophysiological means and used to synthesize and process sound as the game progresses, the question of meaning in sound remains—and therefore the effect of that sound or combination of sounds on player psychophysiology. Context is important—perhaps all-important. The sound of footsteps may not in itself be scary, but the sound of footsteps creeping up on you from behind in the fraught, sepulchral gloom of a survival horror game certainly can be. How then does sound work in context to engender the desired emotions, and what are the minimum constituent elements of that context that must be taken into account to produce the perfectly efficacious fearful sound?

The answers to such questions have relevance not just for computer games as entertainment. Increasingly, computer games and adaptations of their software are used for therapeutic purposes—and one intriguing area is the treatment of post-traumatic stress disorder. Here, patients undergo experiences designed to recreate past events with as much sensorial and emotional realism as possible, but in a controlled environment. In this situation, the production of fearful sounds in particular contexts must be tailored to the individual, but, at the same time, it must be possible to govern the level of fear inducement of such sound as therapy progresses (see, e.g., Rizzo and Kim 2005).

The example cited in the previous paragraph is one way in which the results of studies on the effects of sound on immersion in the gaming environment can be extrapolated to nongaming scenarios and thus to a wider understanding of the perception of sound and our relationship to it in the real world. In every moment of our lives, we are immersed in a sound environment; we are thus an element of an ever-changing acoustic ecology to which we bring our past experiences of sounds, along with their learned meanings, relevance, and related emotions. As the discussion here has demonstrated, computer games can provide a controllable, experimental platform with which to conduct the study of sound perception.

There remains much to be researched and practically achieved—but sonic emotion-eering, even if only within the realm of computer games, promises to deliver a more sustained, immersive, and enjoyable gaming experience than is currently possible. More fun—that, after all, is what gaming is about.

References

Bibliography

Ballas, James A. 1994. Delivery of information through sound. In *Auditory Display: Sonification, Audification, and Auditory Interfaces*, ed. Gregory Kramer, 79–94. Reading, MA: Addison-Wesley.

Barthes, Roland. 1977. *The Grain of the Voice*. Trans. Stephen Heath. London: Image Music Text.

Blattner, Meera M., Denise A. Sumikawa, and Robert M. Greenberg. 1989. Earcons and icons: Their structure and common design principles. *Human-Computer Interaction* 4:11–44.

Breinbjerg, Morten. 2005. The aesthetic experience of sound: Staging of auditory spaces in 3D computer games. http://www.aestheticsofplay.org/breinbjerg.php (accessed March 3, 2013).

Brenton, Harry, Marco Gillies, Daniel Ballin, and David Chatting. 2005. *The Uncanny Valley: Does it exist and is it related to presence?* 19th British HCI Group Annual Conference. Proceedings of the Human-Animated Characters Interaction. Edinburgh.

Chion, Michel. 1994. *Audio-vision: Sound on Screen*. Ed. and trans. C. Gorbman. New York: Columbia University Press.

Collins, Karen. 2008. *Game Sound: An Introduction to the History, Theory, and Practice of Video Game Music and Sound Design*. Cambridge, MA: MIT Press.

Ekman, Inger, and Petri Lankoski. 2009. Hair-raising entertainment: Emotions, sound, and structure in Silent Hill 2 and Fatal Frame. In *Horror Video Games: Essays on the Fusion of Fear and Play*, ed. Bernard Perron, 181–199. Jefferson, NC: McFarland.

Farnell, Andy J. 2007. An introduction to procedural audio and its application in computer games. http://www.obiwannabe.co.uk/html/papers/proc-audio/proc-audio.html (accessed March 3, 2013).

Farnell, Andy J. 2011. Behaviour, structure, and causality in procedural audio. In *Game Sound Technology and Player Interaction: Concepts and Developments*, ed. Mark Grimshaw, 313–339. Hershey, PA: IGI.

Fencott, Clive. 1999. Presence and the content of virtual environments. http://web.onyxnet.co.uk/Fencott-onyxnet.co.uk/pres99/pres99.htm (accessed March 3, 2013).

Garner, Tom, Mark Grimshaw, and Debbie Abdel Nabi. 2010. *A preliminary experiment to assess the fear value of preselected sound parameters in a survival horror game.* Proceedings of Audio Mostly 5th Conference on Interaction with Sound 2010. Piteå, Sweden, September 14–16, 2010. New York: ACM Press.

Garner, Tom, and Mark Grimshaw. 2011. A climate of fear: Considerations for designing an acoustic ecology for fear. Proceedings of Audio Mostly 6th Conference on Interaction with Sound 2011. Coimbra, Portugal, September 7–9, 2011. New York: ACM Press.

Gaver, William W. 1986. Auditory icons: Using sound in computer interfaces. *Human-Computer Interaction* 2:167–177.

Grimshaw, Mark. 2008a. *The Acoustic Ecology of the First-Person Shooter: The Player Experience of Sound in the First-Person Shooter Computer Game.* Saarbrücken: VDM Verlag.

Grimshaw, Mark. 2008b. Sound and immersion in the first-person shooter. *International Journal of Intelligent Games and Simulation* 5:119–124.

Grimshaw, Mark, ed. 2011. *Game Sound Technology and Player Interaction: Concepts and Developments.* Hershey, PA: IGI.

Grimshaw, Mark. 2012. Sound and player immersion in digital games. In *Oxford Handbook of Sound Studies*, ed. Trevor Pinch and Karin Bijsterveld, 347–366. New York: Oxford University Press.

Grimshaw, Mark, Craig Lindley, and Lennart Nacke. 2008. Sound and immersion in the first-person shooter: Mixed measurement of the player's sonic experience. Proceedings of Audio Mostly 3rd Conference on Interaction with Sound 2008. Piteå, Sweden, October 22–23, 2008.

Grimshaw, Mark, Siu-Lan Tan, and Scott D. Lipscomb. 2013. Playing with sound: The role of music and sound effects in gaming. In *Psychology of Music in Multimedia*, ed. Annabel Cohen, Siu-Lan Tan, Roger A. Kendall, and Scott D. Lipscomb, 289–314. New York: Oxford University Press.

Hug, Daniel. 2011. New wine in new skins: Sketching the future of game sound design. In *Game Sound Technology and Player Interaction: Concepts and Developments*, ed. Mark Grimshaw, 314–415. Hershey, PA: IGI.

Jørgensen, Kristine. 2009. *A Comprehensive Study of Sound in Computer Games: How Audio Affects Player Action.*. Queenston: Edwin Mellen Press.

Kramer, Gregory, Bruce Walker, Terri Bonebright, eds. n.d. Sonification report: Status of the field and research agenda. http://www.icad.org/websiteV2.0/References/nsf.html (accessed November 28, 2014).

Nacke, Lennart, Mark Grimshaw, and Craig A. Lindley. 2010. More than a feeling: Measurement of sonic user experience and psychophysiology in a first-person shooter game. *Interacting with Computers* 22:336–343.

Rizzo, Albert A., and Gerard J. Kim. 2005. A SWOT analysis of the field of virtual rehabilitation and therapy. *Presence* 14 (2): 1–28.

Schafer, R. Murray. 1994. *The Soundscape: Our Sonic Environment and the Tuning of the World.* Rochester, VT: Destiny Books.

Slater, Mel. 2007. If you respond as if it were real, then it is Presence. http://www.cs.ucl.ac.uk/research/vr/Projects/PRESENCCIA/Public/001_01-PeachI-MelSlater-June07-STARLAB.pdf (accessed March 3, 2013).

Stockburger, Axel. 2003. The game environment from an auditive perspective. Proceedings of Level Up Digital Games Research Conference. Utrecht Universiteit, November 4–6, 2003.

Thompson, Kristin, and David Bordwell. 2003. *Film History: An Introduction*, 2nd ed. New York: McGraw-Hill.

Truax, Barry. 2001. *Acoustic Communication.* Westport, CT: Ablex.

Ward, Mark. 2010. Voice, videogames, and the technologies of immersion. In *VOICE: Vocal Aesthetics in Digital Arts and Media*, ed. Norie Neumark, Ross Gibson, and Theo van Leeuwen, 267–280. Cambridge, MA: MIT Press.

Waterworth, John L., and Eva L. Waterworth. 2014. Distributed embodiment: Real presence in virtual bodies. In *The Oxford Handbook of Virtuality*, ed. Mark Grimshaw, 589–601. New York: Oxford University Press.

Sound and Media

id Software. 1999. *Quake III Arena.* Software, Activision.

2K Games. 2007. *BioShock.* Software, 2K Games.

Production Processes

31 "Syd's Theme": On the Conceptualization of Audio Production Processes

Franco Fabbri

"Syd's Theme": A Phenomenological Description

"Syd's Theme" is the conventional, unofficial title accepted within the Pink Floyd fan community of the four-note motif heard for the first time on the album *Wish You Were Here* (1975), at about 3:54 from the beginning of the first track, "Shine On You Crazy Diamond, Pt. 1";[1] it marks the beginning of the section that is usually indicated as "Part III" in published scores (which consist of approximate transcriptions of the original recording, as usual in such cases).[2] The four notes of the motif (b_3 flat, f_4, g_3, e_4), played by an electric guitar, can be heard in the foreground, equally spaced as quarter notes on a "free" 3/4 meter (or, depending on transcriptions, 6/4), over a background consisting of a multitracked synthesizer pad, in G minor; the last note (at the beginning of a bar, slightly accented) is usually notated as a longer note (a dotted half note, or a dotted whole note, again depending on transcriptions), tied across bars until the motif is played once more—but this is the result of an inaccurate transcription, as the last note of the motif actually fades out "into" the synth pad, before a new instance of the motif can be heard. The motif is played three times, then a fourth instance is played after a shorter time interval[3] with drums filling in the space at the end of the motif, then again, with a slow rock ballad drum-kit accompaniment starting on the final note, on the downbeat, and then again, before the lead guitar pattern changes. After the drum-kit entry, the motif is clearly perceivable as the song's riff: as such, it will be repeated many times (sixteen occurrences in "Shine On You Crazy Diamond, Pt. 1," including two "reminders" during the baritone sax solo at the end of Part V; the final note is changed in two occurrences in Part III).

Pointing at Sounds, Naming Sounds

If I were sharing the same space with the reader and had a record player at hand, much of the description I wrote in the preceding paragraph would be superfluous. I could simply play the record and *point at* "Syd's Theme." It is what I usually do in lectures, or with my students, or in radio programs. This is also a common form of communication among fans—during collective listening and, especially, among musicians in many different contexts: from members of an amateur band trying to cover a piece, to professionals in a recording studio (Fabbri 2010). In the case of "Syd's Theme," *pointing at* it is even simpler because literally millions of people in the world know it: it would be enough just to sing it (mimicking the sound of an electric guitar, maybe: "dan dan dan daaan") or even to name the title of the song, or the album, or the group, or the guitar player. I think that "Syd's Theme" in its audible form is a largely accepted metonym for all of these (including "David Gilmour," yes, and even "Fender Stratocaster")—and vice versa, of course. So, if a producer or guitarist wants to obtain *that* specific sound, apart from *pointing at* it by simply playing a recording, he or she can use its name or any of its accepted metonyms. In a studio environment, or in a similar context (such as a rehearsal or the preproduction phase of a tour), someone knows how to translate a sound name into the technical procedure to obtain it: a selection of instruments and playing techniques, sound processors, cabling, amplifiers, microphones, and so forth. Often, the relevant competence is (unequally) shared

Figure 31.1
Band members and a sound engineer listening and discussing in a control room, 1971 (photo courtesy of Antonio Zanuso, Milan).

among different people: so, for example, (not exactly in the case of "Syd's Theme," but it could be),[4] a producer or any of the band members may know how to place that sound into the context of the recording (or the live performance), the relevant instrumentalist knows which instrument to choose, which playing technique, which foot-operated effect units and other sound processors, which amplifier—while the sound engineer knows which microphones and transformers, outboard sound processors (especially reverb, compressors and expanders, graphic or parametric equalizers), mixing desk patches, and so forth. So, "dan dan dan daaan" can mean an awful lot of things!

Meaning

But what does it mean *to mean*? In his *Theory of Semiotics* (1976, 67), Eco writes: "What is, then, the meaning of a term? From a semiotic point of view it can only be a *cultural unit*. In every culture, 'a unit … is simply anything that is culturally defined and distinguished as an entity. It may be a person, place, thing, feeling, state of affairs, sense of foreboding, fantasy, hallucination, hope or idea' (Schneider 1968, 2). We shall see later how a cultural unit can be defined semiotically as a semantic unit inserted into a system."

Immediately afterward, Eco introduces another fundamental concept, the *interpretant*, quoting from Peirce. So, a sign is: "anything which determines something else (its *interpretant*) to refer to an object to which itself refers … in the same way, the interpretant becoming in turn a sign, and so on *ad infinitum*" (Peirce 1931–1958, 2, 300, quoted in Eco 1976, 69).

In other terms, a sign (1)—a combination of expression and content, of signifier and signified—may have as its content (signified) another sign (2), the *interpretant*, whose content is the same, or similar, to the content of sign (1) and whose expression is related to the expression of sign (1) by synonymy, metonymy, metaphor, or because it is a representation of the same content in another semiotic system (like a drawing), or because it is simply pointed at, or for other reasons (see Eco 1976, 70). An elementary example of an interpretant is a sign whose expression is a translation into another language, like /dog/ and /Hund/, both related to a cultural unit conventionally representing a "dog," or a "Hund" (the same animal).[5] So each sign points to other signs (interpretants) in an infinite chain, in a process of *infinite semiosis*. Therefore, the meaning of an expression is never defined once and for all, but is constantly approximated and explained by interpretants. We can understand that semiosis works by approximation based on the same example that we have just referred to. Is a "dog" really the same animal as a "Hund"? Is the semantic field around the concept (the cultural unit) "dog" structured in exactly the same way as that around "Hund"? Classic semantic studies provide many examples showing differences in the structuring

Table 31.1

	Baum		arbre
træ			
	Holz		
			bois
skov			
	Wald		
			forêt

of the semantic field around similar (but not identical) cultural units, like the example in table 31.1 (from Hjelmslev 1961, 50).

We can see that different expressions in Danish, German, and French do not correspond to exactly the same content: /træ/ covers the semantic space of "Baum" and "arbre," but also part of the semantic space of "Holz" and "bois"; /bois/ covers the semantic space of "Holz," but also part of "Wald," and so forth. Sign functions, Hjelmslev explains, establish a form—the *content-form* (which is represented by the shape of the diagram, or by the empty diagram, as Eco says). As cultural units, units like "Baum," "arbre," and so forth are the *content-substance* (Eco 1976, 72).

The Semantic Space of the Recording Studio

Similar examples can be cited about the semantic space of the sound qualities conventionally adopted by recording industry professionals (for early accounts on the subject, see Fabbri 1987 and Fabbri 2012b, 28–31; for more recent ones, see Porcello 2004; Fabbri 2010).

The example given in table 31.2 is also useful in order to understand that cultural units are defined both in relation and in opposition to one another. Both in English and in Italian, a sound without effects is "dry," or "secco" (almost the exact equivalent in standard linguistic usage). There is a certain degree of cultural construction here, as the adjective is applied in a typical studio environment where it is customary to have a "dead" response (that is, no natural reverb). So, operationally, a dry sound is the sound one gets when an input channel carrying the signal of a microphone, or the direct line output of an electric instrument, is opened (this is, of course, an interpretant of the /dry/–"dry" sign and explains how concepts like "dry" in studio practice become shorthand for technical procedures and relevant verbal instructions: "leave it dry" is shorter than "do not apply any effect to that signal"). /Wet/ in English means the opposite, that is, a sound treated with reverb or echo or any other effect similarly based on reflections, repetitions, modulations

Table 31.2

English General meaning	As sound concept	Italian As sound concept	General meaning
Dry	No reverb, no FX	No reverb, no FX	Secco
Wet	Reverberated, echoed, with FX	N/A Reverberated	Bagnato Riverberato Alonato
		With FX	Effettato

of the original signal (a chorus, a flanger, but not—to my knowledge—a parametric equalizer). Although there are expressions that suggest an iconic usage of /wet/ (like /drowned in echo/, or other relations to the concept of "moisture," like the usage of /halo/ as a synonym of /reverb/ or /echo/), it seems that the function of "wet" is simply one of opposition with respect to "dry." /Bagnato/ (the equivalent of /wet/ in Italian) has no meaning in studio language. /Riverberato/ and /alonato/ (the only term with a connotation of moisture, a visual one: think of the halo around the moon on a foggy night) refer to a sound treated with reverb or echo, but not with other effects. A sound treated with chorusing or flanging effects is "effettato." In this case, Italian usage is less metaphorical than English.

The sound of David Gilmour's Fender Stratocaster in "Syd's Theme" can be described as "wet." From reports, we know that it was treated with a delay unit and a foot-operated flanger (an early one),[6] but of course it is its audible sound quality that prompts us to refer to that cultural unit to describe it. And that quality is somehow illusory, as the background synth pad provides the kind of prolongation of the guitar sound one usually associates with reverb. Of course, that sound may have other meanings that can be related to its timbral quality or to other musical parameters: a *musematic* analysis, based on the method developed by Philip Tagg, may suggest that "Syd's Theme" can be associated with *paramusical connotations*.[7] *Hypothetical substitution* may confirm that those four notes associated with the synth background and the overall sound are a *museme*: changing the guitar sound (such as using a heavily distorted "metal" sound, or the opaque, "dry" sound of a jazz guitar) would probably change the meaning not only of that short passage, but of the whole song. It may well be that a producer or performer will attempt to suggest a precise, intended meaning in studio work,[8] but codified concepts like "dry," "wet," "fat," "thin," and a rich library of references, such as "David Gilmour's guitar in 'Syd's Theme,'" form a kind of intermediate semantic space, meaningful enough for the purposes of the trade without being too idiosyncratic.

Figure 31.2
This chapter's author as a guitarist with his Fender Stratocaster and foot-operated effects, 1982 (photo courtesy of Maurizio Dugoni, Milan).

Conventions

How are such semantic conventions established? The name should not suggest that a community or any group of people sit down together at a table and agree on rules, although some semantic conventions may indeed be established in that way (by compiling a poetic manifesto, for example, such as the *Manifesto della Musica Futurista*).[9] According to philosopher David K. Lewis, the following could be one definition of *convention*:

A regularity R in the behavior of members of a population P when they are agents in a recurrent situation S is a convention if and only if, in any instance of S among members of P,

(1) everyone conforms to R;
(2) everyone expects everyone else to conform to R;
(3) everyone prefers to conform to R on the condition that the others do, since S is a coordination problem and uniformity to R is a proper coordination equilibrium in S. (Lewis 2002, 42)

In an article about diachronic processes in music genres, I wrote:

In a recurring music event, maximizing the pleasure of each of the participants, or ensuring everyone has the best understanding of what is going on, or minimizing the amount of information that must be processed to obtain pleasure and understanding may be perceived as coordination problems; conforming to genre conventions, to use Lewis's words, is "a proper coordination equilibrium" in that recurring event. (Fabbri 2012a, 185)

I would suggest that a studio production is definitely a similar kind of "recurring music event." Although, of course, studio production is not usually perceived as a genre, it is categorized similarly—namely, as a context (and a set of events) where certain "rules" hold, and as opposed to another context: that of live performance. Performing musicians, producers, and engineers conform to linguistic conventions about sounds and studio procedures; such conventions may be shared generally or be valid within one or more communities (typically, but not exclusively, musicians vs. engineers, with producer mediation). Conceptualizing sounds and sound processes is a way to spare energy, concentrate on the intended results, and maximize pleasure (through feeling competent, for one, and like an accepted member of a professional community, see Porcello 2004).

Like all conventions, linguistic conventions are subject to change: as a consequence, research on sound concepts must encompass synchronic and diachronic aspects, covering not only differences in coding among coexisting communities, but also the changes taking place over time within the same community. Concepts and methods from disciplines such as musicology, semiotics, linguistic anthropology, and historiography[10] are essential for future studies in this field.

Notes

1. Coauthored by Roger Waters, Rick Wright, and David Gilmour, in various combinations.

2. This may sound confusing. On the album, "Shine On You Crazy Diamond" is divided into two tracks, one at the beginning of the track list and the other at the end, named respectively "Shine On You Crazy Diamond, Pt. 1" and "Shine On You Crazy Diamond, Pt. 2." Each is divided formally into parts, nine in total: I–V and VI–IX respectively. "Syd's Theme" is located in "Shine On You Crazy Diamond, Pt. 1," at the beginning of Part III.

3. In scores, the first two occurrences of the motif (including the long tail of the last note) last eight bars (four, if the 6/4 meter is chosen: see, e.g., Waters, Gilmour, and Wright 1991), and the next three last four bars (two, in 6/4). In the recording, the time between the start of the first note in the motif and the start of the corresponding note in the next occurrence (i.e., the total duration of each occurrence) is, respectively, 12.5″, 11.9″, 5.4″, 5.3″, 5.3″. The sixth occurrence (before the pattern change) is shorter.

4. A thorough description of how David Gilmour's guitar sounds in "Shine On You Crazy Diamond" can be found here: http://www.gilmourish.com/?p=287.

5. From now on, I will use the same convention introduced by Eco (1976): words enclosed in slashes (like /dog/) represent the expression of a sign, and words enclosed in double inverted commas (like "dog") represent the sign's content.

6. I was, however, not able to find any reference to the usage of an outboard reverb unit, although the general aspect of the guitar and synthesizer sounds clearly suggests it. In a studio—and definitely in the seventies—reverb is readily available on each channel at the turn of a knob.

7. For a general description of Tagg's musematic analysis, see Tagg 1999.

8. "I need a sound like that of cockroaches running on a marble floor." My mother, who worked in the seventies as production assistant at Italian radio, once interrupted composer Luigi Nono and studio engineer Marino Zuccheri, working at the Studio di Fonologia Musicale, Milan, with this request. According to RAI's rules, they had to oblige.

9. See *Manifesto of Futurist Musicians* at http://www.unknown.nu/futurism/musicians.html.

10. See, e.g., Rick Altman's usage of methods derived from the *École des Annales* for the reconstruction of basic concepts in cinema history (Altman 2004, 6).

References

Bibliography

Altman, Rick. 2004. *Silent Film Sound*. New York: Columbia University Press.

Eco, Umberto. 1976. *A Theory of Semiotics*. Bloomington: Indiana University Press.

Fabbri, Franco. 1987. Cucina elettronica e paesaggi immaginari. Background e prospettive metodologiche per un'estetica del "sound." In *Tempo di rock,* ed. Paolo Prato, 23–31. Florence: Giunti. Repr. in Fabbri, *Il suono in cui viviamo: Saggi sulla popular music,* 3rd. ed. (Milan: Il Saggiatore, 2008), 286–293.

Fabbri, Franco. 2008. *Il suono in cui viviamo: Saggi sulla popular music,* 3rd. ed. Milan: Il Saggiatore.

Fabbri, Franco. 2010. "I'd like my record to sound like this:" Peter Gabriel and audio technology. In *Peter Gabriel: From Genesis to Growing Up,* ed. Michael Drewett, Sarah Hill, and Kimi Kärki, 173–182. Aldershot: Ashgate.

Fabbri, Franco. 2012a. How genres are born, change, die: Conventions, communities and diachronic processes. In *Critical Musicological Reflections,* ed. Stan Hawkins, 179–191. Aldershot: Ashgate.

Fabbri, Franco. 2012b. Genre theories and their applications in the historical and analytical study of popular music: A commentary on my publications. A thesis submitted to the University of Huddersfield in partial fulfillment of the requirements for the degree of Doctor of Philosophy by Published Works. University of Huddersfield. Also available at https://universitaditorino. academia.edu/FrancoFabbri.

Hjelmslev, Louis. 1961. *Prolegomena to a Theory of Language.* Madison: University of Wisconsin.

Lewis, David K. 2002. *Convention: A Philosophical Study.* Oxford: Blackwell.

Peirce, Charles Sanders. 1931–1958. *Collected Papers.* Cambridge, MA: Harvard University Press.

Porcello, Thomas. 2004. Speaking of sound: Language and the professionalization of sound-recording engineers. *Social Studies of Science* 34 (5): 733–758.

Schneider, David M. 1968. *American Kinship: A Cultural Account.* New York: Prentice-Hall.

Tagg, Philip. 1999. Introductory notes to the semiotics of music. http://www.tagg.org/texts.html.

Tagg, Philip. 2009. *Everyday Tonality: Towards a Tonal Theory of What Most People Hear.* New York: The Mass Media Scholars' Press.

Waters, Roger, David Gilmour, and Rick Wright. 1991. Shine On You Crazy Diamond (Part III). In *Pink Floyd: Wish You Were Here; Guitar Tablature Edition,* 37–43. London: Pink Floyd Music Publishers.

Sound and Media

Pink Floyd. 1975. *Wish You Were Here.* LP, Harvest, SHVL 814.

32   Sonic Signatures in Record Production

Toby Seay

In 2012, the Library of Congress of the United States made the following declaration, creating an important relationship between technology and aesthetics: "There must be a systematic and sustained effort to compile and collect information related to legacy recording technology and practices: where it is, how it works, and the characteristics, or 'audio signatures,' of the recordings themselves" (Library of Congress 2012, 18). Although this declaration is aimed at promoting the preservation of sound recordings, it strongly resonates with my research into the record production practices that create sonic signatures. In a recent paper regarding the technical environment of Sigma Sound Studios in Philadelphia, I asked if it was even possible to determine the sonic influence of technology on Sigma's recorded output, and, if it was possible, what is the best way to detect this influence? (Seay 2012) The answer to that question remains elusive. However, having lived and worked as an audio engineer in one city with an inferred signature (Nashville) and now living, teaching, and researching in another (Philadelphia), it is impossible to escape the notion of sonic signatures within the music community. Therefore, I continue to explore how studio technology and workflow might influence the creation of sonic signatures.

I have always held the opinion that identifiable sonic signatures were, for the most part, mythology created by marketers who wished to place a musical community's output into a neat package. For instance, William Ivey describes the Nashville Sound as being the creation of homogenous instrumentation, nonstandard musical notation, and a strategic marketing campaign (1982, 133–134). Furthermore, the Sound of Philadelphia is described by Tony Cummings as "much more than the sound of soul" but "also the sound of cashiers rubber-stamping money drafts" (1975, 8). However, these descriptions tend to focus on the musical and business output rather than the sonic output of each production center. For instance, if the Sound of Philadelphia had instead been called the Music Industry of Philadelphia, there would be little to argue about. In addition, Simon Zagorski-Thomas, in comparing the sonic signatures of the UK vs. the United States, employs a great deal of caution when trying to empirically identify sonic signatures while at the same time noting the proliferation of producers, musicians, and engineers who claim that these signatures do exist (2012, 57).

Although I feel there is a high degree of mythology to the topic of sonic signatures, I also find it necessary to explore the topic by looking at the technology involved in its creation, in addition to the recorded music. A case can be made for a technical influence if one examines the writings of Susan Schmidt-Horning (2004). She describes a technical workforce that is trained on-the-job for the most part, tacitly learning recording skills directly from a senior engineer. I would further argue that most engineers could trace their developmental lineage either back to a prime influential figure, or to a group of mentors, placing them in a school of influence. For instance, at Sigma Sound Studios, the chief engineer and studio owner Joseph Tarsia trained virtually all of Sigma's technicians. If Tarsia did not train them, a senior engineer who was trained by Tarsia trained them. This scenario places all of the engineers within a common technical culture, therefore creating the potential for a local aesthetic.

Along with this local aesthetic is the indelible transformation of sound that takes place in a recording. In the technical sense, as soon as a microphone (or any transduction) is introduced, a translation of reality is underway. Not only will a microphone introduce its own character to the sound, but the engineer's placement of the microphone and signal manipulation will further distance the recording from the original experience. In a performance sense, a recording can be made of individual performances separated by time and space, creating a recorded document that could only exist through the recording process. Therefore, a sound recording has the spatially distant and detached contextual characteristics that John Thompson describes as a "mediated experience" (1995, 227–228). In this sense, the listener is not in the place of performance and does not experience it as a shared event with other listeners. It has been my experience as an engineer that reducing the effects of this mediation and trying to create a recording that is as close to reality as possible is rarely a production goal. Instead, it is my contention that the goal of the recording process is to create a new artistic reality as a new object with its own aesthetic character rather than to document a singular event.

However, this new aesthetic is presented on its own terms as a singular experience that is both created by the record production process and accepted by consumers as a new entity. Paul Théberge (1989) describes this rationalization as "the result of a shift in recording aesthetics away from the 'realistic' documentation of a musical event to the *creation* of one" (104). He goes on to say that the process of modern music production also creates, a "myth of community," in that there is a tacit agreement between creators and consumers as to the perceived reality of a sound recording (ibid., 108). Therefore, sound recordings, especially music productions, are both mediated and rationalized by all participants, creating a larger sound recording system or culture. This culture requires all participants to subscribe to the definition, consciously or not, of what a sound recording represents.

Are sonic signatures mythology, marketing, or reality? It is my view that regardless of any empirical evidence to support reality, it is necessary to examine the relationships between the contents and agents of recorded sound to better paint the picture of mediation and rationalization.

Sound Elements and Agents

To better examine sonic signatures, I have identified the individual elements and agents associated with a sound recording (figure 32.1). However, I caution that this diagram is not meant to microscopically analyze individual parts of a recording as if each element can stand on its own merit; but by identifying these elements and their agents, one can look at relational effects and the sound recording as a whole. Or, as Théberge states: "Recording technology must be understood as a complete 'system' of production involving the organization of musical, social, and technical means" (1997, 193). It is with this systemic approach that the elements and agents in this diagram interact through what I call the *influential domain*, whereby each element has the potential to inform all other elements. This diagram presents an interactive system that deters the study of any one element without considering its context by focusing the discussion toward the influential domain.

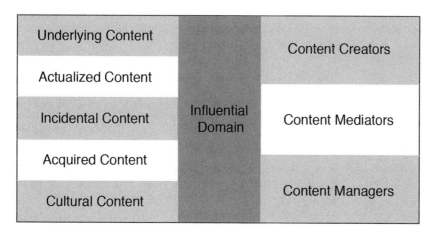

Figure 32.1
Element and agent relationships of a sound recording (diagram courtesy of Toby Seay).

Though most are self-explanatory, the following are definitions for each element and agent:

- Underlying content—unperformed literary and/or musical work
- Actualized content—source sound from a performance of the underlying content
- Incidental content—additional sounds as affect of performance (acoustic properties, sound field manipulation, etc.)
- Acquired content—captured underlying, actualized, and incidental content within a sound recording
- Cultural content—time and place, market, audience (target or not), expectations, reactions, and impact
- Influential domain—abstract confluence where content elements interact
- Content creators—songwriters, performers, and so forth
- Content mediators—engineers, producers, and so forth
- Content managers—marketers, consumers, and so forth

It is important to note that there is no assumption that these elements exist in any order. For instance, an improvised performance will exist as actualized content before the underlying content can be discussed. Furthermore, some acquired content can exist before some actualized content, as is often the case with multitrack recording procedures.

The cultural content is particularly interesting, as none of the other elements can exist without an influencing culture. That culture may be local, such as the studio environment or recording community, or it can be much broader, such as a recording's national or ethnic point of origin. Within a studio environment, the cultural content may be a reflection of common practices that manifest from content mediators.

Sound Element and Agent Relationships

My research in this area has been focused on the Sound of Philadelphia and the influence Sigma Sound Studios had on creating it. To better understand this relationship, former Sigma engineers were asked to participate in a survey regarding sonic signatures, which asked for their opinion on what part technology played in the creation of the Sound of Philadelphia. Interestingly enough, when asked what characteristics best describe the Sound of Philadelphia, every respondent described the music, not the technology. For instance, chief engineer and founder Joe Tarsia (pers. comm., 2013) replied, "large orchestrations, strings, horns, and full rhythm section." Furthermore, staff engineer Carl Paruolo (pers. comm., 2013) said that recordings at Sigma were made "in three different rooms, it couldn't have been the 'room sound,' so I would have to first say the musicianship, followed by the arrangements and production."

These answers seem to reinforce the view that sonic signatures have more to do with genre, performance, and musicological concerns than technology or sound studies concerns.

When asked specifically what sonic characteristics can be heard in a Sound of Philadelphia recording, however, nearly all respondents pointed to what Tarsia (pers. comm., 2013) described as "Sigma's unique use of delayed echo that really opened the stereo landscape." This predelayed echo (created with a tape delay patched before the Electromesstechnik [EMT] 140 plate reverb's input) was a surprisingly common response. Sigma was a well-documented innovator when it came to technology. Tarsia felt his competitive edge was staying ahead of technological trends. Therefore, Sigma was known for being an early adopter of 16-track and 24-track tape machines, Dolby noise reduction, and console automation. Staff engineer Arthur Stoppe (pers. comm., 2012) describes Sigma as being a "world-class studio, frequently on the bleeding edge with such technology," and staff engineer Jay Mark (pers. comm., 2013) commented: "We had very high electronic and mechanical standards. Our consoles were hand built and custom designed by us." However, the methodology of using predelay was the most cited technologically significant influence on the studio's sonic output.

While Paruolo points out that the sound of the room could not have been part of Philadelphia's sonic signature, many Sigma engineers surveyed noted that the sound of an ensemble playing together in the studio provided a sonic signature. For instance, when listening to multitrack recordings from Sigma, Tarsia (pers. comm., 2013) made the comment that some recordings "sounded like Sigma" but others did not. The level of instrument spill (sound from one source bleeding into the microphone intended for another) between tracks was his determining distinction. Paruolo (pers. comm., 2013) describes sessions at Sigma "where the rhythm section, the string section, and the horn section were all recording in the same room at the same time, and sometimes, even the vocalist." When there was a lot of spill, it represented ensemble performances rather than overdubbed performances with greater isolation between instruments. In the 1971 Len Barry recording "Girl You're Too Young," the drum spill into the string microphones is vital to the sound of the drums. However, the 1973 Techniques recording "Shake It, Wake It" was recorded in the same room, but it does not exhibit the same high level of spill because of a more overdubbed approach. Therefore, this sonic signature of live performance came from how the room was used, not its acoustic properties. Porcello (2002) speaks to this same situation when describing the difference between the Austin and Nashville sounds, making a clear reference to recording procedures that controlled the amount of instrument spill as the influencing factor between these sonic signatures rather than the musical content.

Conclusions and Questions

The results of this study display the difficulty in determining fact from fiction in regard to sonic signatures. Multitrack technology and capturing the sound of the room are often cited as influencing factors in the creation of a sonic signature. Schmidt-Horning makes reference to the diminished use over time of natural ambience in recordings by saying "the need to maintain separation of individual tracks in order to better control each instrument in the final mix forced engineers to once again obliterate the sound of the room" (2012, 30–31). Does this mean that sonic signatures are only possible in studios where the sound of the room is incorporated? Does that place sonic signatures in a forgone era? Or, does it simply shift the sonic signature from an organic representation of a live performance to one of manufactured sonic creation that relies on instruments to be, as Kealy describes, "recorded separately and then replayed and edited in minute detail" (1979, 15)? It is my view that it is very difficult to identify the acoustic characteristics of a specific space within a recording, but that it may be easier, and perhaps more relevant, to identify how the recording was made within the space—which places the recording within a locally identifiable culture by maintaining a strong relationship between the content and its agents.

This view is supported by the philosopher George Sabine, who wrote: "Without the individual, as has many times been shown, the universal is a bare identity about which nothing can be predicated. Without the relational aspect of experience, on the other hand, the totality disappears entirely" (1907, 157). This statement points directly to the importance of context. Although it is easy to get bogged down in describing individual components of a recording, it is the element and agency relationships that provide the clues to sonic signatures—similar to a balloon filled with air that, when squeezed, displaces the contained air to another area, creating a new aesthetic with the same contents. Sonically speaking, the myriad means and methods of sound capture have a significant influence on the actualized content.

Finally, I would like to refer to a movie by RCA Victor created to explain the art of recording. In describing how recordings are made, the narrator states: "It is not just the mechanical reproduction of a concert performance, it's the art of creating the *effect* of a concert performance" where "the end object is a planned illusion. The illusion, as you ultimately hear the music played back through your phonograph, that you have the best seat in the concert hall" (*The Sound and the Story*, 1956). Even in 1956, recordings were considered a manufactured reality: one that consumers accepted and subscribed to as a single sonic event. Mediated and rationalized, sonic signatures contain multiple facets of mythology, marketing, and reality.

References

Bibliography

Cummings, Tony. 1975. *The Sound of Philadelphia*. London: Methuen Paperbacks.

Ivey, William. 1982. Commercialization and tradition in the Nashville sound. In *Folk Music and Modern Sound*, ed. William Ferris and Mary L. Hart, 129–140. Jackson: University Press of Mississippi.

Kealy, Edward R. 1979. From craft to art: The case of sound mixers and popular music. *Work and Occupations* 6 (3): 3–29.

Library of Congress. 2012. *National Recording Preservation Plan*. Washington, DC: LOC.

Porcello, Thomas. 2002. Music mediated as live in Austin: Sound, technology, and recording practice. *City & Society* 14 (11): 69–86.

Sabine, George. 1907. The concreteness of thought. *Philosophical Review* 16 (2): 154–169.

Schmidt-Horning, Susan. 2004. Engineering the performance: Recording engineers, tacit knowledge and the art of controlling sound. *Social Studies of Science* 35 (5): 703–731.

Schmidt-Horning, Susan. 2012. The sound of space: Studio as instrument in the era of high fidelity. In *The Art of Record Production: An Introductory Reader for a New Academic Field*, ed. Simon Frith and Simon Zagorski-Thomas, 29–42. Farnham: Ashgate.

Seay, Toby. 2012. Capturing that Philadelphia Sound: A technical exploration of Sigma Sound Studios. *Journal on the Art of Record Production* 6: n.p.

Théberge, Paul. 1989. The "sound" of music. *New Formations* 8 (summer): 99–111.

Théberge, Paul. 1997. *Any Sound You Can Imagine: Making Music/Consuming Technology*. Hanover, NH: Wesleyan University Press.

Thompson, John B. 1995. *The Media and Modernity: A Social Theory of the Media*. Stanford, CA: Stanford University Press.

Zagorski-Thomas, Simon. 2012. The US vs. the UK sound: Meaning in music production in the 1970s. In *The Art of Record Production: An Introductory Reader for a New Academic Field*, ed. Simon Frith and Simon Zagorski-Thomas, 57–76. Farnham: Ashgate.

Sound and Media

Barry, Len. 1971. *Girl You're Too Young*. Seven-inch single, Vanguard Records, unreleased.

The Sound and the Story. 1956. Director: Jam Handy. Documentary, Radio Corporation of America.

Techniques. 1973. *Shake It, Wake It*. Seven-inch single, Jerry Ross Productions, unreleased.

33   Phonographic Work: Reading and Writing Sound

Rolf Großmann

Two styluses have written history in music: the quills of Bach, Mozart, and Beethoven, and the gramophone needle (figures 33.1 and 33.2). Both of them physically document something that has to do with sound, and both serve as media for the transmission of tradition and the creation of art forms. Reading and writing are our most important cultural skills. This applies most of all to written language, but also to other notation practices. In the case of musical notation and phonography, for instance, both the methods used and the semantic label "-graph" suggest that they can be understood as systems of reading and writing. A pointed object inscribes something onto a particular material (paper, wax, tinfoil, etc.), which then exists after this process as an (in-) formed object.[1] As a medium, it represents a third aspect; it can be used, copied, and archived with regard to this new aspect. In contrast to the technical media of photography and cinematography, which took their first steps in the nineteenth century, in the case of phonography and the usage of the gramophone the analogy to writing can almost literally be grasped.[2] Even William Fox Talbot referred to photography as a "pencil of nature" in 1844, though in that case, a photochemical process took place that bears no resemblance to the artisanal activity of writing—which points to an important cultural aspect that lies beyond any concrete technical methods. Photography and phonography have assumed a cultural function, that of painting and musical notation, which were formerly assigned to brush and pencil. Instead of the painter, it is now the camera that is doing the "painting," and the music is engraved onto a phonograph cylinder.

By the time of tape recording, with its "read and write heads," the procedure had nothing to do with writing performed by humans. The process of writing and reading is invisible and cannot be carried out without being mediated by technology. In the installation *Random Access* (Nam June Paik, 1963), a sound head is guided across tape fragments that have been stuck onto a surface: this only uses the metaphor of reading, of exploring an unknown media text. At the same time, in its materiality and referentiality, it demonstrates a difference between technical and manual notation.

Figure 33.1
A classical writer—of music. Statue of Beethoven at the Münsterplatz, Bonn
(Germany) (photo courtesy of Rolf Großmann).

Figure 33.2
A divine writer. The Writing Angel trademark, record label (1900–1907), the Gramophone
& Typewriter Company, London.

Friedrich Kittler (1985) explored this transformation in *Aufschreibesysteme 1800/1900* ("systems of inscription")[3] as the media-technological core of cultural change and demonstrated an important point for policies on methodology, as well as a new direction for research in the humanities in particular: the cultural techniques of writing can be detected in the cultural practice of the technical media and are subject to continuous transformation. However, Kittler also speaks of the fundamental differences among the technical functions of media, which are of considerable importance regarding sound media. To him, the gramophone is the medium of "the real,"[4] with its object being the physically formed "precultural" world of sonic waves: "Thus, the real—especially in the talking cure known as psychoanalysis—has the status of phonography" (1999, 16).

Turning Kittler's psychologizing view into a cultural perspective, the difference between gramophone writing and musical notation becomes clear. The reference of musical notation is a cultural practice that began with Guido d'Arezzo in the eleventh century and simultaneously founded the composition techniques of European art music and proceeded to develop along with it. Here, the meaning of the notes is generated by the notion of the pitches of tones that are shaped by their respective musical practice and temporally meaningful arrangement. So this notation is related to a cultural construction, an idealized system of notes and rhythmic arrangements (the *tone universe* of Western art music). Only against the background of this construction does any handling of this writing, whether it be for the purpose of remembering, performing, or composing, make sense. It is therefore not only the ability to write it down manually but also the knowledge of its complex cultural context of meaning that is part of this particular media literacy.

In contrast, the phonograph's technical reading and writing down of the sound initially requires nothing other than the technical apparatus itself. Its media-semiotic reference is the vibration and not a culturally shaped notion of tones based on systems of tonality and scales. Although the usage of this machine in its respective environment is just as little independent of culture as the range and the perception of the noises and sounds it engraves, as Kittler rightly points out, technological writing relates to a reality before its cultural forming (the *universe of sonic vibrations*). In addition, the apparatus itself is capable of creating sounds, an ability that was previously attributed to humans and musical instruments. Its culturalization as a system of writing—comparable to the development of musical notation from the Guidonian hand to neumes to score notation—therefore progresses on a different basis than that of the idealized and rationalized notation systems of European art music.

"Classical" Thinking—Composition

Entering into a new system of media reference has consequences. When sound notation, along with its apparatuses, became the medium of working with music in the
twentieth century, what characterized this work, and can one continue to refer to it as
composing? In art music, composing is a process of providing musical structure—and
not any specific sound—with a form. Although it is related to performing in a complementary manner, composing is nevertheless distinct. According to the established
conception of composition, it is tied to working with the musical material of notation,
to concepts of melody, harmony, rhythm, and arrangement. The object of this work
is a network of established concepts, methods, and rules, or, as Eduard Hanslick puts
it: "*geistfähiges Material.*"[5] His "classical" foundations are the methods and forms of
handling melodies and polyphonies, described as contrapuntal work, functional harmony, or motivic-thematic work. Even where extended—for example, graphic—scores
are used, the separation of structural representation and realization as sound continues
to exist.[6]

In contrast, phonography, as a system of writing, has only a little to do with these
traditional musical parameters. It does not notate pitches and rhythms, but instead
progressions of vibrations; at the same time, its apparatuses have the character of a
(musical) instrument and link the process of technical reading to a "performance"
of that which has been written down. With regard to the referenced musical material—and thus from the perspective of "classical" composition work—this medium is
inferior to the former system of music notation: the musical structure must first be
filtered out of the background noises and the manifested sound; as an instrument, it
is a second-degree generator of sound that merely reproduces the musical instruments
of a preceding presentation with qualitative losses. If work is performed against this
background at the level of phonographic signals, it is evident that the aim should not
be that of composing but of improving quality—or of a manipulation that creates the
impression of an optimally authentic reproduction. The result is "a recording aesthetic
of 'concert hall realism' and 'high fidelity,'" as Edward R. Kealy (1979, 9) describes the
"craft-union mode" of record production. Paradoxically, even in this phase of what
appears to be phonorealism, the stated goal of the production is illusion: "The listener's illusion that he was sitting in Philharmonic Hall rather than in his living room"
(ibid., 210–211).

In fact, the concept of *composition* is actually rarely used in the context of technological media, the emphasis being more on *design, effect,* and *manipulation.* These are
terms of accidence, and they presume a primarily existing essence that is merely modified through phonographic media. The sound (the sonic materiality) of the music is
its necessary precondition, although it remains subordinate to its structure as a performance variable. The term *composition* connotes a differentiated structural organization

and would thus have a claim to reflection and value, which have so far been ascribed solely to art music. Regardless of all the criticism of the dominant methods for an analysis of art music, the field of popular music studies is also associated with this manner of thought.[7] Therefore, if there is any mention of composition in the context of phonographic work, far-reaching changes in structure and form are presupposed. This condition is first observed in the discourse on popular music studies in the context of early overdubbing and subsequently multitrack recording. "Les Paul's approach to recording was, without doubt, a form of composition, a literal 'putting together' of the music" (Théberge 1989, 105). Chris Cutler, who was among the first to describe the significance of the "media of electronic transformation and recording" for musical innovations in the popular music of the twentieth century, stresses the option to turn performances in multitrack recordings into the object of collective work and composing. To him, recording is "a medium of composition for performers" and, with a view to the avant-garde of *musique concrète* and electronic music, "a medium of performance for composers" (Cutler 1984, 286–287). I would like to address this perspective in more depth in the following section and focus the discussion on the medium as a writing system.

Working with Phonographic Material

Here, the term *phonographic work* has three points of reference: as an object (work of art), as process (working), and as an abstract term for a common practice that includes a specific knowledge and methodology. I will concentrate on the third meaning, which is deduced from the "trademarks" of written composition in classical music, such as the already mentioned motivic-thematic work—and which clearly demands comparable appreciation as a compositional process.

However, in the interaction between objectification and working with the objects, this type of conceptualization always incorporates the two aforementioned aspects of meaning as well. In focusing on phonography as it is related to performing, the character of the record as an elaborated artifact itself is eclipsed. Since we "are used to treating records as musical events *in themselves*" (Frith 1988, 21; original emphasis) in this era of advanced hi-fi, records are able to become independent of their performance and construct a sound reality of their own. The medium leads us to "new forms of *creative art*—forms in which the capturing of performances ... becomes not an end in itself but a gathering of raw material which can be treated in various ways: sped up, slowed down, chopped about, mixed, distorted and so on, as part of considered composition" (Clarke 1983, 199–200). It is obvious that such methods of phonographic composition had to lead to a profound crisis in the authentic representation of performances in rock music. This contradiction is only resolved when "the art of record production" (Frith and Zagorski-Thomas 2012) is viewed from the perspective of a new reference system of composition and performance: in forming artifacts at the level of vibrations and samples and the independent media-technical performance of these artifacts. In order to gain a better understanding of the impact of phonographic notation and the electronic and digital transformations of phonographic signals associated with it, two fundamental extensions of the perspective are necessary:

• Crossing the boundaries of the "popular" to gain access to the experiments of the avant-garde of the twentieth century, and
• Including musical cultures whose practice is not primarily shaped by the value standards of Western art music

This broadening of the perspective relates first and foremost to all forms of the development and sedimentation of the practical knowledge of creating music with this new notation of sound. It applies to performative practice, to the development of new instruments, and to working with phonographic material in the studio or with

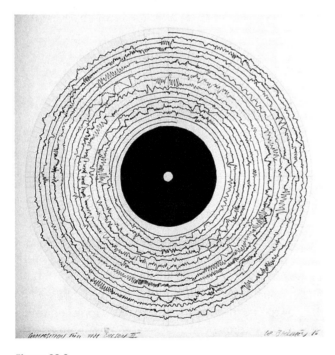

Figure 33.3
Moholy-Nagy's dream as a visual artwork. *Komposition für Tim Wilson II*, KP Brehmer, 1986 (photo courtesy of Ursula Block). Ink, tempera paint on graph paper, 33.0 × 33.0 cm. Courtesy gelbe MUSIK/Ursula Block, © VG Bild-Kunst Bonn 2015. Tim Wilson was a familiar figure on talk shows who was able to "read" unlabeled records and assign them to the correct recording artist.

a notebook. Thus one of the pioneers of phonographic thinking can be found at the intersection of visual art and synesthetic ideas in the 1920s. Bauhaus artist Lázsló Moholy-Nagy, taking the idea of writing with a phonographic needle literally in the twenties, attempted to create a universal instrument with the aid of an alphabet of etched patterns ("Ritzschrift-ABC" ["groove-script alphabet"]; Moholy-Nagy 1926, 363)—but that was based on a misconception. The direct, manual engraving of sounds in the grooves of the record proved impossible. The process of immediate reading and writing had to remain a technical one. However—and this is where his vision really was forward-looking—the mechanics and the mechanisms really can be *played*. Christian Marclay, who aptly refers to himself as a "record player,"[8] or Kid Koala, who creates new melodies with the pitch control slider, are contemporary examples of performers playing directly with the turntable's reading mechanisms. In fact, the performative acquisition of reading and writing processes is a long-standing tradition in the twentieth century that is based on the key functions of altering the playing speed and layering or mixing several signals. The experimental instruments of Pierre Schaeffer (built by Jacques Poullin), the *morphophone* and the *phonogène*, emerge in parallel with developments such as Harry Chamberlin's tape-replay home organ (which became the *mellotron*) and King Tubby's delay and filter devices in Jamaican dub, which have since found their way into the everyday culture of popular music via sampling and hip hop. Finally, playing with phonography has established itself with instruments ranging from the DJ setup, the mellotron, the sampler, with tape delays such as the Echoplex or the Watkins Copicat to live sequencers, virtual instruments, and software plugins connected to new, sensory interfaces.

In this case, working in a studio perhaps most closely resembles the traditional role of the composer, who forms musical material in the absence of instruments and performances. The cutting, mixing, and processing of recorded material presupposes not only a certain distance from the performative act and technological skills, but also the ability to work extensively with *musical material*—that is, the aesthetically significant, culturally sedimented phenomena and methods. That this composing process was not understood as such results from the classical music-based thinking explained above and the notions of authorship, originality, works, and performance that it implies. For, in addition to the self-recorded original material by, for example, Pierre Schaeffer, Jimi Hendrix, and Brian Wilson, since the 1980s the raw material of musical work has also been material shared with others (dub versions, for example) or completely external material extracted from other recordings on vinyl, tape, or as a sample (as used by Grandmaster Flash, DJ Shadow, or Public Enemy, among others). If the new material—such as Pierre Schaeffer's *objets sonores*—and their processing provoked a dispute over the classical concept of composing, then working with archives of specific performances of already "finished" music—that is, out of the DJ's crates, the loops, and the chopped beats—lies completely outside the concept of Western art music.

In a traditional sense, an author's composition ideas are manifested in a successful performance. The audibly perceptible result is the end of a chain of material, composition, and notation in which the musical structures are defined. Any adoption of parts of the notation or performance from another work of art or performance is regarded as plagiarism. However, in phonographic notation, the process begins with the performance. It is reperformed in the medium as a second-degree performance or further processed as phonographic material—or both. The respective performances are now entering phonography as "secondary orality" (Ong 1982), and this orality—and along with it structures of the previous work—becomes the object of instrumental playing, composing work, and the construction of new artifacts that now emerge as phonographic work. Here, the modes of production and individual styles are combined to form hybridizations from diverse origins and selections and are represented in the charts, as is the case with Madonna's "Music," Kanye West's *808s & Heartbreak*, or Daft Punk's "Get Lucky." Thus, an aesthetic historiography develops on the basis of technological media storage and transformation that is often misunderstood as an aesthetics of repetition or as recycling. In fact, it is no more and no less than the results of working with phonographic media as systems of notation and cultural archives.

Magic Science

In order for such a practice to develop outside of the concept of Western art music in the first place, alongside the experiments launched by the European avant-garde and its popular culture adaptations, a new, active approach to audio-technology and recording was required as a driving force. Originally, the technology of mass media had the negative connotation of serving as an instrument of cultural industry and leading to a passive, alienated culture of music. This position can be summed up as follows: "Pop is a classic case of alienation: something human is taken from us and returned in the form of a commodity. Songs and singers are fetishized, made magical, and we can only reclaim them through possession, via a cash transaction in the market place" (Frith 1988, 12).

Simon Frith puts forward this standard argument in order to then subject it to a detailed critique. Here, it serves a similar purpose. This intentionally exaggerated critique of cultural industry contains two key aspects: *alienation* and *magic*. In order to make them productive, we will look at the active roles of two groups of actors in the pop universe for whom the acquisition of pop as a commodity is certainly not of central interest. As extreme positions, these include both the artists themselves and those excluded from the established commodity cycles for lack of purchasing power. Replacing *commodity* with *(media-)technology* yields new options to handle pop and pop recording: by adopting audio-technology and using it creatively. Using Jimi Hendrix's own characterization, Paul Clarke refers to a "magic science" (1983, 195–198) as a new, positive mode of applying technology in the studio that is at the same time located outside live performance: "This is where Hendrix's term 'magic science' comes in useful, in that it insists we approach rock songs with an ear not only for the 'magic' of our experience of them, but also for the 'science'—the technological processes—through which the music is created" (ibid., 196).

Jimi Hendrix is an early prototype of the alienated inhabitants of the "Black Atlantic,"[9] who sought to adopt technology in ways that were beyond its inherent rationality. These include not only theatrical and experimental live feedback activities, but also, and in particular, a new way of handling studio technology. Their use of the studio bears the hybrid traits of simultaneously magical and thoughtful work with material whose emphasis on sound and whose orality and expression initially appear to reject all processing into a written medium. Here, technology does not serve the purpose of some calculated improvement of a musical product's commodity character; instead, technology is used to enhance its aesthetic intensity.

However, the crucial step in developing a literacy of recording required an even more radical breaking away from Western concepts of music and was also taken in the context of the "Black Atlantic." The social and societal conditions in Jamaica—a country that was hardly able to participate in the circulation of cultural industry commodities and whose production facilities functioned in early capitalist conditions—provided a fitting framework for this. Here, records were produced in the early 1960s, but more to be used in sound systems than to be sold to the public (Manuel and Marshall 2006, 449). So the aim of production was to make something that would have been regarded as having little sense in the Western world: the use of phonographic media as a media-instrumental part of a widespread, established performance practice. Partly alienated from its function and partly modified, studio equipment such as filters, reverbs, delays, and mixers served as a creative set of instruments in a new type of media performance instead of being used to create the illusion of authentic reproduction.[10] Thus the recording had now also overcome its role as a reproducing medium in the everyday practice of popular music that had only been assigned a meaning with reference to an original performance. The consequences of this change of roles are not only new phonographically montaged live genres such as dub, dancehall, or hip hop, but also, with "versioning" and "riddim," new concepts of a dynamized musical identity with a shared authorship (ibid.). At the same time, attention is shifted from melodic and harmonious forms of development to the sound and rhythm of repetitive patterns—that is, to what appears to be of secondary importance in regard to the composition of Western art music. In this context, and only a little later, with "Breakbeat Science," a central and highly differentiated field of "magic science" or, as Kodwo Eshun puts it, referring to the 1960s Sun Ra Arkestra, "Mythscience" emerges (Eshun 1998, 00[–004]): "For Flash [Grandmaster Flash, RG] in '81, going to the lab means approaching the studio as a research centre for the breaking down of the beat. In the lab, the

Breakbeat is isolated and replicated, to become the DNA of rhythmic psychedelia" (ibid., 02[013]). The sound laboratories and sound systems of dub are an elementary instrument-based and conceptual ecosystem for the transcultural aesthetic dissolution of boundaries in phonographic work and for its development. The scopes for action created here relate to all areas of sound and temporal structure and have an impact regarding its usage in composition, ranging from sound effects, multitrack recording, mixing and remixing to the digital world of the twenty-first century. They are characteristic and prototypical of a new link between humans and media technology that has ultimately become established in all studios performing artistic work. Against this background, the modes outlined by Kealy in 1979 on the way from "craft-union mode" to "art mode" represent the first variants of generalized media literality flowing into a broad practice of phonographic work. Only the establishment of new relations between technical notation, its reading and writing apparatuses, and cultural practice constitutes a reference of phonography differing from that of acoustic vibrations. The *Aufschreibesystem* of the acoustically real (see above) thus becomes the overarching field of activities for performing and composing music, and phonographic work transforms musical performance into a new type of musical material.

Translated by Mike Gardner

Notes

1. A detailed investigation of the "Genealogics of Acoustic Inscription" can be found in Levin 2006, 49.

2. Cf. Adorno (1934) 1990, 56: "It is covered with curves, a delicately scribbled, utterly illegible writing…"

3. Regrettably, this central aspect of inscription (word for word, the title translates as: "Systems of Inscription") was not adopted in the title of the English edition: "Discourse Networks 1800/1900" (Stanford, 1990).

4. And yet this aspect of the "real" is already reflected by Kittler: "Media 'define what really is'; they are always already beyond aesthetics" (1999, 3).

5. See Großmann and Hanáček, "Sound as Musical Material," in this vol.

6. Chris Cutler gives a detailed description of these "innate qualities of notation" against the background of rationalization and industrialization (1984, 284–286).

7. Thus while Philip Tagg's advanced method of analysis provides for a consideration of "electro-musical and mechanical aspects," "compositional texture and method" is assigned to the "aspects of tonality and texture" (1982, 48). Finally, in an almost traditional manner, his subsequent

model analysis of the most minute musical elements (the "musemes") refers to melodic phrases and motifs.

8. Exhibition Christian Marclay, *Record Player*, 1984, New York, Galerie Paula Cooper and *Christian Marclay—Record Player,* DVD video documentary directed by Luc Peter, Switzerland, 2000.

9. Drawing on Richard Middleton, Clarke describes aspects of a cultural hybridization that Paul Gilroy (1993) explores in detail using the term "Black Atlantic."

10. For details, see Williams 2012. The remix in 1970s New York disco represents a similar development of media performance, although it lacks the radicalness and the desire for experimentation characteristic of dub.

References

Bibliography

Adorno, Theodor W. (1934) 1990. The form of the phonograph record. Trans. Thomas Y. Levin. *October* 55:56–61. Original version (German): Die Form der Schallplatte. *23: Eine Wiener Musikzeitschrift* 17–19 (December 15, 1934), 35–39 (signed under the pseudonym "Hektor Rottweiler").

Chanan, Michael. 1995. *Repeated Takes: A Short History of Recording and Its Effects on Music.* London: Verso.

Clarke, Paul. 1983. "A magic science": Rock music as a recording art. *Popular Music* 3:195–213.

Cutler, Chris. 1984a. Technology, politics, and contemporary music: Necessity and choice in musical forms. *Popular Music* 4:279–300.

Cutler, Chris. 1984b. *File Under Popular.* London: November Books.

Eshun, Kodwo. 1998. *More Brilliant Than the Sun: Adventures in Sonic Fiction.* London: Quartet Books.

Frith, Simon. 1988. *Music for Pleasure: Essays in the Sociology of Pop.* New York: Routledge.

Frith, Simon, and Simon Zagorski-Thomas, eds. 2012. *The Art of Record Production: An Introductory Reader for a New Academic Field.* Farnham: Ashgate.

Gilroy, Paul. 1993. *The Black Atlantic: Modernity and Double Consciousness.* Cambridge, MA: Harvard University Press.

Kealy, Edward R. 1979. From craft to art: The case of sound mixers and popular music. *Work and Occupations* 6:3–29.

Kittler, Friedrich. 1985. *Aufschreibesysteme 1800/1900.* Munich: Fink. Trans. Michael Metteer, with Chris Cullens as *Discourse Networks 1800/1900* (Stanford, CA: Stanford University Press, 1990).

Kittler, Friedrich. 1999. *Gramophone, Film, Typewriter*. Trans. Geoffrey Winthrop-Young, and Michael Wutz. Stanford, CA: Stanford University Press. German Edition: *Grammophon, Film, Typewriter* (Berlin: Brinkmann & Bose, 1986).

Levin, Thomas. 2006. "Tones from out of nowhere": Rudolph Pfenninger and the archaeology of synthetic sound. In *New Media, Old Media: A History and Theory Reader*, ed. Hui Kyong Chun Wendy and Thomas W. Keenan, 45–82. New York: Routledge.

Manuel, Peter, and Wayne Marshall. 2006. The riddim method: Aesthetics, practice, and ownership in Jamaican dancehall. *Popular Music* 25 (3): 447–470.

Moholy-Nagy, László. 1926. Musico-mechanico, mechanico-optico: Geradlinigkeit des Geistes—Umwege der Technik. Special Issue "Musik und Maschine." *Musikblätter des Anbruch* 8/9:363–367.

Ong, Walter. 1982. *Orality and Technology: The Technologizing of the Word*. London: Methuen.

Tagg, Philip. 1982. Analyzing popular music: Theory, method, and practice. *Popular Music* 2:37–67.

Talbot, William Henry Fox. 1844. *The Pencil of Nature*. London: Longman, Brown, Green, and Longmans.

Théberge, Paul. 1989. The "sound" of music: Technological rationalization and the production of popular music. *New Formations* 8 (2): 99–111.

Williams, Sean. 2012. Tubby's dub style: The live art of record production. In *The Art of Record Production: An Introductory Reader for a New Academic Field*, ed. Simon Frith and Simon Zagorski-Thomas, 235–246. Farnham: Ashgate.

Sound and Media

Christian Marclay—Record Player. 2000. Director: Luc Peter. Video documentary, Intermezzo Films.

Daft Punk. 2013. "Get Lucky (Daft Punk Remix)." Twelve-inch single, Columbia, 88883746911.

Madonna. 2000. "Music." Seven-inch single, Maverick, Warner Bros. Records, 716826.

Sun Ra and His Myth Science Arkestra. 1967. *Cosmic Tones for Mental Therapy*. LP, Saturn Research, SR-408.

West, Kanye. 2008. *808s & Heartbreak*. CD, Roc-A-Fella Records, B0012198-02.

34   Listener Orientation

Mark J. Butler

I think the main difference is, in electronic music, there's a lot of ways to create something that runs—that is static, but nevertheless, it's creating something. Take a drum computer: you turn it on and it plays a pattern. And you cannot turn on a drummer. A drummer always has to do something in order to work. And the drum computer, you turn it on and the pattern is there, but the action of the person who is playing the drum computer is *changing* the pattern.

—Robert Henke (Monolake), interview with the author, Berlin, Germany, July 22, 2005

In the above statement, Robert Henke contrasts electronic music with styles played on traditional instruments. He makes a distinction between continuously "doing something" in order to produce a sound, as an acoustic drummer would, and initiating a process that runs on its own once set in motion. He also articulates a tension between a process moving inexorably forward through time and the "static," almost physical sense of presence that a pattern evokes when it is just "there." In speaking about his musical creations, he describes independently functioning sounds from an external perspective—almost as if he were listening to someone else's music—rather than as finished compositional products to which authorship might be ascribed.

The sites of contrast in Henke's remarks consistently express a divided perspective: on the one hand, he speaks as an agent of sonic genesis; on the other, as someone who stands outside the event he has initiated, hears it, and evaluates it. In the latter role, he acts particularly as a *listener*. I describe this kind of perspective on sound as *listener orientation*. This term captures a set of attitudes that is widespread within electronic dance music. A DJ or laptop set characterized by listener orientation is simultaneously performance-based and interpretive; it encompasses both the production and consumption of sound. The musician's attitude is reflective and characterized by a dual consciousness. I elaborate these qualities in the following article. Although I will be speaking primarily of performance, listener orientation is equally characteristic of electronic dance music's compositional processes; it can best be understood as a way of perceiving an unfolding sonic event.

DJing Between Listening and Doing

In dialogue with Henke's remarks, I would like to share a personal anecdote. Although I had begun to notice some of the attitudes I describe here after interviews with musicians, I first became keenly aware of listener orientation as I was learning to DJ, especially during a period in which I frequently practiced classical piano and DJing in immediate succession. The former was a discipline I had been practicing for most of my life, while the latter was completely new to me. The resulting gaps in my experience and knowledge drew the discrepancies and points of contact between the two activities into sharp relief. After practicing, I often recorded my thoughts immediately in journal entries.

While DJing, I was struck by the way in which I could step outside a sound as I was making it happen. In fact, this oscillation between listening and doing is an essential part of this kind of musical performance: DJs are continuously evaluating the current configuration of sounds; determining if, when, and how it should change; and thinking about what sort of sound or record should follow next. While this is happening, the record continues to turn, and loops and sequences continue to repeat, until the next sonic action is undertaken. Referring back to Henke's description, the sound is always "there"; it is always running. This sonic presence arises through the combined effects of repetition and an endless flow of sound, two of the most distinctive features of this musical style.

By contrast, classical pieces offered me rests, chances to breathe, breaks between movements. Yet they also required me to be continually involved in the production of the sound; if I became too enraptured by what I was playing—if I became too much of a *listener*—the piece fell apart. At the same time, I noticed that I have often been drawn to pieces featuring constant rhythmic motion, such as Chopin's etudes or Schubert's *Impromptu in E flat*, D. 899. These works also offer an uninterrupted flow of sound and cyclical temporal organization. Structuring time in recurring cycles fosters a certain reflectivity; one can begin to stand outside a temporal unit and perceive it even as it unfolds.

As a classical performer, I sought to bolster my abilities by practicing listening outside of a performance context. While pursuing a graduate degree in piano, I began the discipline of "silent practice." Often advocated by piano teachers, but rarely (I suspect) carried out, this mode of exercise involves running a composition silently through one's head, away from one's instrument. I turned to the technique in desperation, during a period in which my teacher was pushing me to learn large amounts of repertoire in a very short time. I was having particular problems memorizing the finale of a late Schubert sonata, a rondo movement in which the initial theme appears in a seemingly endless series of keys and textures. I could imagine the piece in my head in a loose, fuzzy way, but when I tried to really make myself hear *every note*, I was startled to discover how vague my knowledge was. Even more surprising, I made the same mistakes *mentally* that I did physically! Practicing silently became a helpful discipline for me, something that freed me from the rut of pointless physical routines and allowed me to refine my inner hearing in new ways.

At the same time, there is a strangeness to the idea of silent practice, a curious separation of music from the body. It reveals the extent to which an understanding of sound as recorded object has permeated our consciousness, even with regard to styles whose origins predate recording. By silently practicing, I slowly and surely produced a full recording in my memory, which in turn enabled me to play it back for others. And—to invert the sequence of cause and effect—one might also ask whether I needed to know a piece this well *because* I was so accustomed to experiencing music through recordings. For me (and presumably for many other classical performers) there was often a huge disjuncture between my primary activity as a musician—practicing—and the performances for which I was supposed to prepare. One required many hours of daily maintenance, centered in particular around *repetition*; the other proceeded irrevocably forward through time and was over in an instant. A bad performance was like a recording that one could not stop; a good one, like the playback I had inscribed in my mind.

These reflections highlight some of the effects of recording practices on musical epistemologies in contemporary classical performance, a context in which the pervasiveness of these practices is rarely admitted or discussed. In electronic dance music, conceptions of sound as recorded object are more immediately apparent. What is perhaps surprising, however, is their incursion into the domain of performance. In characterizing their performances, for instance, laptop musicians frequently use the language of discovery: they "find" a perfect combination between two loops; they "realize" a new way in which a track might be arranged; they "hit upon" previously unknown sonic possibilities. What almost escapes notice is the serendipitous, externally oriented attitude they display—projected, after all, onto music *they themselves* wrote. Similar emphases appear often in descriptions of DJing, as in the following characterization by DJ Shiva:

What's really fun is when it's completely off the cuff, just accidental. You were playing this song, and you're flipping through your crate, thinking, thinking, thinking, "What's gonna go? OK, that!" And you put it down on there, and you cue it in, and you're like, "Holy shit! I didn't know that!" Because it has a sound that's exactly the same as the sound in that record, only it's on another beat. (Lisa Smith [DJ Shiva], interview with the author and Cornelia Fales, December 3, 2001)

In the context of a DJ set, this attitude of discovery is less surprising, since DJing is in fact about finding novel combinations of other musicians' records. The more significant quality evident in both descriptions is the aforementioned view of musical sound as recorded object. Notably, this perspective holds even when the sounds in question are self-generated. And it continues to interact with more traditional concerns about music's technical qualities. DJ Shiva, for instance, pinpoints timbral and rhythmic qualities as the key to the discovery she describes: two sounds that are "exactly the same," with one "on another beat." The constituents of this musical play, however, are entire recordings.

The Epistemology of Performing with Recordings

The attitudes that I describe as listener orientation are thus inextricably intertwined with the penetration of modern media practices into all forms of music. In electronic dance music, this interpenetration reveals itself in a particularly striking fashion, as performances themselves are crafted from the alteration and reconfiguration of recorded elements. The physical technologies of its production and performance also promote listener orientation. By shifting some of the responsibility for sound production from the performer to the machine, electronic dance music technologies cultivate the emergence of a distinctively interpretive role *during performance*. Whereas a conventional musician "always has to do something in order to work," the DJ or laptop performer can turn on a device such as a drum computer, and "the pattern is there." The performer's agency does not disappear, of course, for the ever-present pattern is only raw material for a musical improvisation. Freedom from continuous concentration on producing "the notes," however, enables the performer to be a better listener, for he or she can step outside the performance and evaluate it while it is ongoing. It is in this way that the technologies of electronically mediated performance support the quality of listener orientation that I have described as reflective; they make it possible for the performer to create and experience a musical event at the same time.[1]

Scholarly and journalistic writing on technology in electronic dance music, however, has not addressed such effects; instead, it has generally emphasized one or both of two main themes. The first is the low cost of the technology involved in production, which in the mid-1980s allowed house and techno musicians to make records with cheap analog devices, such as the Roland TR-808 and TR-303, and more recently has enabled the production of internationally distributed recordings from relatively simple home studios. The second recurring theme is the low degree of technical training required for composition. In time, the technologies involved can be mastered on one's own, and most electronic dance musicians are self-taught.[2]

These emphases, although not irrelevant, do not tell the whole story, for they fail to address the experiential possibilities fostered by the technologies of electronic dance music.[3] Indeed, very little critical attention has been devoted to the actual experience of performing, despite the rise of performance studies, music-theoretical literature on performance and analysis, and other areas of inquiry. The same gap exists in fields in which performance is a primary activity. After writing the previously mentioned narratives of my experiences as a classical performer, I was struck by their unfamiliarity; from my twenty years of formal piano instruction, I can recall no extended discussion of what a performance should feel like for the musician(s) or how it *does* feel. Instead, discourse around performance focused almost exclusively on issues of mastery and interpretation.

For contemporary audiences and musicians, the dual perspective afforded by listener orientation likely derives from the rise of the recording as the primary format through which we encounter music. Prior to these infinitely repeatable, rewindable, pausable objects, performance was more strongly characterized by evanescence. Now, although musical events continue to move forward through time as always, there is also a novel sense in which we can *hear*, or understand in an auditory manner, the disruptive effects of recording on the linear flow of time.[4] And perhaps we also desire this kind of experience, this enhanced ability to reflect on musical sound as it unfolds, precisely because the epistemology of recording practices has so thoroughly pervaded our consciousness.

Acknowledgments

This essay is based on material that appears in my 2014 book *Playing with Something That Runs: Technology, Improvisation, and Composition in DJ and Laptop Performance*. I thank Oxford University Press for permission to include it. The book takes as its intellectual starting point the fact that electronic dance music performances—namely, DJ and laptop sets—are fluid, contingent, improvised *events*, which are at the same time formed entirely from prerecorded *objects*. In exploring the tensions between these modalities, the project aims to problematize several key dichotomies that have informed discourse on musical performance and recorded sound. These include fluid/fixed, process/product, improvised/composed, and live/prerecorded.

Notes

1. One modality through which musicians may further perform a response to recorded sound is dance. When musicians embody their responses through dance, they are performing listener orientation in a visible manner. This is most evident when they let go of their interfaces and dance with their whole body, something that would not be feasible without technological support, of course.

2. I deliberately use the word *training* rather than *skill*. Although the latter does appear frequently in arguments of this sort, it is problematic. Electronic dance music requires substantial skill in both composition and performance, and respected musicians spend years developing their abilities.

3. An exception to this lacuna appears in a 2002 master's thesis by John Bowers, in which the author briefly describes machine-facilitated listener orientation (without using this term) in a section on "variable engagement." "Different forms of engagement have different *phenomenologies* associated with them," he writes. "Listening can take different forms depending on whether one is listening as one is bodily engaged or listening to an independent machine production … Algorithmic material or the playback of recordings may 'buy time' for the set-up and initiation of other materials" (Bowers 2002, 46–47). As an ethnographically informed and self-reflexive account of interactive electronic performance, Bowers's thesis is unique.

4. The way in which recording "stops time" is a recurring theme in the literature on sound technologies; see especially Katz 2004.

References

Bowers, John. 2002. Improvising machines: Ethnographically informed design for improvised electro-acoustic music. MA thesis, University of East Anglia.

Butler, Mark J. 2005. Unpublished interview with Robert Henke [Monolake], July 22.

Butler, Mark J. 2014. *Playing with Something That Runs: Technology, Improvisation, and Composition in DJ and Laptop Performance*. New York: Oxford University Press.

Butler, Mark J., and Cornelia Fales. 2001. Unpublished interview with Lisa Smith [DJ Shiva], December 3.

Katz, Mark. 2004. *Capturing Sound: How Technology Has Changed Music*. Berkeley: University of California Press.

35 (Re-)Mastering Sonic Media History

Jens Gerrit Papenburg

Every time has its sound. The study of historic sounds and soundscapes that do not have to be aestheticized as music has become an established research field in contemporary humanities scholarship. In contrast to that kind of sound history, sonic media history deals with the media, forms, and formats that make the sounds of history audible in the first place and that make them a possible object of study. Media history remains a sonic history decisively by asking if the media of the sounds of history have a specific sound themselves, an implicit *idio-sound*.[1] Or, in other words, sonic media history analyzes the opacity of sonic media—or rather the thresholds between the shimmering through, the *sounding through* on the one hand, and sonic media's becoming-imperceptible on the other hand—as a historical phenomenon. For that endeavor, sonic media history combines media theory with sound studies and musicology.

Media's Idio-Sound between Contingency and Design

Media's idio-sound coincides with media's materiality, physicality, and—in the case of digital media—their codification. As a mere technical subject, it is primarily modeled in relation to frequency ranges and dynamic ranges. But these ranges, again, correlate with narratives such as the grand narrative of increasing "fidelity" and with discourses such as the one on "cold" CD sound and "warm" vinyl sound, which combine aesthetic judgments, nostalgic references to the past, practices of listening, and selected technologies. In this regard, idio-sound includes "the diffuse and atmospheric comfort of the small but bright gramophone sound" that—according to young Adorno in 1927—"corresponds to the humming gaslight" but also encompasses the "shimmering of the 18 kilohertz frequency range of a 128k mp3 encoding" (Kösch 2005). Moreover, idio-sound refers to the aging of media materialities that have become audible: a yellowing of visual media conforms to the so-called *copy effect*, hearable as a pre-echo of magnetic tape that has long gone unheard, to the crackles and pops of scratched vinyl, to the skipping of a CD.

However, idio-sound is also one that is designed in the space between aesthetics and technology. In this chapter I propose that the analysis of the sonic dimension of popular music's release practices provides a fruitful case study to access the design of media's idio-sound—because it has become common practice to release (and rerelease) *one single* music production in the form of *several* different media and formats over a period of several decades. The various releases of one single production thereby compete with each other through their sound as well, through subtle, minor sonic differences. As a consequence, we do not have to catch up with the soundtracks of avant-gardist "cracked media" artists (Kelly 2009) or listen to the nostalgic simulations of vinylizer plugins in order to analyze crucial aspects of the design of idio-sound. Instead, we can listen closely to the work of a specific type of engineer whose efforts are still heavily understudied—is it only because of his or her allegedly nerdy appearance?—but who offers a small but important contribution to how popular music sounds today and to how it sounded yesterday: the mastering engineer.[2] Mastering engineers attempt to master media's idio-sound. In their remastering of old record productions, they are interested in making the old sound new. In doing so, he or she (notably, the industry is still primarily dominated by male mastering engineers, although there are exceptions) contributes to a paradoxical historicization of the sounds of popular music.

Mastering as Media and Format Transposition

With the advent of magnetic tape in recording studios in the late 1940s, music was produced with tape, but it was released as a record. That is where mastering came in: the mastering engineer turned tapes into records. Thus, we can start the history of mastering in a narrower sense in the late 1940s, when mastering emerged as a conversion of one medium into another, or, more precisely, as a "media-technological transposition."[3] Although magnetic tape became the primary medium for producing music in the recording studio, record labels released music as records. In the beginning, that form of transposition was often referred to as a mere technical transfer, which became known as transferring or even as rerecording (see Uecke 1956). Also, in the 1950s recording studios began to install separate mastering rooms with disk-cutting lathes for phonograph records (see Putnam 1960). The spatial differentiation between recording, production, and post-production processes correlated with the distinction between recording engineer and mastering engineer. In the late 1960s, the first independent mastering studios appeared in the United States—such as Doug Sax's The Mastering Lab in Hollywood, Frankford/Wayne Mastering Lab in Philadelphia, or Sterling Sound in New York.

With the digitalization of music production and the introduction of the CD, mastering stopped processing the media transposition and started to process a format transposition, something which still exists today when music gets mastered for particular popular media players—indicated by specific slogans such as "Mastered for iTunes" (see Katz 2013). However, mastering in the digital era has continued to process media transpositions as the marketing concept "digitally remastered," which followed the CD's commercial launch in 1982, made obvious. Under this slogan, Columbia Records was able to sell the label's old "Jazz Masterpieces" for a second time—not in the form of the LP again, but in the form of the new CD. In addition, there are continuous examples of mastering as a form of media transposition, demonstrated by the periodic rereleases of popular music's "classic" albums, such as a remastered version of The Beatles' catalog—in fact, while writing this essay, Capitol Records rereleased that catalog as a fourteen-LP boxed edition—in mono.

Mastering, as media or format transposition, marks a mediation between the production and the release of music as vinyl record, CD, or sound file. The mastering engineer fine-tunes the sound of a production and mediates between technology and aesthetics. Thus, according to mastering guru Bob Ludwig, mastering is "the last creative step and the first manufacturing step in the record-making process" (cited in Sutherland 1998). For a mastering engineer, sound does not exist "in and of itself" but in relation to listening practices and media technologies. That is, the mastering engineer is "finishing a piece of music or any auditive work for a certain purpose," as Berlin-based engineer Rashad Becker puts it (Becker and Henke 2008). In addition, he or she is involved in the creation of the general sound of a release—and that can be tricky, especially in the case of soundtracks or compilation albums that include sound material from different decades and genres.

The function of the mastering engineer is comparable to the function of the editor in book production. Just as the latter manages the process from the author's manuscript to the page proof, the former manages the process from the producer's final mixdown to the matrix that is used for pressing. In regard to the book, the historians Guglielmo Cavallo and Roger Chartier point out that a text never exists "in itself," that is, "outside of the physical support that offers it for reading (or hearing)" (1999, 5). The same point also applies to popular music's releases, with the difference that the physical supports and media of popular music are complying with a set of industry standards. Those standards, according to Friedrich Kittler, "determine how media reach our senses" (2010, 36). The goal of those standards is for the sonic media themselves to go unheard industry-wide, for the media to become—to make use of a media theory concept—"anesthetic" (cf. Ernst 2004; Vogl 2007): we should listen to the sounds of music and not, say, to the sounds of the vinyl records themselves, that is, not directly to the media's idio-sound. But we can add here that these two sonic domains—the sound of music and the idio-sound of media—are not clearly distinguishable once music exists primarily in the form of a record. Moreover, the industry standards for sonic media are not always well-defined. That became particularly clear during so-called "format wars" such as the "war of speeds" between RCA Victor's comparatively fast-spinning single record and CBS Columbia's slow-spinning LP in the late 1940s, as well as in the debates about the standardization of the recording curve—the Recording Industry Association of American (RIAA) recording curve became the industry standard in 1954 (see Tang 2004, 147–152).[4] Hence, fine-tuning the sound of music based on the particular physics and standards of media is one function of the mastering engineer. The engineer thereby constructs a relationship between sound and technology through which the sound is made to meet the requirements of mass reproduction. This kind of fine-tuning is not merely a matter of executing a set of technical instructions. As early as the 1950s, a lot of engineers who were not impressed by the standardization of the recording curve were primarily interested in making "hot" records, which were supposed to "stand out when played on jukeboxes" (Schmidt-Horning 2013, 189). Sun Records owner Sam Phillips points out that he added "intended distortions" during mastering to achieve "proper loudness" (cited in Broven 2009, 152). In this case, rather than going unheard, the media's idio-sound is being designed, resulting in a kind of 1950s proto-loudness war (see also Devine 2013). Thus, the mastering engineer assures that the sound of music functions efficiently as a part of concrete listening practices, such as playlist listening, which is, again, organized in relation to the listening technologies and devices, such as jukeboxes, through which music is heard.

Moreover, the establishment of mastering in the music production process around 1950 coincided with the particular organization of the music economy as an economy dominated by the mass production of records (see Attali 1985; Frith 1992; Anderson 2006; Wicke 2009). In Jacques Attali's bold description of this economy, he defines a particular subject that can assist in more precisely situating the mastering engineer on a theoretical level. Attali designates this subject as the "molder," which he understands as someone "who creates the mold," which, in turn, "allows the mass reproduction of an original" (1985, 128). Tim Anderson specifies Attali's mold as the "initial master recording" (Anderson 2006, xxxv). We can once again add to Anderson's remarks that the mold serving as a model for mass (re-)production is produced by the mastering engineer who masters a production's mixdown. In short, we can identify Attali's molder as a mastering engineer.

Anderson shows that a "recording-based media economy" (ibid., 22) is driven by an exploration of entertainment groups' and labels' back catalogs and archives. From a sonic media history perspective, we can note here that this exploration resonated strongly with the release of material that had already been released in different media formats.[5] In popular music studies, those release practices have primarily been analyzed as an economic issue—as a strategy to minimize risk for record labels in a turbulent music market (see Keightley 2004; Wikström and Burnett 2009). But the release of a particular music production in different and changing media formats over the years coincides with minor audible differences as well—and these differences are the product of mastering.

In the following section, I will discuss how mastering is involved in the fine-tuning of media transpositions and how mastering mediates between sound, technologies, and listening practices on the basis of two case studies.

Case Study I. Rock 'n' Roll—"Long Tall Sally": Remastering Changing Listening Practices

Because 1950s rock 'n' roll productions have been released as a series of different and changing media and formats over a period of about sixty years now, they provide a pertinent case study for an analysis of the exploration of back catalogs as well as of media's designed idio-sound. The selected rock 'n' roll hit production by Little Richard— "Long Tall Sally"—will be instrumental here in pointing out crucial aspects of the epistemological potential of a sonic media history. Five different releases of that production will be addressed in the following: (1) a seven-inch single release (Specialty 1956); (2) a shellac single release (London 1956); (3) a release as part of the compilation LP *Alan Freed's "Golden Pics"* (End 1961); (4) a release—a so-called "master" version—of the production on the triple CD *The Specialty Sessions* (Specialty 1989); and (5) a "hi-end" rerelease of Little Richard's first LP, which contains "Long Tall Sally" as the B-side's opening track (MoFi 2008).

In February 1956, Robert "Bumps" Blackwell produced "Long Tall Sally" with Little Richard in Cosimo Matassa's New Orleans J&M Studio for Art Rupe's label Specialty Records. One of the most bizarre rereleases of this production also acts as an illuminating object for a sonic media history: the aforementioned "hi-end" rerelease of Little Richard's first LP *Here's Little Richard*. It was put out by an audiophile label as a 180-gram vinyl pressing relying on a specific—half-speed—mastering technique, which is meant to enable better resolution of the high frequency range. However, an audiophile release of a 1950s independent rock 'n' roll production might clash with the sound aesthetics of this production because indie labels such as Specialty "allowed more noise into a record's grooves than the majors would tolerate" (Zak 2010, 163). Coinciding with fans' increasing ages and social advancement, the productions of Little Richard entered into the low-noise, middle-class living room as part of a hi-end listening culture.

The releases of Matassa's production of "Long Tall Sally" articulate minor sonic differences at a very general level: LP releases are cut much more quietly than single releases, the surface noises of shellac sound differently from those of vinyl, and CDs are comparatively low-noise. Moreover, the high frequencies are more present in the vinyl than in the shellac releases, as we can hear that the ride cymbal during the saxophone solo is comparatively accentuated throughout the audiophile release and also throughout the seven-inch single. But what makes the 1956 seven-inch single release stand out is that it sounds very loud because of its being cut at the margin of distortion. Moreover, it includes a characteristic reverb missing in the other releases.

We can assume that the 1956 seven-inch release was targeting listening practices in which records from different labels are listened to in a series through a device that is made for the serialization of records, such as a jukebox or an automatic record changer. The single's reverbed, mid-accentuated, heavy-compressed sound at the margin of distortion might contribute to the production working as a part of that kind of listening practice. In that case, the record's sound would be designed in relation to a listening practice in which the sound of records does not relate to the sound of a live performance, but rather to the sound of other records of other studios and labels with which this single is played in a series and with which this single has to compete.

Moreover, the single release's distortions resonate with the distortions that Specialty had already included in the record production, and that was also due to Little Richard's vocal technique: "Richard screamed so hard. His dynamic range was so terrific. Richard would be singing like this [whispers] and then all of a sudden BOW!! The needles would just go off the dial" (Blackwell, cited in White 1994, 75). Those distortions are also a part of "Long Tall Sally."[6] They are accentuated through the single release and toned down through the LP releases. In the case of the compilation LP *Alan Freed's "Golden Pics,"* the sound of "Long Tall Sally" has to work as part of a fixed, compiled series of "acoustic snapshots," which the fallen DJ Freed put together in an attempt to revive rock 'n' roll after it had become outmoded around 1960. Although the next tune in a jukebox series might be unknown, in the LP promoted by Freed it was already determined.

Even though the mastering of "Long Tall Sally" clearly demonstrates a set of sonic differences, it remains an anonymous procedure that is not explicitly mentioned in the release practices. The audiophile release and the 1960s LP release are, however, exceptions in this case. The latter displays a stamp in the run-out groove area that authorizes the mastering as a product of Bell Sound—a recording studio based in New York City. Similar practices of mastering authorization should become widespread over the next decades.

Case Study II. Techno and House: Mastering the Physicality of the Twelve-Inch Single

Electronic dance music—and especially early Chicago house and Detroit techno releases from the 1980s and 1990s—created another pertinent case study for a sonic media history, in part because the early evolution of these music forms is closely connected to a specific record format—the twelve-inch single. This format combines the diameter of the LP with the speed of the small 45-single and is itself a byproduct of the 1970s mastering of disco (see Papenburg 2014). The twelve-inch single enables not only the release of long tracks as singles, but its physics allows for the cutting of records with a wide cut. The design of the bass frequencies benefits from this additional space in particular. Moreover, the twelve-inch single's higher speed makes a sophisticated resolution of high frequencies possible. The louder a record is cut, the more it develops a specific idio-sound—and twelve-inch discs with only about ten minutes of music on each side make a loud cut possible.

1970s disco culture organized music production as a highly professionalized process based on a strict division of functionality; in early house and techno culture music, production is organized as an integrated process, pushed forward by young amateur producers working in cheap tape-based home studios at a do-it-yourself (DIY) level. For instance, one of the first Chicago house twelve-inch singles, "On and On" (Jes Say 1984), is "produced, arranged & performed by Jesse Saunders"—as the label sticker informs us. But, in the end, the early house producers were not in a position to produce records on their own. Thus, Saunders's record was produced in Chicago's record pressing plant Musical Products, owned by Larry Sherman (see Bidder 2001, 25–34). Sherman, who also cofounded the legendary house label Trax Records, was notorious for using recycled vinyl with poor sound. Thus, we can suppose that the anonymous mastering process was at best aimed at simply transferring the recording from tape to disc. Also, in his 1986 article about Chicago's early house scene, music journalist Barry Walters reported that the twelve-inch singles "often snap, crackle and pop louder than the percussion" (1986, 60).

In contrast to Chicago house, mastering acquired a distinctive status early on in Detroit techno. As early as the 1970s, disco mastering became increasingly personalized by specifying the mastering engineer on the record sleeve or by "vinyl graffiti" on the record's dead wax. On the record's inside margin, mastering engineers carved individual identification codes or little pictograms (see Miller 1998). Such strategies toward the authorization of vinyl mastering were heavily explored in Detroit techno and were combined with the exploitation of vinyl's idio-sound.[7] That was also up to Ron Murphy and his company NSC. Starting in the late 1980s, Murphy mastered some of the most vital techno releases of the first and second generation of Detroit techno producers.[8] At the beginning, he adapted the sound of amateur home producers such as Kevin Saunderson, Jeff Mills, Mike Banks, and Richie Hawtin to professional standards: "They mostly recorded their stuff on tape at home and the quality was really terrible" (Murphy, cited in von Thülen 2006).[9] Thus, mastering could be critical, especially for the "classical one-man home recording approach" of record production, as was common in early electronic dance music (Becker and Henke 2008). By adapting the sound to that of a professional level, Murphy's post-production made the sound competitive with the sound of other twelve-inch singles and improved its effectiveness as a part of dance clubs' sound systems. As a result, Murphy was mythologized by techno's historiography: "This sound—the minute difference in the way its records are mastered—may be part of the Detroit techno mystique" (Sicko 1999, 159).

Murphy (trade-)marked the releases with his studio logo, by scratching "NSC" in the run-out groove—and sometimes also on the extended space between the tracks. He exploited the record's physicality by cutting locked grooves, records with two grooves, and records that run from the inside out. Murphy does not only transfer sound to vinyl, he also takes the physics of the twelve-inch vinyl into consideration. For instance, in the case of *Rings of Saturn*, a double twelve-inch album by X-102 that Murphy mastered, the "rings" are not solely in reference to an arbitrary cosmic appellation of the record; they can be found as several endless or locked grooves on the two discs as well. Murphy was interested in a loud record cut (see Murphy cited in Morales 2000). Thus, a bass line that sounds comparatively thin in a record production can be transformed into an overtone-heavy sound during mastering.

Research Perspectives

Upon examination, a sonic media history becomes especially obvious in the field of popular music, although it is not reducible to that field alone. When sound history deals with sound as a material phenomenon, instead of only with descriptions of sound, it is supplemented by a sonic media history that enables an analysis of how sounds from the past are presented to the senses. As demonstrated in the case studies, whether the sounds of history are presented in the form of twelve-inch singles or of, say, LPs or MP3s makes a difference for a sound history and for sound research. Rather than simply going unheard, media's idio-sound is also another possible level of design—one that links listening practices, business models, and technical standards. Although the constitutive function of media for record production has become a heavily studied subject over the last few years, media's idio-sound is ready for further development as a part of a field of sound and music studies that listens in-depth.

Notes

1. I use the concept of idio-sound here in critical reference to organology's idiophones. Compared to the idiophones that are part of a musical instrument taxonomy based on different modes of sound production (see Hornbostel and Sachs 1914), idio-sound combines sound production with sound mediation. Thanks to Veit Erlmann for this reference.

2. A couple of practical handbooks on mastering have been published (e.g., Katz 2007, 2013). Although the study of mediated sound and produced music has become an increasingly popular subject for sound studies and musicology over recent years (Sterne 2012 and, e.g., Frith and Zagorski-Thomas 2012), there is still a gap in the scholarship regarding a cultural and historical analysis of mastering—with a few exceptions (see Hodgson 2010, 189–230; Papenburg 2011; Schmidt-Horning 2013, 188–191; Nardi 2014).

3. For a discussion of the "untranslatability" of media, Friedrich Kittler (1990) introduced the concept of "media transposition." Although Kittler used the concept primarily in relation to

information theory, we can follow here the musical sense of the concept and add that, in our context, the concept is not referring to a music of notes but to a music of sounds and unsounds.

4. This curve defines the degree of emphasis and de-emphasis of frequencies during the cutting of a record and the playing back of a record.

5. Moreover, the issue of rereleasing still existed in the late 1940s when CBS Columbia brought the LP to the market as a new medium and rereleased a collection of 78 shellac discs as LPs, such as Frank Sinatra's *The Voice of Frank Sinatra* (Columbia, USA, 1948) (see Osborne 2012, 98–100).

6. We can hear these distortions in the release of the master tape of "Long Tall Sally" (Specialty 1989), which is, of course, itself remastered.

7. This practice was also picked up in Chicago house, but in a much more low-key way, where twelve-inch releases were authorized by a discretely etched "BP," (standing for the engineer Bud Pressner) in the run-out groove.

8. For a generation-based historiography of Detroit techno, see May 2006.

9. Alongside the media transposition from tape to twelve-inch single, there is another media transposition that is significant for house and techno: the conversion from tape and twelve-inch to LP. This contributed to the fact that a local music scene was exploited at a global level after the British label London Records released the first house compilation LP (1986) and after the British label Virgin Records released the first techno compilation LP (1988). However, I have focused in particular on the mastering of the media conversion from tape to twelve-inch single in the part of the chapter about early house and techno.

References

Bibliography

Adorno, Theodor. (1927/1965) 2002. The curves of the needle. In *Essays on Music*, ed. Richard Leppert, trans. Susan H. Gillespie et al., 271–276. Berkeley: University of California Press.

Anderson, Tim J. 2006. *Making Easy Listening: Material Culture and Postwar American Recording.* Minneapolis: University of Minnesota Press.

Attali, Jacques. 1985. *Noise: The Political Economy of Music.* Trans. Brian Massumi. Minneapolis: University of Minnesota Press.

Becker, Rashad, and Robert Henke. 2008. *Mastering.* http://www.monolake.de/interviews/mastering.html (accessed November 26, 2014).

Bidder, Sean. 2001. *Pump Up the Volume: A History of House.* London: Channel 4 Books.

Broven, John. 2009. *Record Makers and Breakers: Voices of the Independent Rock 'n' Roll Pioneers.* Urbana: University of Illinois Press.

Cavallo, Guglielmo, and Roger Chartier, eds. 1999. *A History of Reading in the West*. Trans. L. G. Cochrane. Cambridge: Polity Press.

Devine, Kyle. 2013. Imperfect sound forever: Loudness wars, listening formations, and the history of sound reproduction. *Popular Music* 32 (2): 159–176.

Ernst, Wolfgang. 2004. Der anästhetische Blick? Wahrnehmung durch Medien. In *Ästhetik: Aufgabe(n) einer Wissenschaftsdisziplin*, ed. Karin Hirdina and Renate Reschke, 65–79. Freiburg: Rombach.

Frith, Simon. 1992. The industrialization of popular music. In *Popular Music and Communication*, ed. James Lull, 49–74. Newbury Park: Sage.

Frith, Simon, and Simon Zagorski-Thomas, eds. 2012. *The Art of Record Production: An Introductory Reader for a New Academic Field*. Surrey: Ashgate.

Hodgson, Jay. 2010. *Understanding Records: A Field Guide to Recording Practice*. New York: Continuum.

Hornbostel, Erich M. von, and Curt Sachs. 1914. Systematik der Musikinstrumente: Ein Versuch. *Zeitschrift fur Ethnologie* 46 (4–5): 553–590.

Katz, Bob. 2007. *Mastering Audio: The Art and the Science*, 2nd ed. Burlington, MA: Focal Press.

Katz, Bob. 2013. *iTunes Music: Mastering High Resolution Audio Delivery*. Burlington, MA: Focal Press.

Keightley, Keir. 2004. Long play: Adult-oriented popular music and the temporal logics of the post-war sound recording industry in the USA. *Media Culture & Society* 26 (3): 375–391.

Kelly, Caleb. 2009. *Cracked Media: The Sound of Malfunction*. Cambridge, MA: MIT Press.

Kittler, Friedrich. 1990. *Discourse Networks, 1800/1900*. Trans. Michael Metteer, with Chris Cullens. Stanford: Stanford University Press.

Kittler, Friedrich. 2010. *Optical Media*. Trans. Anthony Enns. Cambridge: Polity.

Kösch, Sascha. 2005. Die Zukunft neu formatieren: Spekulationen über künftige Seinsweisen von Musik. In *Techno-Visionen: Neue Sounds, neue Bildräume*, ed. Sandro Droschl, Christian Höller, and Harald A. Wiltsche, 55–61. Vienna: Folio.

May, Beverly. 2006. Techno. In: *African American Music. An Introduction*, ed. Mellonee V. Burnim and Portia K. Maultsby, 331–352. New York: Routledge.

Miller, Chuck. 1998. Herbie was here: Stories of the mysterious messages in the runoff grooves. *Goldmine* 24 (7): 461.

Morales, Mizie. 2000. Dubplate Special: Detroitstory. http://de-bug.de/mag/dubplate-special -detroitstory/ (accessed November 26, 2014).

Nardi, Carlo. 2014. Gateway of sound: Reassessing the role of audio mastering in the art of record production. *Dancecult* 6 (1): 8–25.

Osborne, Richard. 2012. *Vinyl: A History of the Analogue Record*. Surrey: Ashgate.

Papenburg, Jens Gerrit. 2011. Hörgeräte: Technisierung der Wahrnehmung durch Rock- und Popmusik. PhD diss., Humboldt-Universität zu Berlin.

Papenburg, Jens Gerrit. 2014. "A great idea after the fact": Das (Er-)Finden der Maxisingle in der New Yorker Diskokultur der 1970er Jahre. In *Popgeschichte*, vol. 2, *Zeithistorische Fallstudien, 1958–1988*, ed. Bodo Mrozek, Alexa Geisthövel, and Jürgen Danyel, 179–200. Bielefeld: transcript.

Putnam, Milton T. 1960. Recording studio and control room facilities of advance design. *Journal of the Audio Engineering Society* 8 (2): 111–119.

Schmidt-Horning, Susan. 2013. *Chasing Sound: Technology, Culture, and the Art of Studio Recording from Edison to LP*. Baltimore, MD: The Johns Hopkins University Press.

Sicko, Dan. 1999. *Techno Rebels: The Renegades of Electronic Funk*. New York: Billboard Books.

Sterne, Jonathan. 2012. *MP3: The Meaning of a Format*. Durham, NC: Duke University Press.

Sutherland, Scott. 1998. In a Maine outpost, a master of sound has built a mecca. *New York Times*, March 28.

Tang, Jeffrey D. 2004. Sound decisions: System, standards, and consumers in American audio technology, 1945–1975. PhD diss., University of Pennsylvania.

Uecke, Edward H. 1956. The control of quality in phonograph records. *Journal of the Audio Engineering Society* 4 (4): 159–162.

Vogl, Joseph. 2007. Becoming-media: Galileo's telescope. Trans. Brian Hanrahan. *Grey Room* 29 (4): 14–25.

von Thülen, Sven. 2006. Detroit: Ron Murphy. http://de-bug.de/mag/detroit-ron-murphy/ (accessed November 26, 2014).

Walters, Barry. 1986. Burning down the house. *Spin* 2 (8): 60–63.

White, Charles. 1994. *The Life and Times of Little Richard: The Quasar of Rock*. Cambridge, MA: Da Capo Press.

Wicke, Peter. 2009. Der Tonträger als Medium der Musik. In *Handbuch Musik und Medien*, ed. Holger Schramm, 49–88. Konstanz: UVK.

Wikström, Patrik, and Robert Burnett. 2009. Same songs, different wrapping: The rise of the compilation album. *Popular Music and Society* 32 (4): 507–522.

Zak, Albin. 2010. *I Don't Sound Like Nobody: Remaking Music in 1950s America*. Ann Arbor: University of Michigan Press.

Sound and Media

Little Richard. 1956. *Long Tall Sally.* Seven-inch single, Specialty, XSP-572-45 (45-XSP-572).

Little Richard. 1956. *Long Tall Sally.* Ten-inch shellac, London/Specialty, MSC 1443-1.

Little Richard. 1989. *Little Richard. The Specialty Sessions.* 3x CD, Specialty, SPCD 8508.

Little Richard. 2008. *Here's Little Richard.* LP, MoFi 1–287 (MFSL-1227–B1).

Saunders, Jesse. 1984. *On and On.* Twelve-inch single, Jes Say, JS9999.

Sinatra, Frank. 1948. *The Voice of Frank Sinatra.* LP, Columbia CL:6001.

Various Artists. 1961. *Alan Freed's "Golden Pics."* LP, End, LP 313 (LP 313A).

Various Artists. 1986. *House Sound of Chicago.* LP, DJ International Records/London, LONLP 22.

Various Artists. 1988. *Techno! The New Dance Sound of Detroit.* 2x LP, 10 Records/Virgin Records, DIX G 75, UK.

X-102. 1992. *Discovers the Rings of Saturn.* 2x EP, Underground Resistance/Tresor. GER.

Cultural Sound Practices

36 Distorted Voices, Afrofuturism, and the Aesthetic Experience of the Self as Other

Jochen Bonz

The 1998 dance pop song *Believe* did not only signify an amazing comeback for Cher—it was also the first appearance of a new vocal sound. Years later, Jens-Christian Rabe would describe this sound as "robot-like sawing, gurgling, fluttering" in his enthusiastic review of the Kanye West album *808s & Heartbreak*, an album strongly influenced by this sound (Rabe 2008, 13). The singing voice's fascinating "grain" (cf. Barthes (1972) 1990; Dolar 2006), located at the border of body and language, had now been digitally potentiated in a simultaneously abrupt and subtle "overturning" of the voice in digital polish. This digitality of sound, practically put *on display*, betrayed a fact that quickly spread through the public: the striking vocal sound was created by software, that is, by the usage of extreme settings on music production software that was actually developed for pitch correction. This phenomenon is therefore called *Auto-Tune*, or *Auto-Tune effect*, like the software.

Following Cher's hit, the digital fluttering of the voice enjoyed further success; although it initially fell silent, it ended up becoming omnipresent. One of its early successes was Daft Punk's single *One More Time*, released in 2000. *One More Time* is a piece of music that is based on the house music style that had become known as "French house" some years earlier, a style that had itself been significantly influenced by the same French producer duo. House is characterized by bass runs, which are reminiscent of 1970s disco and funk, but which can become breakingly thin and slippery to the point of abrasiveness, only to then burst with volume, providing a rich sound in the next moment, as massive digital noise filters and sound compressors are applied. Just as the bass runs are almost continuously expanding and narrowing, blowing up big and then becoming small, the Auto-Tune effect is also used to digitally vary the voice of the guest singer, Romanthony—to insiders known as a deep house singer, with an appropriately voluminous voice for that genre. Here, his voice is overturning and overflowing in digital iridescence, constantly collapsing into itself and breaking loose from what would have once been conventionally considered to be the "natural" voice-body.

This effect came back onto the stage in 2007/2008, when it influenced the single hit *Sensual Seduction/Sexual Eruption* by Snoop Dogg and Lil Wayne's *Lollipop*, as well as the above-mentioned album *808s & Heartbreak* by Kanye West. Snoop Dogg, Lil Wayne, and Kanye West had all become famous as rappers, and at least Dogg and West were already able to look back at long and successful careers. In that year, they transformed from rappers into singers. Their form of expression changed from speaking in a musical way to formulating lines of melody. This vocal-sound-like sonic shape dazzles: it shines and shatters, mutters and flutters. The situation was such that this other speak-singing was once again perceived as sounding new and unknown at this point in time, and that accounts for some of the comments in Rabe's glowing review, in which he talks about his perception of the modulated voice sounds as follows: "One has to imagine metallic sawing and disruptive swirling, occasionally turning over on itself. Not human, but humanoid. It really penetrates the listener, and also sounds pleasantly other-worldly, delirious" (2008, 13).

Since its reappearance in 2007/2008, the Auto-Tune effect has become one of the optional stylistic devices; beyond that, it has turned into a pop music sound signature of the first decade of the twenty-first century (see Reynolds 2011). Today, the Auto-Tune sound has a place in many different music aesthetics, such as in the case of the London dubstep project Burial and its vast, distorted, massive bassline-infused sound spaces; and in the U.S. American independent rock music band Poliça, which is influenced by electronic dance music and R&B.

If you are wondering where the digitally quivering voice went between the hits of the late 1990s and the successful establishment of the sound ten years later, then here is the answer: it was in contemporary soul music, sometimes called hyper soul or simply R&B in general. An example of this is the piece "You Don't Know" by 702, a vocalist group paradigmatic for R&B at that time. This song was released on their self-titled album in 1999. Missy Elliot—whose sound aesthetic represented the state of the art in black music at the time—was one of their producers. 702 only uses the digital quivering of the voice as a stylistic device periodically; that is not the case with the rapper T-Pain. Since 2005, he has made the Auto-Tune distortion into his trademark. He has been a guest singer on many R&B hits, and most likely prepared the path for the renewed appearance of the effect in 2008. He is now selling an iPhone app called *I am T-Pain*, which can be used to distort the user's own voice. As Alexander Weheliye (2002) in *Posthuman Voices in Contemporary Black Popular Music* stated (using 702 as an example), both T-Pain and 702 use the digital quivering "without any decisive connection to the signification of the lyrics" (37). Although the vocoder synthesizer and talk box instruments used in 1970s and 1980s funk and soul productions brought about comparable sound effects, they sounded more analog; moreover, they were used to emphasize certain parts of the lyrics. Today, there is no such semantic connection between sound and statement. The implementation of the Auto-Tune effect thus has a contingent effect; the digital quivering is apparently just there, meaningless.

The Desire for the Unknown

The Auto-Tune effect represents one of the most powerful sounds in pop music and is therefore an aesthetic phenomenon of probably worldwide—but certainly Western—relevance. Therefore, the scope of the interpretive examination of this sound's appeal must be correspondingly broad. The point at which this interpretation should begin, however, must be chosen more specifically. For, as Weheliye's writings demonstrate and the example of 702 and T-Pain make clear, the Auto-Tune effect is located within a tradition of an estrangement of the singing voice, which itself constitutes a theme in the history of the African-American pop aesthetic. This theme can be understood as a component of a comprehensive line of tradition, for which Mark Dery (1993) coined the term "Afrofuturism." Dery subsumes a variety of phenomena under this term, including Ralph Ellison's *Invisible Man*, images by Jean-Michel Basquiat, films such as John Sayle's *The Brother from Another Planet* and Jimi Hendrix's album *Electric Ladyland*. These phenomena's inherent Afrofuturistic moment can be specifically understood as an expression of the hope for less racist political relations. Taking up a formulation by Dery, Alondra Nelson describes the political project of Afrofuturism more generally: "African American voices with other stories to tell about culture, technology and things to come" (Nelson 2002, 9). What should be understood by "other stories"? Stories by others? Stories of others? Stories of the Other? Stories of Othering? David Toop (1985) answers that question in *The Rap Attack*, one of the reference texts for Afrofuturism studies. Although the answer is not explicitly expressed, it is implicitly presented in the form of a recurrent theme that traverses a series of topics, which Toop puts into context on just five of the book's pages.

That series begins with an explanation of the relevance that the drum computer had for early hip hop, as well as for earlier funk pioneers such as Stevie Wonder and Sly Stone. Toop then cites the legendary hip hop DJ and producer Grandmaster Flash, in order to provide an explanation for how he came up with his name. He describes how "Flash" was a play on comic super heroes, which was linked then to "Grandmaster," a title that had become familiar to Bronx youth through martial arts films. Using a quote from film scholar Thomas Cripp, Toop explains how kung fu films supplanted the popularity of films with black heroes, the so-called Blaxploitation films: "Black youth, then, recoiled from fantasies of lust and power, choosing instead symbols from another culture that provided metaphors for Afro-American experience despite their Oriental settings. Martial arts films offered blacks comic strips of pure vengeance dramatized in a choreography of violence unobtainable within the literal context of American social realism" (Toop 1985, 129). Toop moves immediately from this citation to another object of fascination for black youth, the video arcades: "Video games have had a big influence on latter day hip hop—the arcades are bleeping, pulsing, 24-hour refuges for the obsessive vidkids with nowhere else to go" (ibid.). From video games, Toop again jumps to a mix by DJ Afrika Bambaataa that includes a version of Yellow Magic Orchestra's piece, which had originally been written by Martin Denny—a white American from Hawaii who specialized in "exotic easy-listening music" (ibid.). On the album by Yellow Magic Orchestra that contains the cover version, there is also a piece entitled "Computer Games": "a maddening simulation of video-machine beeps, rumbles and banal tunes" (ibid.). At this point, Toop comes to the influence that the German krautrock band Kraftwerk had on Bambaataa. Toop analyzes hip hop youth's fascination in 1980 with Kraftwerk's robotlike, ultracontrolled performance, which evoked a broad semantic field of "Germanness," as being fascination for the extreme opposite of what George Clinton's P-Funk empire was offering, with his extroverted combination of sex, science fiction, and comics.

Kraftwerk were the most unlikely group to create such an effect among young blacks. Four be-suited showroom dummies who barely moved a muscle when they played, they were nonetheless the first group using pure electronics to achieve anything like the rhythmic sophistication of quality black dance music. ... The George Clinton funk empire and its theatre of excesses had taken sex, sci-fi and comic-book abandonment about as far it could go on stage; four Aryan robots pressing buttons was a joke at the other extreme. (ibid., 130)

Contrary to what one would expect from the term "Afrofuturism," the overall theme of this series of topics does not concern any utopic future. The link between Kraftwerk, kung fu films, superhero comics, Martin Denny's fantastic exotica, and so forth, is much more about relative Otherness, foreignness, the unknown. This reference to something utterly different thus appears to constitute a major moment in Afrofuturism; a desire for something that behaves differently from that which is already known, in particular through the usage of technical means. Relating that point to the Auto-Tune effect, one could say that the Afrofuturistic unknown has now taken the place of the voice. The foreign and the different has therefore become the Subject that makes statements. This Subject vexingly disappears "behind that which was perhaps her/his voice at one time, but which is now definitely far outside of her/himself" (Rabe 2008, 13).

The Mimetic Experience of the Self as Other

In order to conclude with a proposal for how the aesthetics of the Auto-Tune effect could be interpreted, taking the qualities into consideration that were worked out based on Toop's writings, it is necessary to allow for two assumptions. First, the conception of music listening as a mimetic experience, in which aesthetic qualities appear as a fantasy experience of one's own Self (see Eggebrecht 1997). In this case, the aesthetic experience consists of the Subject *her/himself* feeling unknown and sensing the presence of technology. There are a variety of interpretations concerning the desirability of this experience. The second assumption that I am proposing here is to conceive of the necessary interpretative framework as being broader than the Afro-American culture—however that culture may be defined. The Auto-Tune effect's worldwide success is a demonstration of why that is required. I would like to sketch out this framework based on one, albeit fundamental, point. It simply consists of the cultural medium's loss of effectiveness: the symbolic order, in the poststructuralist sense (see Bonz 2011). One of the consequences of this cultural change is that the subjective experience of reality is no longer a result of the Subject's long-term ties to values, to specific categories of perception, to permanent objects of desire. As Slavoj Žižek writes, the dimension of the symbolic order is "reduced to the status of floating islands of signifiers, albuminous *îles flottantes*, basking in a sea of yolky enjoyment" (1997, 70). The surroundings of these islands possess a precarious ontological status, which can be understood by using Lacan's notion of the Real (Waltz 2007). It is the place where this situation's Subject is located; a place where one cannot really be. For that reason, living under these conditions means having to move between different symbolically structured fields, without ever being in an actual home. For the Subject, that involves a requirement to integrate different worlds with their respective objects, values, motives, and so forth. The changes from one world to another must also be designed, and their forever-perceptible contingency must be endured. That means that the Subject of these cultural relations must tolerate the repeated entering and exiting of worlds with local validity, as well as the principle absence of any cultural order that could claim comprehensive validity. What must be tolerated here consists mainly of the experience of the unknown: upon reentrance, every world articulated by a symbolic order becomes foreign once again—even if it is just in the moment of stepping in or out. For the Subject, repeated entry into local worlds is moreover linked to the experience that not only her/his environment is subjected to radical transformations. The Self also becomes repeatedly foreign during this process. Indeed, dependent on the particular order, the Subject her/himself is a respective Other—a different Subject.

For the Subject of the late modern Western culture, this transformation experience undoubtedly represents a great challenge both in principle and in daily life. And in the Real of the expressive culture of music? There, this transformation is experienced in a playful way. To quote Norbert Elias, listening to pop music belongs to one of the "class of mimetic or play activities" (Elias and Dunning [1969] 1986, 69). "Most leisure events arouse emotions which are related to those which people experience in other spheres: they arouse fear and compassion, or jealousy and hatred in sympathy with others but in a manner which is not seriously perturbing and dangerous as is often the case in real life. In the mimetic sphere they are, as it were, transposed into a different key. They lose their sting" (ibid. 80). Emotions that are difficult to endure in daily life can be enjoyed during play, as they are blended with a "'kind of delight'" (ibid. 80) in this case. *In that sense, the attraction of the digital fluttering of the voice can be understood as a playful, mimetic experience of the transformation of the Self.* As a demand of late modern daily reality, the experience of the Self as Other is enjoyed through the playing of pop music. The aspect of foreignness has its place in this interpretation of the Auto-Tune effect. What about the technology aspect, the presence of digital flatness and abruptness? The link that merges the unknown in the Auto-Tune effect with the technological has its own internal logic against the background of these considerations: the symbolic order is also a medium, a technology, a machine. In a passage from the 1954/1955 seminar *The Ego in Freud's Theory and in the Technique of Psychoanalysis*, Lacan leaves no doubt:

> I am explaining to you that it is in as much as he is committed to a play of symbols, to a symbolic world, that man is a decentred subject. Well, it is with this same play, this same world, that the machine is built. The most complicated machines are made only with words. Speech is first and foremost that object of exchange whereby we are recognized, and because you have said the password, we don't break each other's necks, etc. That is how the circulation of speech begins, and it swells to the point of constituting the world of the symbol which makes algebraic calculations possible. The machine is the structure detached from the activity of the subject. The symbolic world is the world of the machine. (Lacan 1991, 47)

Friedrich Kittler, among others, later pointed out that that comment was not just about any machine, it was about the computer (see Kittler [1989] 1993).[1] Lacan's seminar in that year stemmed from an examination of cybernetics and led Lacan to a definition of language and symbolic orders as digital media: For "a series of characters can always be traced back to a sequence of 0 or 1" (ibid., 385).

One possible reason why the mimetic play with the late-modern experience of transformations of the Self should include the presence of the digital, in its flatness and abruptness, lies therefore in the digitality of the manner in which the medium of the symbolic order functions. While entering and exiting these worlds, is it not experienceable by the Subject in its flatness and in the abruptness with which the worlds created by it are validated? With the Auto-Tune effect, that experience is transposed into mimetic play.

Translated by Jessica Ring

Note

1. Kittler recognized the possible application of Lacanian psychoanalysis to media theory and discusses this realization in depth (Kittler [1989] 1993, 385).

References

Bibliography

Barthes, Roland. (1972) 1990. The grain of the voice. In *On Record: Rock, Pop, and the Written Word*, ed. Simon Frith and Andrew Goodwin, 293–300. London: Routledge.

Bonz, Jochen. 2011. *Das Kulturelle*. Paderborn: Fink.

Dery, Mark. 1993. Black to the future: Interviews with Samuel R. Delany, Greg Tate, and Tricia Rose. *South Atlantic Quarterly* 94 (4): 735–777.

Dolar, Mladen. 2006. *A Voice and Nothing More*. Cambridge, MA: MIT Press.

Eggebrecht, Hans Heinrich. 1997. *Die Musik und das Schöne*. Munich: Piper.

Elias, Norbert, and Eric Dunning. (1969) 1986. The quest for excitement. In *Quest For Excitement: Sport and Leisure in the Civilizing Process*, 63–90. Oxford: Blackwell.

Kittler, Friedrich. (1989) 1993. Die Welt des Symbolischen: Eine Welt der Maschine. In *Draculas Vermächtnis: Technische Schriften*, 58–80. Leipzig: Reclam.

Lacan, Jacques. 1991. *The Seminar, Book II. The Ego in Freud's Theory and in the Technique of Psychoanalysis, 1954–1955*, ed. Jacques-Alain Miller, trans. Sylvana Tomaselli. New York: W. W. Norton.

Nelson, Alondra. 2002. Future texts. Introduction. *Social Text* 71, 20 (2): 1–15.

Rabe, Jens-Christian. 2008. Das große Flattern: Tune, was du nicht lassen kannst; Digitale Tonhöhenkorrektur, Kanye West und das irre kalte Monster Pop. *Süddeutsche Zeitung*, December 18, 13.

Reynolds, Simon. 2011. The songs of now sound a lot like then. *New York Times*, July 15. http://www.nytimes.com/2011/07/17/arts/music/new-pop-music-sounds-like-its-predecessors.html.

Toop, David. 1985. *The Rap Attack: African Jive to New York Hip Hop*. London: Pluto Press.

Waltz, Matthias. 2007. Das Reale in der zeitgenössischen Kultur. In *Verschränkungen von Symbolischem und Realem: Zur Aktualität von Lacans Denken in den Kulturwissenschaften*, ed. Jochen Bonz, Gisela Febel, and Insa Härtel, 29–55. Berlin: Kadmos.

Weheliye, Alexander G. 2002. "Feenin": Posthuman voices in contemporary black popular music. *Social Text* 20 (2): 21–44.

Žižek, Slavoj. 1992. *Mehr-Genießen*. Vienna: Turia + Kant.

Sound and Media

Cher. 1999. *Believe*. CD, Warner Bros, B00000F1D3.

Daft Punk. 2000. *One More Time*. CD, Virgin Records, B0000544BO.

Lil Wayne. 2008. *Lollipop*. CD, Cash Money, Universal Motown, B00180OSI6.

702. 1999. *702*. CD, Motown, Universal Music Group, B00000JBFJ.

Snoop Dogg. 2007. *Sensual Seduction/Sexual Eruption*. CD, Doggystyle, Geffen, B00153ZKVA.

West, Kanye. 2008. *808s & Heartbreak*. CD, Roc-A-Fella, Def Jam, B001FBIPFA.

37 Critical Listening

Carlo Nardi

Picture three different scenarios: a sound engineer mastering an audio recording in a secluded, spartan studio in a peaceful city park;[1] a freelance professional listening to an Internet radio station while working on a translation project; and a student forming an opinion on a particular subject by listening to an audio lecture. All these situations involve the adoption of specific listening habits and strategies, but they differ in the way these habits and strategies have been learned, in their purpose, as well as in other aspects that are worth a closer look. In this chapter, I will succinctly discuss those three case studies to illustrate just as many prerequisites of critical listening. At the same time, by underscoring the epistemological, emancipating, and empowering potential of critical listening, I will demonstrate how its ultimate goal is the dismantling of ideological discourses. While drawing extensively on recent theorizations within sensory studies, I will outline a concept and a practice that are embedded more specifically in Marxist theory.

First Scenario: Objective Techniques of Listening in Audio Mastering

Audio mastering is the last stage of record production: it is the moment in which a sound recording is given its final form before it is ready for the market. For that reason, this process finds itself at a crucial intersection between aesthetics and marketing, thus demanding the simultaneous negotiation of creativity and technique, production and listening, intellectual work and manufacturing. The process of audio mastering requires various listening skills, which are sometimes divided by recording engineers into two broad categories: "analytical listening" and "critical listening" (see Moylan 2007, 90; Corey 2010, 5), with the first involving competence about music genres and music aesthetics and the second requiring the application of objective techniques informed by a knowledge of acoustics. Regarding the latter, sound engineers apply what Sterne (2003), in his consideration of how the ethos of Western rationalization permeated the cultural construction of the sense of hearing through the use of the stethoscope and the telegraph, defined as *"audile techniques"*: "A set of practices of listening that were articulated to science, reason, and instrumentality and that encouraged the coding and rationalization of what was heard" (ibid., 23). These techniques are applied by sound engineers; for instance, in the case at hand, audile techniques include strategies to neutralize features of the environment and to ensure detachment from human instability in the perception of sound. Through the application of selected procedures that permit, for instance, an empirical differentiation between sound and noise or the elimination of unwanted resonances, it is possible to gain an "objective" perspective on sound or, in other words, to obtain predictable and replicable results.

Audile techniques such as those employed by mastering engineers definitely demonstrate an acute capacity to analyze sound details, paired with an equally sophisticated ability to synthesize acoustic information. However, based on long-standing social norms, such skills are disengaged from the critique process since the position of sound engineers within the structure of labor usually relegates them to a technical role. In short, they are hired to apply the techniques rather than to question those techniques or the aesthetics that they implicitly serve. This position, moreover, prevents sound engineers from liberating techniques from the socioeconomic constraints that restrict their application to a particular context. As a consequence, "objective" techniques of listening are a necessary but not a sufficient condition for critique.

Second Scenario: Background Listening as a Critical Competence

In the second scenario, a professional—let us say a translator writing a German version of an English manual for a pressure transmitter—selects a radio station that meets his or her musical taste while providing a suitable background to the main activity. Few would describe this instance as one of critical listening, but the reason I am considering it is twofold. First of all, music, and especially recorded music, is ubiquitous as a constitutive part of media—including films, television, telephone hold, games, and the Internet, as well as in the urban environment in places such as parking lots, waiting rooms, stores, and malls. Second, and related to this, not only is background music everywhere; background listening has also become a predominant mode of listening (Kassabian 2013, 2002). One important implication of that is that ubiquitous listening often involves an element of choice—choosing to use instrumental music as a background for watching that film because you know that the director set scenes to classic Stax soul, or going to that particular pub because they do not play techno music—and this also means that inattentiveness does not necessarily imply passivity. Background music can improve performance in certain situations—and even when that is not the case, it is common practice among a significant section of the population, which means that people have become accustomed to playing music while performing other tasks. The fact that the music is in the background does not mean that it is unheard, nor that a person's attention cannot shift, bringing music in and out of attention. While intellectual and manual faculties are conditioned by the listening mode, with effects on both the way a task is performed and how the music is experienced, it is important to recognize that this relationship is reciprocal. This means that listening as a nonprimary task or, better, listening as a concurrent activity, can just be the "adequate" mode of listening in certain situations (see Stockfelt 1997). Therefore, the critical listening proposed here diverges substantially from Adorno's (1962) concept of structural listening, as far as the latter advocates an abstraction of the work of art from any conditioning (see Subotnik 1996).

To sum up, what could provisionally be called background listening expresses specific, albeit implicit, musical competence, as well as a more general cultural competence, owing to the normality of this mode of listening, which can even superimpose itself on products that have been conceived for attentive listening. Nonetheless, that competence has a weak spot that points to an inherent risk of background listening. In the example at the beginning of this section, it is not the functional usage of music in and of itself that should be blamed—as arguably all music performs some kind of function—but the fact that music, either by improving performance or by making a task more pleasing, is employed to maximize the output of work. In this way, music competence, listening skills, and aesthetic values are made subservient to the ideology of work and, hence, to the reproduction of the relations of production.[2] Figuratively speaking, in order to be critical, listening has to be placed in a new perspective in which it no longer constitutes the background of the primary social framework.

Third Scenario: Audio Books and the Textual Paradigm of Critical Listening

Audio lectures can be used for learning purposes when commuting or when performing activities that do not require primary attention. Audio lectures and, more generally, audiobooks do not simply offer an aural alternative to the written word—they actually provide a different kind of experience, with significant implications for the way a text is approached, whether that text preexists the spoken lecture or not. In particular, owing to the "specific time-boundedness of the auditory object" (Wittkower 2011, 217), the speaker dictates the pace of the reading, forcing the listener to maintain constant concentration while simultaneously relieving him or her from the laborious effort of reading. The speaker also provides interpretive clues through the adoption of styles of enunciation and prosodic variations, making the audiobook comparable to a theatrical performance (see Rubery 2011).

I have chosen the example of audio lectures because they require a special disposition, and for that reason they cannot be entirely equated with listening to a radio drama or an audio novel. Just as in a lecture hall, listeners are expected to apply their critical faculties to their full extent. As a matter of fact, critical listening—sometimes also called active listening or perceptive listening—is a concept commonly used in education, where it is considered a prerequisite of critical thinking. The latter, according to most textbooks (see Rudinow and Barry 2008; Cohen 2009; Lau 2011), consists of taking reasonable decisions after systematic thinking; it involves the adoption of consequential and logical strategies whose purpose is to avoid fallacies of thinking, thus striving to judge things objectively. It rejects dogmatic knowledge, stereotypes, and similar cognitive shortcuts, as well as affective interference. Finally, it is self-reflective: "Thinking critically is thinking about your thinking so that you can clarify and improve it" (Chaffee 2012, 52). Somewhat reductively, *The Greenwood Dictionary of Education* defines critical listening as follows: "Thorough listening to a presentation with the intent to analyze what is being heard, not to argue with it" (Wright 2011, 112). This description, however, is insufficient in many ways. In the first place, it considers listening merely a matter of responsiveness; in the second place, it does not seem to contemplate the possibility that listening can also be self-reflexive—meditating on how we listen, listening to how we speak, and so forth.

More generally speaking, as the proposed example suggests, any definition that disregards the specificity of auditory communication is, in and of itself, inadequate and would jeopardize any critical intent. In fact, an audio lecture articulates various forms of mediations, some of which can be likened to similar processes in publishing, such as editing; others, which are particular to the acoustic medium, such as ambient sounds or the presence of a voice, bear with them both nonverbal signs and information about the speaker's gender, ethnicity, class, and so forth. In other words, even though they may go unnoticed, other elements are being conveyed while the attention is focused on what the speaker says and, therefore, on the verbal level of communication. The problem is that the concept of critical listening as it is commonly used in education and philosophy is mainly derived from a textual paradigm; as a consequence, it is unable to grasp the broader auditory dimension of listening since it neglects the various mediating elements that contribute to that dimension.

A Comprehensive Critical Listening Perspective

So, what does listening need in order to be critical? First, a critical listening perspective must include the types of competence listed in the previous sections and, at the same time, it should include a transformative purpose. More precisely, critical listening requires technical skills that would allow the subject to gain "objective" information about the acoustic environment and human communication—that is, knowledge that is verifiable and that can be falsified (Popper 1962) through similar observations. Second, familiarity with a range of listening modes, including background listening, is paramount for at least two reasons: to be able to take the perspective of other actors and to inscribe the act of listening within actual frames of social interaction. It is only in this way that sound will not be relegated to an abstract domain, but instead considered as part of a larger field of symbolic transactions. Third, the analytical skills used in critical thinking are also relevant, provided that they are applied to the entire auditory domain.

Furthermore, these skills should be aimed at revealing the contradictions of reality. In particular, that translates into siding with the exploited and the disadvantaged to show how the current reality is unacceptable to them. Knowledge has to be aimed at transcending the contradictions of the contemporary world, hence an element of praxis is inescapable. As Boltanski argues, "unlike 'traditional theory,' 'critical theory' possesses the objective of reflexivity. It can or even must … grasp the discontents of actors, explicitly consider them in the very labour of theorization, in such a way as to alter their relationship to social reality and, thereby, that social reality itself, in the direction of *emancipation*" ([2009] 2011, 5). All elements involved—theory, skills, praxis—are interrelated: there can be no real change if theory is disengaged from either reality or social transformation. If normative listening is imbued with the same worldview that ideologically justifies it, critical listening can transcend ideology based on its transformative power.

But what are the specifics of sound and hearing in relation to a critique of the Real? In other words, what can critical hearing say about reality and how can it foster its change? Here, the term critical hearing is intended in at least two ways: in the first place, as a means of knowledge; second and not less importantly, as auditory knowledge itself—that is, as a way to reveal the biases of sensory knowledge. According to this second denotation, we need to assume that the various senses function in different ways in a specific historical time and in a specific place, up to a certain degree. Accordingly, critical listening can be used as a tool to reveal contradictions that are inscribed within the normal functioning of the other senses, and, owing to its undeniable prominence, the sense of sight in particular.

These observations prompt further epistemological questions. How can critical listening be effective in disclosing and dismantling structural inequality and formations of power? How can we apply techniques of listening in an emancipatory way? Needless to say, there are no general rules that apply to every situation, because, just as critique is linked to the real, listening is also historically and socially defined. In other words, borrowing Mannheim's ([1929] 1936) theory of ideology, both listening capacity and the faculty of hearing itself are shaped by a particular ideology that varies across society and is the expression of particular interests, along with a total ideology that is typical of a certain society or a social class at a certain time (49–53). According to their class, gender, race, and so forth, different groups will be taught, encouraged, rewarded, and/or forbidden to listen in certain ways. Moreover, at the same time, they will (consciously or not) listen in specific ways in certain situations—not only because that is what is expected from them, but also because it allows them to pursue the particular interests of their group or, in the case of subordinate groups, because it prevents them from threatening the hegemony of the dominant groups. Listening capacity is therefore linked to the habit of selecting certain auditory data from the environment while ignoring others, together with more specific comprehension skills and strategies, with obvious implications for the process of knowledge. On a broader level, just as all societies share a certain worldview, language, and structure of thought, they also share ways of sensing the world.[3] In this regard, Howes underscores the close interdependence between the senses and social organization: "Sensory models not only affect how people perceive the world, they affect how they relate to each other: sensory relations are social relations" (2004, 55).

By applying methods that ensure objectivity and consistency of evaluation, critical listening makes it possible to overcome particular ideologies. On the other hand, emancipation from a total ideology of the senses is more complex to pursue because the instruments used to disclose ideological discourses are informed by the same structure of thinking. Thus, how can we be sure that our actions are actually critical and not ideological? Mannheim, once again, comes to the rescue. In a simplified manner, we can say that, although those actions that reproduce the status quo are ideological, those actions that effectively manage to change reality—real structures of power, exploitation, and inequality—are what we can call *utopian,* following Mannheim ([1929] 1936, 173). This, nevertheless, bears a degree of uncertainty that is inherent to the nature of social change: change is, in fact, a possibility and a necessity simultaneously, but its real forms cannot be anticipated with certainty. Only ex post, that is, through empirical inquiry, is it possible to ascertain whether an action was actually utopian or, instead, ideological. Although any inherent uncertainty of critique should not inhibit its pursuit, it can at least free it from the aura of predictability and inevitability with which critical theory is sometimes imbued, especially in the orthodox versions of its philosophy of history. Consequently, if we do not want to transform critique itself into an instrument of exploitation, we must be aware of this inherent uncertainty. In the meantime, fully aware of this uncertainty, few could possibly disagree with the opinion that the democratization of auditory knowledge is a worthy aim to pursue.

Notes

1. In this example, I refer to the 2004 ethnographic study I conducted at Calyx Mastering, in Viktoriapark, Berlin (Nardi 2007, 2005). On the process of audio mastering, see also Nardi 2014 and Papenburg 2011.

2. On the Taylorist use of alternating segments of music and silence to break up the workflow and improve workers' productivity, see Tagg 1997.

3. Gell (1995), Stoller (1989), Feld (1982), and Seeger (1981), among others, have brilliantly illustrated the variety of sensory ideologies with their studies on, respectively, Umeda-speaking

people in Papua New Guinea, the Songhay of western Africa, the Kaluli of Papua New Guinea, and the Suyà of Mato Grosso.

References

Adorno, Theodor W. 1962. *Einleitung in die Musiksoziologie: Zwölf theoretische Vorlesungen.* Frankfurt am Main: Suhrkamp.

Boltanski, Luc. (2009) 2011. *On Critique: A Sociology of Emancipation.* Trans. Gregory Elliott. Cambridge: Polity Press.

Chaffee, John. 2012. *Thinking Critically*, 10th ed. Boston: Wadsworth.

Cohen, Elliot D. 2009. *Critical Thinking Unleashed.* Plymouth: Rowman & Littlefield.

Corey, Jason. 2010. *Audio Production and Critical Listening: Technical Ear Training.* Burlington, VT: Focal Press.

Feld, Stephen. 1982. *Sound and Sentiment: Birds, Weeping, Poetics, and Song in Kaluli Expression.* Philadelphia: University of Pennsylvania Press.

Gell, Alfred. 1995. The language of the forest: Landscape and phonological iconism in Umeda. In *The Anthropology of Landscape: Perspectives on Place and Space*, ed. Ed Hirsch and Michael O'Hanlon, 232–254. Oxford: Clarendon Press.

Howes, David. 2003. *Sensual Relations: Engaging the Senses in Culture and Social Theory.* Ann Harbor: University of Michigan Press.

Kassabian, Anahid. 2002. Ubiquitous listening. In *Popular Music Studies*, ed. David Hesmondhalgh and Keith Negus, 131–142. New York: Bloomsbury.

Kassabian, Anahid. 2013. *Ubiquitous Listening: Affect, Attention, and Distributed Subjectivity.* Berkeley: University of California Press.

Lau, Joe Y. F. 2011. *An Introduction to Critical Thinking and Creativity: Think More, Think Better.* Hoboken, NJ: Wiley.

Mannheim, Karl. (1929) 1936. *Ideology and Utopia: An Introduction to the Sociology of Knowledge.* Trans. Louis Wirth and Edward Shils. London: Routledge & Kegan Paul.

Moylan, William. 2007. *Understanding and Crafting the Mix: The Art of Recording.* 2nd ed. Burlington, VT: Focal Press.

Nardi, Carlo. 2005. Zen in the art of sound engineering. In *The Proceedings of the 2005 Art of Record Production Conference.* London College of Music & Media, London, September 17–18.

Nardi, Carlo. 2007. Fare musica: Un processo intersensoriale. *Critica Sociologica XLI* 162 (summer): 79–93.

Nardi, Carlo. 2014. Gateway of sound: Reassessing the role of audio mastering in the art of record production. *Dancecult* 6 (1): 8–25.

Papenburg, Jens Gerrit. 2011. Hörgeräte: Technisierung der Wahrnehmung durch Rock- und Popmusik. PhD dissertation, Humboldt-Universität zu Berlin.

Popper, Karl R. 1962. *Conjectures and Refutations: The Growth of Scientific Knowledge.* New York: Basic Books.

Rubery, Matthew, ed. 2011. *Audiobooks, Literature, and Sound Studies.* New York: Routledge.

Rudinow, Joel, and Vincent E. Barry. 2008. *Invitation to Critical Thinking,* 4th ed. Belmont, CA: Thomson Wadsworth.

Seeger, Anthony. 1981. *Nature and Society in Central Brazil: The Suyà Indians of Mato Grosso.* Cambridge, MA: Harvard University Press.

Sterne, Jonathan. 2003. *The Audible Past: Cultural Origins of Sound Reproduction.* Durham, NC: Duke University Press.

Stockfelt, Ola. 1997. Adequate modes of listening. In *Keeping Score: Music, Disciplinarity, Culture,* ed. David Schwarz, Anahid Kassabian, and Lawrence Siegel, 129–146. Charlottesville: University Press of Virginia.

Stoller, Paul. 1989. *The Taste of Ethnographic Things: The Senses in Anthropology.* Philadelphia: University of Pennsylvania Press.

Subotnik, Rose Rosengard. 1996. *Deconstructive Variations: Music and Reason in Western Society.* Minneapolis: University of Minnesota Press.

Tagg, Philip. 1997. Understanding musical time sense: Concepts, sketches, and consequences. http://www.tagg.org/articles/timesens.html. Revised and expanded version in *31 artiklar om musik: Festskrift till Jan Ling,* 21–43. Göteborg: Skrifter från Musikvetenskapliga institutionen, 9, 1984.

Wittkower, D. E. 2011. A preliminary phenomenology of the audiobook. In *Audiobooks, Literature, and Sound Studies,* ed. Matthew Rubery, 216–231. New York: Routledge.

Wright, Jeffrey M. 2011. Critical listening. In *The Greenwood Dictionary of Education,* 2nd ed., ed. John W. Collins III and Nancy Patricia O'Brien, 112. Westport, CT: Greenwood Press.

38 Sonic Cartoons

Simon Zagorski-Thomas

The history of recorded music is often characterized as a story of improved "realism"—a story of the development of high fidelity. Yet, in parallel with the technical challenge of achieving full frequency response and full dynamic range in a recording, studio designers and sound engineers have sought to reduce the muddiness caused by low-frequency reverberation and to increase perceived clarity by using multiple microphones in close proximity to instruments and spreading them across the stereo image—all of which creates a simplified or schematic version of reality: a cartoon of sound. In the 1940s and 1950s, a great deal of effort went into studio design that provided architectural solutions to these problems, such as bass "traps" and parabolic reflectors (see Schmidt-Horning 2012 for a detailed description of this period). In later years, analogue and digital signal processing have added to the techniques available to both enhance clarity and create artificial sound worlds. My experience of recording began in the late 1970s as a musician and continued through the 1980s and 1990s as a sound engineer, producer, and composer. Although I realized in a pragmatic way that notions of realism and quality were both applied differently and valued differently in different musical genres,[1] it wasn't until I became an academic in 2003 that I thought about this in any great detail. Thus, an important subtext here is the issue of conscious intention and the nature of social phenomena: I am not suggesting that recording professionals think in terms of sonic cartoons when they are working, but that this is the result of a long-term social, historical, and technological process.

Examples

Miles Davis's *Kind of Blue* (1959) album cover includes the Columbia Records logo with the tagline "Guaranteed High Fidelity." While the quality of the recording may be superb, the notion that it is a faithful reproduction of what was heard in the studio on those two days in Columbia's 30th Street Studios in 1959 is problematic. The only place it sounded like that was in the control room as it came through the speakers—and even that is not quite true, as that was later altered by the various mastering processes that have been applied to the different releases. Looking at photographs of the session (see figure 38.1), there is a wall of acoustic screens about three meters high arranged around the musicians to reduce the amount of reverberation being picked up. This way the long, rich reverberation of 30th Street Studios[2] is relegated to the background and only becomes more noticeable when the dynamics get louder. The sound of the saxophones and trumpet were "sweetened" with short, artificial reverberation from the studio's echo chamber (Frank Laico, sound engineer on *Kind of Blue*: see Kahn 2001, 102). From the photographs and interviews in Kahn's (2001) book, it appears that there were seven microphones used:[3] one for each instrument and two on the drum kit that were then mixed down (including the echo chamber) to the three-track tape machine so that the "tenor saxophone and piano shared the left track, trumpet and bass the center, and alto saxophone and drums the right" (ibid., 102). The *invariants* at play here—the way that high frequency sound content can suggest proximity and the ways that reverberation both suggests space and can affect our sense of both the strength and power of a sound source—have been highly manipulated. The result is a totally artificial combination of intimacy and space,

Figure 38.1
Photo from the *Kind of Blue* recording session (photo: Don Hunstein © Sony Music Entertainment).

and yet the clarity of all the individual sounds (owing to the relatively close micro-phone placement), the convincing (albeit technologically constructed) stereo image, and the subtle reverberation combine together to create a cartoon of an intimate jazz club acoustic experience. In fact, this kind of cartoon staging for jazz recordings has become a template for the sound that jazz clubs have come to emulate through the use of discrete microphones and public address systems.[4]

That example hopefully serves to explain some of the issues. Even classical record-ings that still retain the notion of replicating the sound of the concert hall very rarely do so with a single pair of microphones in a coincident pair—that is, like a pair of ears in the concert hall. They use multiple "spot" microphones as well as artificially created, wide stereo images through the usage of various placement techniques. In short, they exaggerate or simplify. At the other extreme, as in the case of abstract visual art, the abstract audioscapes of dub reggae and contemporary electronica have used textures and spatial imagery that are entirely imagined—yet they rely on the fact that they are drawing on some facet of "reality," some *invariant*, to create meaning. For example, the types of echo used may not be realistic, but the very notion of an echo is inextricably bound to enclosed space, and even if the echoes build in volume and become distorted rather than fading away, we hear it as a spatial effect—albeit an imaginary or impos-sible one.

Introduction to the Theoretical Side

These schematized representations or *sonic cartoons* are an inherent part of the way that recording works. Just as looking at a photograph or a film of a desert island is not the same as being there, a "realistic"[5] recording is not the same as a concert. But, more to the point, the vast majority of recordings are complex spatial and temporal distortions or, perhaps more accurately, constructions. And just as visual art can range from highly figurative work—from photography and realist painters to impressionist and abstract painting, for example—so can recorded sound, by staging a musical performance in ways that make it sound like a concert hall experience or like nothing on earth.

I insist that what the draughtsman, beginner or expert, actually does is not replicate, to print, or to copy in any sense of the term but to mark the surface in such a way as to display invariants and record an awareness. Drawing is never copying. It is impossible to copy a piece of the environ-ment. Only another drawing can be copied. We have been misled for too long by the fallacy that a picture is similar to what it depicts, a *likeness*, or an *imitation* of it. A picture supplies some of the information for what it depicts, but that does not imply that it is in projective correspondence with what it depicts. (Gibson 1979, 279)

This quote provides a clue as to the theoretical basis for this article: Gibson's ecologi-cal approach to visual perception, which has been applied to auditory perception and

music by Eric Clarke (2005), among others. The notion of invariants, general characteristics rather than specific auditory structures, is to be further understood through
Lakoff and Johnson's term *image schema* (Johnson 1990; Lakoff 1990) and Fauconnier
and Turner's work on *mental spaces* and *conceptual blends* (Fauconnier and Turner 2003).
The following quote from Gibson illustrates this point: "Whatever the artist may do
… he cannot avoid showing his surface *in the midst of other surfaces of an environment.
A picture can only be seen in a context of other nonpictorial surfaces*" (Gibson 1979, 272;
emphasis in original).

By and large, the same is true of recorded music: there are very few moments when
we mistake "live" performance for recorded music and vice versa. And this is despite
the fact that popular music concert practice is now mediated to emulate the sound
of recordings: multiple microphones reduced down to a stereo image reproduced
through left and right speaker arrays in a concert hall. Despite the claims of high fidelity, recorded music is not a copy of "a piece of the environment." Even in a formal-
ized acoustic concert hall situation, when we are in a relatively static position, small
head movements produce spatial information that cannot be reproduced in recording,
our choice of which element to watch affects what we hear,[6] and nonmusical environmental noise (audience members coughing and moving in their seats) is subject to the
same ambiance as the music.

The Perception of Recorded Sound

Even if the aural experience of the concert hall could be accurately reproduced—
which it cannot—our perception is both multimodal and informed by context. Most
importantly, perception is an activity that involves an active engagement with the
environment, and the illusion of recording realism is based on a passive relationship
between the listener and a pair, or any other array, of speakers. Although binaural
information is important in spatial hearing, we also use bodily movement (head movement in particular) to build a more accurate, multimodal interpretation by checking
how the binaural information changes in relation to changes in our bodily position. In
an environment with multiple players arranged across a stage, these head movements,
along with the visual stimulus, reinforce our interpretation of this spatial arrangement.
If this arrangement is represented by the relative volumes of these players in two or
more speakers, we not only have no visual clues, but our head movements reveal the
illusory nature of the stereo (or surround) image.

As far as the multimodal nature of perception is concerned, there is work that
suggests that the somatic perception of ultrahigh and ultralow frequency sound, the
effect of vibration on the skin and internal organs, is a significant contributor in
our perception of sound (see, e.g., Oohashi et al. 2000). However, when it comes to

perceiving sounds that we have not made ourselves, visual perception is probably the most important other mode. Not only does seeing a mouth move affect how we hear the sound it appears to be making, but the pairing of a visually perceived gesture and an aurally perceived sound (e.g., how a bow moves across strings or how fingers touch piano keys) work together from the lowest levels of neural response to improve perception (Kayser et al. 2010). In effect, when we look at an individual in an ensemble, we perceive their instrumental contribution more clearly. This obviously has important implications for the difference between being at a concert and hearing recorded sound with no visual representation of the performance; but, even with video, we return to the idea of perception being proactive. With a video, the editor or director determines the point of view—what kinds of close-ups or tracking shots and so forth will be presented and when; with a concert, the listener directs their gaze at will.

The third issue is that of nonmusical environmental noise. If, at a concert hall, there is coughing, shuffling, traffic noise, or any other environmental sound, it occurs in the same space as the musical performance and is subject to the same acoustic ambiance. If I am listening to a recording of a concert in a living room, car, or on headphones, the environmental noise is both different in character and will have a different form of ambiance than the recorded sound. Going back to Gibson, if a *"picture can only be seen in a context of other nonpictorial surfaces"* (Gibson 1979, 272), the sound of a recorded performance and its studio environment can only be heard in a context of other nonmusical and different environmental sounds.

Sonic Cartoons

The proactive and embodied nature of our perceptual and interpretive processes ensures that we are always aware of the differences between reality and representations. In addition to this, representational works do not have to be representations of reality; they can also be representations of imagined realities or even of the impossible. Thus, Salvador Dalí can create a visual representation of "soft" clocks in *La persistencia de la memoria* (1931) and Prince can create a track ("When Doves Cry," 1984) where the kick and the snare from the drum kit are staged in different-sounding spaces and several overdubs of his voice are in different spaces again.[7] Throughout the history of recording, alongside the aspiration to create a realistic representation of the concert hall, there has been a parallel aspiration to create novel, imaginary representations of performances.

The point about these representations is that they draw on our previous experience and knowledge of spatial sound, but do so schematically: by representing some of the auditory characteristics of space, but not others, or by creating something that is reminiscent of these characteristics (see Zagorski-Thomas 2010, 2012). When I describe these types of representation as *sonic cartoons*, I use the term not just to mean that certain features are exaggerated, but also that certain features might be simplified to achieve greater clarity.

Conclusion

In conclusion then, we often think of video and sound recording as having the property of somehow "capturing" reality. However, as Gibson (1979, 279) points out, these media simply create a representation of some of the characteristics or *invariants* that we use to create interpretations of what is going on around us. The proactive nature of our perceptual system combined with the context of playback means that we will become aware of their representational nature even when the utmost effort is made to recreate the experience of that reality. Recorded music, despite its aspirations to high fidelity, has embraced the artifice of its representational nature and turned it to its advantage in a number of ways. Even before the quality of the technology was able to match the acoustic specifications of our hearing system in terms of frequency and dynamic range, multiple microphones and close placement were being used to create cartoon representations of clarity: for example, where the listener is impossibly close to all the instruments in an ensemble simultaneously. And this idea of clarity in particular, and the various techniques that have been developed over the last century to achieve it, is central to the notion of sonic cartoons. Record production is continually looking for strategies to allow it to clean up the messiness of reality: to thin out reverberation in order to create the sense of space without the muddiness, to suggest the low-frequency power of a large space without the loss of definition, to exaggerate the intimacy of closeness with microphone and processing techniques, or to create artificially spread out and wide stereo images of ensembles. Just as visual cartoons remove unnecessary detail and exaggerate important features so as to create distorted and simplified representations of some aspect of reality, so do sonic cartoons in recorded music. And these strategies for creating simpler and clearer representations of musical performances have also become the basis for the creation of impossible and imaginary stagings of performances—the more abstract end of the sonic cartoon world.

The notion of *sonic cartoons* is part of a larger project to incorporate the production of recorded music into the mainstream of musicology (Zagorski-Thomas, 2014). My plans for future research in this area involve both a wider range of case studies and a more detailed examination of how specific forms of schematic sound can encourage particular interpretations in a listener.

Notes

1. During this period I was recording a wide range of music: historical and contemporary art music (classical, for want of a better word), jazz, various styles of rock, dance music, reggae, hip hop, South and West African music, salsa, and music for television, ads, and films.

2. Columbia's 30th Street Studio was converted from a large Greek Orthodox church of over ten thousand square feet (nine hundred square meters) with a high, vaulted wooden ceiling (Kahn 2001, 75).

3. From the photos and Laico's interview, some, and possibly all, of the microphones were Telefunken U-47 large diaphragm condenser microphones (Kahn 2001).

4. Jazz clubs in London such as Ronnie Scott's and the 606 create similar cartoons of intimacy and "high fidelity" by exaggerating the high-frequency content of their sound reinforcement systems. This notion of live sound emulating recorded sound is an area that is ripe for further research.

5. By "realistic," I mean binaural stereo—a recording made with two microphones to simulate the positioning of the two ears of a listener.

6. For evidence of this, see McGurk and MacDonald 1976. An illustration of the McGurk effect can be seen at http://www.youtube.com/watch?v=G-lN8vWm3m0 (accessed August 27, 2012).

7. The bass drum has a long but artificially curtailed reverb reminiscent of a largish bright hall, and the snare has a shorter, brighter reverb like a studio chamber. The voices range from very close and unreverberant to a larger airy space, but also include processing reminiscent of telephone voices at some points.

References

Bibliography

Clarke, Eric F. 2005. *Ways of Listening: An Ecological Approach to the Perception of Musical Meaning.* New York: Oxford University Press.

Fauconnier, Gilles, and Mark Turner. 2003. *The Way We Think: Conceptual Blending and the Mind's Hidden Complexities.* New York: Basic Books.

Gibson, James J. 1979. *The Ecological Approach to Visual Perception.* Hillsdale, NJ: Psychology Press.

Johnson, Mark. 1990. *The Body in the Mind: The Bodily Basis of Meaning, Imagination, and Reason.* Chicago: University of Chicago Press.

Kahn, Ashley. 2001. *Kind of Blue: The Making of the Miles Davis Masterpiece.* New York: Da Capo Press.

Kayser, Christoph, Nikos K. Logothetis, and Stefano Panzeri. 2010. Visual enhancement of the information representation in auditory cortex. *Current Biology* 20 (1): 19–24.

Lakoff, George. 1990. *Women, Fire, and Dangerous Things: What Categories Reveal about the Mind*. Chicago: University of Chicago Press.

McGurk, Harry, and John MacDonald. 1976. Hearing lips and seeing voices. *Nature* 264 (5588): 746–748.

Oohashi, Tsutomu, Emi Nishina, Manabu Honda, Yoshiharu Yonekura, Yoshitaka Fuwamoto, Norie Kawai, Tadao Maekawa, Satoshi Nakamura, Hidenao Fukuyama, and Hiroshi Shibasaki. 2000. Inaudible high-frequency sounds affect brain activity: Hypersonic effect. *Journal of Neurophysiology* 83 (6): 3548–3558.

Schmidt-Horning, Susan. 2012. The sounds of space: Studio as instrument in the era of high fidelity. In *The Art of Record Production: An Introductory Reader to a New Academic Field*, ed. Simon Frith and Simon Zagorski-Thomas, 29–42. Farnham: Ashgate.

Zagorski-Thomas, Simon. 2010. The stadium in your bedroom: Functional staging, authenticity and the audience led aesthetic in record production. *Popular Music* 29 (2): 251–266.

Zagorski-Thomas, Simon. 2012. Musical meaning and the musicology of record production. *Beiträge zur Popularmusikforschung* 38:135–148.

Zagorski-Thomas, Simon. 2014. *The Musicology of Record Production*. Cambridge: Cambridge University Press.

Sound and Media

Davis, Miles. (1959) 1997. *Kind of Blue*. CD, Columbia Records, CK 64935.

Prince. 1984. "When Doves Cry." Seven-inch single, Warner Brothers Records, W9286.

Contributors

Karin Bijsterveld is a historian and professor of science, technology, and modern culture at Maastricht University. Among her publications are *Mechanical Sound* (MIT Press, 2008) and *The Oxford Handbook of Sound Studies* (with Trevor Pinch; Oxford University Press, 2012). She is currently coordinating the project Sonic Skills: http://fasos-research.nl/sonic-skills/.

Susanne Binas-Preisendörfer, PhD, was born in 1964 in Berlin. She was active in the East German music scene "die anderen bands," and she founded the Berlin situated-listening gallery singuhr-hörgalerie: http://www.singuhr.de. Binas-Preisendörfer has been a professor of music and media studies since 2005; she currently teaches and conducts research at the University of Oldenburg, focusing on the history and aesthetics of mediatized musical forms, music and globalization, music business, youth cultures and popular music, arts and cultural policies. Since 2013, she has served as chair of the German-speaking branch of the International Association for the Study of Popular Music (IASPM): http://www.iaspm-dach.net.

Carolyn Birdsall is assistant professor in media studies at the University of Amsterdam. Her monograph *Nazi Soundscapes: Sound, Technology, and Urban Space in Germany, 1933–1945* was published by Amsterdam University Press in 2012 (also available through open access: http://www.oapen.org). Her current research focuses on practices of sound recording and archiving in German broadcasting (1930–1960), with a related interest in emerging concepts of documentary sound in European radio and sound film. Homepage: http://home.medewerker.uva.nl/c.j.birdsall.

Jochen Bonz has conducted extensive research on the aesthetics of pop music and the ways in which they are related to processes of identification and the shaping of listeners' subjectivities. His PhD thesis deals with the culture of house and techno music (2008, "Subjekte des Tracks [The Subjects of Tracks]"). In that work, as well as in his second book (2011, *Das Kulturelle* [The Sphere of Culture]), he combines an ethnographic approach with concepts from Lacanian psychoanalysis. His recent research project is concerned with the fans of soccer/football. Methodologically speaking, it is based in soundscape studies as well as in ethnopsychoanalysis. Bonz was based at the University of Bremen, Germany, for many years. He currently holds the position of assistant professor in the Department of History and Cultural Anthropology at Leopold-Franzens-University Innsbruck, Austria.

Michael Bull is professor of sound studies at the University of Sussex. He is the founding editor of the *Senses and Society Journal* published by Bloomsbury and the founding editor (along with Veit Erlmann) of the forthcoming *Sound Studies Journal* (Bloomsbury). He has published widely in the field of sound studies, most recently putting out a four-volume edition on sound studies for Routledge (2013). Bull is currently writing a monograph on sirens and editing an expanded second edition of *The Auditory Culture Reader* (with Les Back).

Thomas Burkhalter is an ethnomusicologist, music journalist, and cultural producer from Bern, Switzerland. He published the book *Local Music Scenes and Globalization: Transnational Platforms in Beirut* with Routledge (2013) and coedited the book *The Arab Avant-Garde: Music, Politics, Modernity* for Wesleyan University Press (November 2013). Burkhalter is the founder and editor in chief of Norient—Network for Local and Global Sounds and Media Culture, and the artistic director of the annual Norient Musikfilm Festival. http://www.norient.com.

Mark J. Butler is a music theorist whose research addresses popular music, rhythm, gender and sexuality, and technologically mediated performance. He is an associate professor and coordinator of the program in music theory and cognition at Northwestern University. He is the author of *Unlocking the Groove* (Indiana University Press, 2006) and the editor of *Electronica, Dance, and Club Music* (Ashgate, 2012). His most recently published book is *Playing with Something That Runs: Technology, Improvisation, and Composition in DJ and Laptop Performance* (Oxford University Press, 2014). It is based on extensive fieldwork with internationally active DJs and laptop musicians based in Berlin.

Diedrich Diederichsen was editor and/or publisher of music journals in Hamburg and Cologne (*Sounds, Spex*) throughout the 1980s. He has taught at a variety of universities since the early 1990s, and has been a professor of theory, practice, and communication of contemporary art at the Academy of Fine Arts Vienna since 2006. His most recent books are *Über Pop-Musik* (Kiepenheuer & Witsch, 2014, in German); *The Whole Earth: California and the Disappearance of the Outside* (co-edited with Anselm Franke; Sternberg Press, 2013, in German/English); and *Psicodelia y Ready-Made* (Edición Adriana Hidalgo, 2010, in Spanish). Diederichsen is continuously working on pop music, contemporary art, modern composition, cinema, theater, design, and politics. For a complete list of publications, see http://diedrich-diederichsen.de. He lives in Berlin and Vienna.

Veit Erlmann is an anthropologist/ethnomusicologist and the endowed chair of music history at the University of Texas at Austin. He has won numerous prizes, including the Alan P. Merriam award for the best English monograph in ethnomusicology and the Mercator Prize from the German Research Foundation DFG. Erlmann frequently presents his work at leading institutions around the world. He has also published widely on music and popular culture in South Africa, including *African Stars* (University of Chicago Press, 1991), *Nightsong* (University of Chicago Press, 1996), and *Music, Modernity, and the Global Imagination* (Oxford University Press, 1999). His most recent publication is *Reason and Resonance: A History of Modern Aurality* (New York, 2010). Currently, he is working on a book on intellectual property law in South Africa that will be published by Duke University Press. A pioneer of sound studies and editor of the classic volume *Hearing Cultures* (Bloomsbury, 2004), he also became founding coeditor of the journal *Sound Studies* (Bloomsbury, London) in 2014. For more on Veit Erlmann's activities, visit his website at http://www.veiterlmann.net.

Franco Fabbri is a musician and musicologist; he teaches popular music and sound studies at the University of Turin. His main interests are in the fields of genre theories and music typologies, the impact of media and technology across genres and musical cultures, and the history of popular music. He has served twice as chair of the International Association for the Study of Popular Music. He is coeditor (with Goffredo Plastino) of the *Routledge Global Popular Music Series* (http://www.globalpopularmusic. net). Among his publications are *Elettronica e musica* (Fratelli Fabbri, 1984), *Il suono in cui viviamo* (Feltrinelli, 1996, 3rd edition il Saggiatore, 2008), *L'ascoltotabù* (il Saggiatore, 2005), *Around the Clock: Una breve storiadella popular music* (Utet, 2008), and *Made in Italy: Studies in Popular Music* (coedited with Goffredo Plastino, Routledge, 2014). For a full list of his publications in various languages, see https://universitaditorino. academia.edu/FrancoFabbri.

Golo Föllmer's audio practice has focused on radio, sound art, and electroacoustic music. He conducts research on sound installation, contemporary music, audio media, and the aesthetics of radio. In 2002, he completed his PhD on networked music. Following a junior professorship, he is now lecturer in the media department at the University of Halle. He does curatorial work and conference organization for *sonambiente* (Berlin, 1996 and 2006), *net_condition* (Karlsruhe, 2000), *RadioREVOLTEN* (Halle, 2006), and *Multisensual Digital Radio Culture* (Berlin, 2013). He is the founder of the master's degree program *Online Radio* and project leader of the European research project *Transnational Radio Encounters*.

Marta García Quiñones is a PhD candidate at the University of Barcelona, where she is finishing a thesis on historical models of music listening and theories of audition. She is also interested in the transformations of listening in the contemporary mediascape, and she has written on radio programming, mobile listening to portable digital players, listening to music in low-attention contexts, and on the return of some forms of collective listening. She edited the collection *La música que no se escucha: Aproximaciones a la escucha ambiental* (L'Orquestra del Caos, 2008) and coedited, with Anahid Kassabian and Elena Boschi, *Ubiquitous Musics: The Everyday Sounds That We Don't Always Notice* (Ashgate, 2013). From 2009 to 2011, she was head of Public Programmes at the Museum of Contemporary Art, Barcelona. She was a member of the international research network "Sound in Media Culture" (2010–2013), funded by the German Research Foundation, and a coinvestigator of the AHRC-funded expert workshop on "Culture, Value, and Attention at Home" (2013).

Mark Grimshaw is the Obel professor of music at Aalborg University, Denmark. He has published over sixty works across subjects as diverse as sound, virtuality, the Uncanny Valley, and IT systems. His last two books were an anthology on computer game audio published in 2011 and *The Oxford Handbook of Virtuality* for Oxford University Press (2014). With coauthor Tom Garner, a monograph entitled *Sonic Virtuality* is due out in 2015 from Oxford University Press.

Rolf Großmann studied musicology, German literature, philosophy, and physics, and completed a dissertation on "music as communication"; he was a freelancer and jazz musician for many years and is currently working at the Leuphana University of Lueneburg (Germany) as professor of "digitale Medien und auditive Gestaltung" and director of the Center of Competence "((audio)) Aesthetic Strategies." He is a founding member of the Institute for Culture and Aesthetics of Digital Media (ICAM), Lueneburg. Research interests: technoculture and the aesthetics of music. Further information and publications: http://audio.uni-lueneburg.de.

Maria Hanáček (†) received her MA in popular music studies from the University of Liverpool in 2007. She was a doctoral candidate at Humboldt University Berlin, where she also taught seminars on music technology and studio production and worked as a research associate and lecturer. Her PhD thesis examines where the creative subject is discursively situated in the process of record production by analyzing the so-called "making of" videos. She was a cofounder of the research network "Sound in Media Culture," one of the main initiators of the refounded German speaking IASPM-branch (IASPM D-A-CH), and a member of the Association for the Study of the Art of Record Production (ASARP).

Thomas Hecken is a professor of new German literature at the University of Siegen. He is the author of numerous books, including *Gestalten des Eros: Die schöne Literatur und der sexuelle Akt* (1997), *Avantgarde und Terrorismus: Rhetorik der Intensität und Programme der Revolte von den Futuristen bis zur RAF* (2006), *Gegenkultur und Avantgarde: Situationisten, Beatniks, 68er* (2006), *Theorien der Populärkultur: Dreißig Positionen von Schiller bis zu den Cultural Studies* (2007), *Pop: Geschichte eines Konzepts* (2009), and

Avant-Pop: Von Susan Sontag über Prada und Sonic Youth bis Lady Gaga und zurück (2012), and the editor of the journal *Pop: Kultur und Kritik* and the blog pop-zeitschrift.de.

Anahid Kassabian is the James and Constance Alsop Chair of Music at the University of Liverpool. Her research and teaching work focuses on ubiquitous music; music, sound, and moving images; listening; disciplinarity; and music and new technologies, especially games, virtual worlds, and pervasive computing. She also works with feminist and postcolonial theories in her writings on the Armenian diaspora and the art and music from the MENA region. She is the author of *Ubiquitous Listening* (University of California Press, 2013) and *Hearing Film* (Routledge, 2001). She coedited the volumes *Ubiquitous Musics* (Ashgate, 2013) and *Keeping Score: Music, Disciplinarity, Culture* (University of Virginia Press, 1997).

Carla J. Maier (née Müller-Schulzke) is a postdoctoral researcher in the base project *Analog Storage Media* at the Cluster of Excellence *Image Knowledge Gestaltung* at the Humboldt University Berlin. She received a PhD from Goethe University Frankfurt/Main for her thesis "Transcultural Sound Practices: Urban Dance Music in the UK." She has taught cultural studies, postcolonial studies, and sound studies. Her research interests include transcultural popular music, cultural theory of everyday sounds, and postcolonial Anglophone literature.

Andrea Mihm is a cultural scientist and art teacher. She studied at the universities of Marburg, Frankfurt/Main, and Turku/Finland. She wrote her thesis on the mechanical/sociocultural interspace of the escalator, published a cultural history of the suitcase, and worked alongside others at the University of Hamburg and for German Lufthansa. At the moment, she is working at a secondary school in Frankfurt am Main, where she lives with her husband and three children.

Bodo Mrozek is a historian based in Berlin. After studying history in Berlin and Amsterdam, he worked as a journalist and editor for the *Frankfurter Allgemeine Zeitung* and *Die Zeit*. He held scholarships from the German Historical Institutes in Washington, London, and Paris. He was a doctoral fellow of the research group "Felt Communities? Emotions in European Music Performances" at the Max-Planck-Institute for Human Development in Berlin, and he is currently an associated scholar of the Center for Contemporary History in Potsdam (ZZF). Mrozek is the coeditor of two volumes *Popgeschichte* (Bielefeld: Transcript, 2014) and the founder of the multilingual blog pophistory.hypotheses.org. He is currently finishing his PhD project on the emergence of a transnational popular youth culture in the 1950s and 1960s.

Carlo Nardi is a research associate at Rhodes University and teaches at CDM—Centro Didattico Musica Teatro Danza. He received his PhD in sciences of music from the University of Trento in 2005. Between 2011 and 2013, he was the general secretary of the International Association for the Study of Popular Music (IASPM). His work has focused on the use of technology from a sensory perspective, authorship in relation to technological change, the organization of labor in music-making, and sound for the moving image.

Jens Gerrit Papenburg is a lecturer and research associate in popular music history and theory at Humboldt University Berlin. He is the sound review editor of the journal *Sound Studies* (Bloomsbury, London) and cofounder of the international research network "Sound in Media Culture" (funded by the German Research Foundation, 2010–2016). He received his PhD from Humboldt University Berlin for the thesis "Hörgeräte: Technisierung der Wahrnehmungdurch Rock- und Popmusik [Listening Devices: Technologization of Perception through Rock and Pop Music]." He is currently working on a postdoctoral research project about a cultural and media history of the "para-auditive" dimension of popular music's sound.

Thomas Schopp studied musicology and history at Humboldt University, Berlin. He worked as a research assistant at the Institute of Music, University of Oldenburg. In his dissertation, Schopp developed a sound history of the disc jockey show in American radio between 1930 and 1970. He now works as a program director for the German National Academic Foundation.

Holger Schulze is a professor of musicology and principal investigator at the Sound Studies Lab at the University of Copenhagen. He serves as a curator for the House of World Cultures Berlin and as a founding editor of the book series *Sound Studies* (in German). He is a founding member and vice chair of the European Sound Studies Association as well as coeditor of the international journal for historical anthropology *Paragrana*. He was a cofounder and the first department head of the new master's program in sound studies at the Berlin University of the Arts, as well as a the director and cofounder of the international research network Sound in Media Culture.

Toby Seay is an associate professor of recording arts and music production at Drexel University. He has had a long career in the music industry as a musician, recording engineer, technical consultant, and audio preservationist. As a recording engineer, Toby recorded artists such as Dolly Parton, Randy Travis, Delbert McClinton, Ringo Starr, David Wilcox, Kirk Whalum, and many others. Toby has worked on numerous Gold and Platinum Certified recordings, as well as eight Grammy winners. Toby's research interests include audio preservation practices and standards, specializing in multitrack audio formats and how these resources can be put to use in audio production education and research. Toby deconstructs record production techniques, examining social and participant dynamics within the recording studio environment and the affects of workflow and recording techniques on musical production outcomes and sonic signatures.

Jacob Smith is associate professor in the radio-television-film department at Northwestern University. In addition to writing the books *Vocal Tracks: Performance and Sound Media* (2008), *Spoken Word: Postwar American Phonograph Cultures* (2011), and *The Thrill Makers: Celebrity, Masculinity, and Stunt Performance* (2012, all University of California Press), he has published articles on media history, sound, and performance.

Paul Théberge is a professor cross-appointed to the Institute for Comparative Studies in Literature, Art, and Culture and the School for Studies in Art and Culture (Music), at Carleton University, Ottawa. He has published widely on issues concerning media, technology, and music and is the author of *Any Sound You Can Imagine: Making Music/ Consuming Technology* (Wesleyan University Press, 1997) and the producer of *Glenn Gould: The Acoustic Orchestrations* (Sony Classical, 2012). He is also coeditor of the forthcoming volume *Living Stereo: Histories and Cultures of Multichannel Sound* (Bloomsbury Academic, 2015).

Peter Wicke is a professor of the history and theory of popular music and the director of the Centre for Popular Music Research at Humboldt University, Berlin. He is a member of the editorial board of the academic journals *Popular Music* (Cambridge University Press) and *Popular Music History* (Equinox Publishing), of the advisory panel of *The Journal of the Royal Musicological Association* (Routledge), of the advisory board of the *Norsk Tidsskrift for Musikkforskning* (University of Agder, Norway), and of the advisory board of the International Institute for Popular Culture at the University Turku (Finland). He has published numerous articles and books on popular music that have been translated into more than fifteen languages. In English, he has published the monographs *Rock Music: Culture, Aesthetics, Sociology* (Cambridge University Press, 1990) and (with John Shepherd) *Music and Cultural Theory* (Cambridge University Press, 1997). Currently, he is working on a book dealing with the history of the technologies of music production (with a strong focus on the nineteenth century).

Simon Zagorski-Thomas is a reader at the London College of Music, University of West London. He is the founder and codirector of the annual Art of Record Production Conference, a cofounder of the *Journal on the Art of Record Production*, and cochairman of the Association for the Study of the Art of Record Production (http://www. artofrecordproduction.com). His publications include *The Musicology of Record Production* (Cambridge University Press, 2014) and *The Art of Record Production* (coedited with Simon Frith, 2012). Before becoming an academic, he worked for twenty-five years as a composer, sound engineer, and producer with artists as varied as Phil Collins, Mica Paris, London Community Gospel Choir, Bill Bruford, the Mock Turtles, Courtney Pine, and the Balanescu Quartet. He continues to compose and record music and is currently conducting research into the musicology of record production, popular music analysis, and performance practice in the recording process.

Index